I0037031

Magnetic Oxides and Composites II

Edited by

Rajshree B. Jotania[1], Sami H. Mahmood[2]

[1]Department of Physics, Electronics and Space science, University school of sciences, Gujarat University, Ahmedabad 380 009, India

[2]Department of Physics, The University of Jordan, Amman-11942, Jordan

Copyright © 2020 by the authors

Published by **Materials Research Forum LLC**
Millersville, PA 17551, USA

All rights reserved. No part of the contents of this book may be reproduced or transmitted in any form or by any means without the written permission of the publisher.

Published as part of the book series
Materials Research Foundations
Volume 83 (2020)
ISSN 2471-8890 (Print)
ISSN 2471-8904 (Online)

Print ISBN 978-1-64490-096-3
eBook ISBN 978-1-64490-097-0

This book contains information obtained from authentic and highly regarded sources. Reasonable efforts have been made to publish reliable data and information, but the author and publisher cannot assume responsibility for the validity of all materials or the consequences of their use. The authors and publishers have attempted to trace the copyright holders of all material reproduced in this publication and apologize to copyright holders if permission to publish in this form has not been obtained. If any copyright material has not been acknowledged please write and let us know so we may rectify this in any future reprints.

Distributed worldwide by

Materials Research Forum LLC
105 Springdale Lane
Millersville, PA 17551
USA
https://www.mrforum.com

Manufactured in the United States of America
10 9 8 7 6 5 4 3 2 1

Table of Contents

Preface

Magnetic oxides continue to take the lead in the magnetic materials world market due to the wide range of their applications, stability, simplicity of production and characterization, and cost effectiveness. The existence of magnetic oxides with different crystallographic structures, and tunable magnetic properties arising from different types of magnetic interactions between magnetic ions within the lattice, paved the way for successful employment of these oxides in a wide diversity of applications. These include, and not limited to, permanent magnets, microwave devices, magnetic refrigeration, sensors, catalysis, and health sector. This special topic volume is an extension of the previous "Magnetic Oxides and Composites, Vol. 31, 2018" published by Materials Research Forum LLC, USA under the umbrella of the book series "Materials Research Foundations". The chapters in this book focus mainly on topics that were not adequately covered by the previous volume, in an effort to highlight the broader spectrum of applications of magnetic oxides and their composites. A total of ten chapters in this volume were dedicated to address various aspects related to synthesis, characterization, and applications of magnetic oxides and composites in different fields.

Chapter one is focused on structural, magnetic, electric and magnetocaloric properties of $Pr_{0.67}Ba_{0.22}Sr_{0.11}Mn_{1-x}Fe_xO_3$ ($x = 0.00$ and $x = 0.05$) perovskites synthesized using solid state method. The investigated perovskites exhibited properties suitable for magnetic refrigeration.

In chapter two, the structural and magnetic properties of Al -substituted $Er_3Fe_{5-x}Al_xO_{12}$ ($0.0 \leq x \leq 0.8$) garnets prepared by conventional ceramic method were investigated. The site selectivity of Al^{3+} ions, and the effect of their concentration (x) were addressed.

Chapter three is concerned with the structural, magnetic, critical behavior, and magnetocaloric effect of $La_{0.67}Ba_{0.22}Sr_{0.11}Mn_{1-x}Fe_xO_3$ ($x = 0.0, 0.1, 0.2, 0.3$ and 1.0) prepared by the conventional ceramic method. The samples exhibited a change in the universality class, and potential for magnetic refrigeration applications at low iron substitution levels.

Chapter four is focused on eco-friendly, biosynthesis of carbon-based metal oxide nanocomposites using green reducers. Applications of graphene-metal oxide nanocomposites in electrochemistry, heterogeneous photo catalysis, and antibacterial activity were discussed.

In chapter five, synthesis of Co-substituted Ni-Cd spinel ferrites ($Co_xNi_{0.5}Cd_{0.5-x}Fe_2O_4$) using citrate gel auto combustion method was described. The antimicrobial activity of the ferrites was examined against Gram negative (Escherichia coli, Klebsiella pneumonia) and Gram positive (Bacillus subtilis, Staphylococcus aureus) bacterial strains, and the results indicated an improvement of the activity by doping the ferrite with cobalt.

Chapter six is based on the applications of metal/metal oxides nanoparticles in organic transformations. The authors have emphasised on the fundamentals of different nanocatalysts in chemical reactions. They have also enlightened the catalytic performance of metal oxide nanocatalyst for high reaction activity, long-term stability, and selectivity to target products for industrial applications.

In chapter seven, the effect of Ca substitution on the physical properties of Co-spinel ferrites ($Co_{1-x}Ca_xFe_2O_4$) prepared by citrate gel method was investigated. The experimental results showed that partial replacement of cobalt by calcium resulted in a decrease of the switching field distribution (SFD), making the samples suitable for high density recording media applications.

In chapter eight, a comparative study of the structural and magnetic properties of lithium ferrite nanoparticles prepared by solid state reaction (SSR) and sol-gel (SG) method was conducted. The study revealed that a single ferrite phase was formed in the SG method at a significantly lower temperature compared with SSR method. The presence of both, ordered and disordered $LiFe_5O_8$ crystal forms were confirmed by Raman spectroscopy. Also, the study revealed higher magnetic parameters for the ferrites prepared by SG method.

Chapter nine reports the synthesis, behavior and biomedical applications of multifunctional ferrites. Synthesis, characterization and hyperthermia measurements of $Cd_xNi_{1-x}Fe_2O_4$ ferrites prepared using the sol-gel process were described and discussed.

In chapter ten, the structural, electrical and magnetic properties of $Ni_{1-x}Cu_xFe_2O_4$ nanoferrites prepared by citrate-gel auto-combustion technique were reported. The effects of Cu substitution on the magnetic properties, dielectric properties, Seebeck coefficient, activation energy, as well as structural parameters were discussed.

Sami H. Mahmood

Rajshree B. Jotania

Magnetic Oxides and Composites II
Materials Research Foundations 83 (2020) 1-20

Materials Research Forum LLC
https://doi.org/10.21741/9781644900970-1

Chapter 1

Investigation on Structural, Electrical, Magnetic and Magnetocaloric Properties of $Pr_{0.67}Ba_{0.22}Sr_{0.11}Mn_{1-x}Fe_xO_3$ Perovskites

K. Snini[1a], M. Ellouze[1a], E.K. Hlil[2b], K. Khirouni[3c]

[1]University of Sfax, LaMMa, B.P.1171, 3000 Sfax, Tunisia

[2] Institute of Néel, CNRS and Joseph Fourier University, BP 166, F-38042 Grenoble Cedex 9, France

[3]Physics Department, The University of Gabes, Tunisian 6079 Gabes

[a]sninikhaled2@gmail.com, [a]mohamed.ellouze@fss.rnu.tn,
[b]El-Kebir.Hlil@neel.cnrs.fr, [c]kamel.khirouni@fsg.rnu.tn

Abstract

The structural, electrical and magnetic properties of $Pr_{0.67}Ba_{0.22}Sr_{0.11}Mn_{1-x}Fe_xO_3$ ($x = 0.00$ and $x = 0.05$) perovskites have been investigated. All samples were prepared using solid state method and characterized by XRD, DC-AC conductivity, Magnetic, Magnetocalric measurements. The variation of conductivity with temperature shows a metal-semiconductor transition and $Tc \sim 220$ K, 90 K for $x = 0.00$ and 0.05 respectively. A paramagnetic to ferromagnetic transitions was found with decreasing temperature. The magnetocaloric effect study indicates that the investigated compounds have the appropriate properties to be suitable candidates to be used as refrigerants.

Keywords

Manganites, X-Ray Diffraction, Rietveld Refinement, Magnetization, Relative Cooling Power

Contents

1.   Introduction

Over the last twenty years, the study of physical properties in doped manganite oxides with general formula $RE_{1-x}AE_xMnO_3$, where RE is a rare earth (RE = Pr, La, Sm, Nd…) ion and AE is a divalent element (AE = Sr, Ca, Ba…), have attracted a great interest. Such studies have been achieved since the discovery of the magnetocaloric effect (MCE) phenomena [1-4] and colossal magnetoresistance (CMR) [5-7]. In order to explain these new properties, many studies in structural, magnetic, magnetocaloric and transport properties of manganites have been investigated. It has been illustrated that they are the consequence of the combination of the double-exchange interaction between ferromagnetically coupled Mn^{3+} and Mn^{4+} ions [8] and the strong electron-phonon interaction known as the Jahn–Teller effect [9]. Many factors like the variation in A-site ions [10] and B-site ions [11] influence the physical properties of manganites. The electric properties of these compounds depend on several factors such as the ratio of the divalent ion, the ionic radii of the metal ion, the elaboration method of the samples, etc. Thus, manganites doped with iron in Mn-site are interesting for several reasons. Those compounds are known to exhibit interesting electronic properties, which are explained by many mechanisms. To optimize the properties of this material a systematic study of the electrical and dielectric properties should be conducted over a wide temperature range and with different doping levels. For this purpose, impedance spectroscopy is an effective method to study not only the properties of intragranular and interfacial regions and their inter-relations, but also their temperature and frequency dependences in order to separate the individual contributions from the total impedance and finally their interfaces with electronically conducting electrodes. This technique is very useful for the investigation of bound or mobile charge in the bulk or interfacial regions of any kind of solid or material such as ceramics [12], semiconductors [13], polymer [14], polycrystalline materials [15], insulators (dielectrics) [16] and superconductors [17] etc. In the last few years, extensive studies by various researchers have been carried out on the effect of doping at both A and

Mn-sites on electrical properties of manganites. Until now, the dependence of electrical and magnetic properties on doping in Mn-site by metal ions has not been well investigated.

The purpose of this work is to study the structural, electrical, magnetic and magnetocaloric properties of $Pr_{0.67}Ba_{0.22}Sr_{0.11}Mn_{1-x}Fe_xO_3$ (x = 0.0 and 0.05) manganite oxides prepared by the conventional solid-state reaction method at high temperature.

2. Experimental procedure

$Pr_{0.67}Ba_{0.22}Sr_{0.11}Mn_{1-x}O_3$ compounds were synthesized using the conventional solid-state reaction method by mixing Pr_6O_{11}, BaO_2, SrO_2, MnO_2 and Fe_2O_3 with purities higher than 99.0 % in the desired proportion according to the reaction:

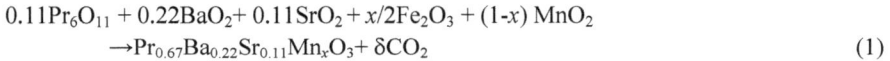

$$0.11Pr_6O_{11} + 0.22BaO_2 + 0.11SrO_2 + x/2Fe_2O_3 + (1-x) MnO_2$$
$$\rightarrow Pr_{0.67}Ba_{0.22}Sr_{0.11}Mn_xO_3 + \delta CO_2 \tag{1}$$

The stoichiometric amounts of metal oxides were initially mixed in an agate mortar. Then the powders were pressed into pellets (about 1 mm in thickness) and sintered at 800, 1000, 1200 and 1300 °C for 24 h with intermediate regrinding and re-pelletizing to ensure a better crystallization. Finally, those pellets were slowly cooled to room temperature in open air. For all heated samples, phase purity and cell parameters were determined by powder X-ray diffraction, recorded at room temperature on a Panalytical X'PERT Pro diffractometer, using $\theta/2\theta$ Bragg Brentano geometry with diffracted beam monochromatized Cu-K_α radiation. The diffraction patterns were collected by steps of 0.017 over the angle range 10–70°. Structural analyses were checked by the Rietveld method using the FullProf program [18, 19]. For electrical measurements, a thin silver film (20 nm thick) was deposited on both sides of the pellet through a circular mask of 6 mm of diameter. Then, we obtained a configuration of a plate capacitor which is used to measure both the conductance and the capacitance. The sample is mounted in a cryostat to vary the temperature from 80 to 340 K. Under vacuum and in dark, measurements are conducted with an Agilent 4294 analyzer using a signal amplitude of 20 mV. Magnetization measurements versus temperature, $M(T)$, in the range 5-400 K and versus magnetic applied field, $M(\mu_0H)$, were carried out using a Vibrating Sample Magnetometer (VSM Oxford model). Magnetocaloric effect (MCE) was estimated based on the magnetization measurements versus magnetic applied field up to 5T at several temperatures.

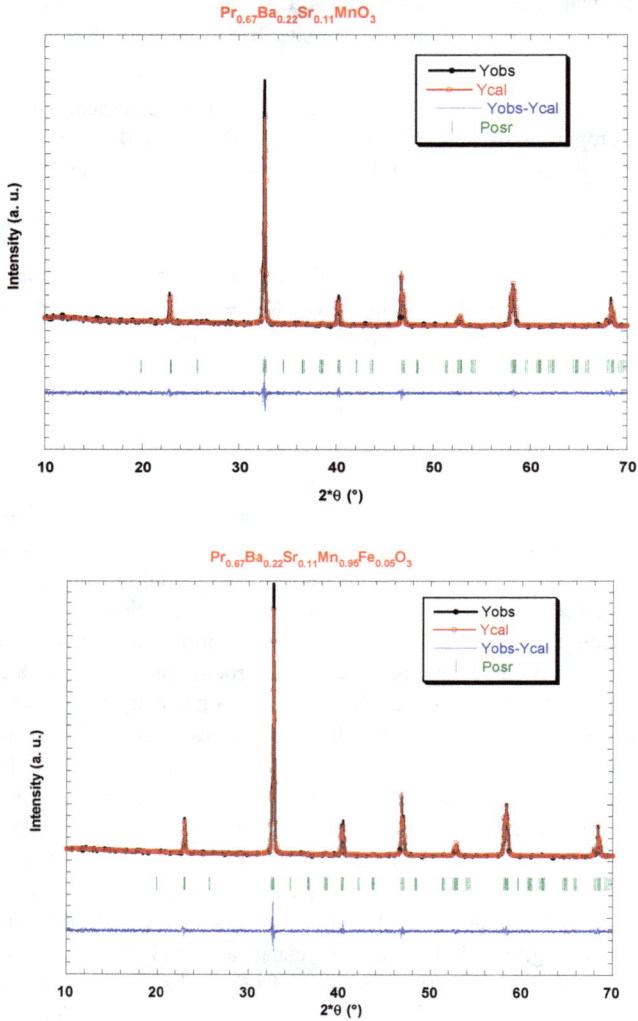

Fig. 1. Observed and calculated X-ray diffraction data and Rietveld refinement for $Pr_{0.67}Ba_{0.22}Sr_{0.11}Mn_{1-x}Fe_xO_3$ for x = 0.00 (a), and x = 0.05 (b). Vertical bars are the Bragg reflections for the space group P_{nma}. The difference pattern between the observed data and fits is shown at the bottom.

Table 1. Values of lattice parameter, unit cell volume (V), Mn–O–Mn bond angle, Mn–O bond length, profile factor (R_p), tolerance factor (t), weighted profile factor (R_{wp}), global chi-square (χ^2), Bragg factor (R_{Bragg}) and crystallographic factor (RF) for $Pr_{0.67}Ba_{0.22}Sr_{0.11}Mn_{1-x}Fe_xO_3$ manganites with x=0.00 and 0.05.

Parameters	x = 0.00	x = 0.05
Symmetry	orthorhombic	orthorhombic
Space group	P_{nma}	P_{nma}
a (Å)	5.5178(2)	5.5191(1)
b (Å)	7.7434(4)	7.7455(3)
$b/\sqrt{2}$ (Å)	5.4754(1)	5.476(4)
c (Å)	5.4824(3)	5.4839 (2)
V (Å³)	234.243	234.425
$<d_{Mn-O}>$ (Å)	2.056(5)	2.028(5)
$<Mn–O–Mn>$ (°)	165.750(1)	165.150(3)
t	0.9168	0.9168
$<r_A>$(Å)	1.2574	1.2574
R_p	7.57	7.54
R_{wp}	9.57	9.44
R_{exp}	8.05	7.69
R_{Bragg}-Factor	3.76	4.13
R_F-Factor	8.57	7.73
χ^2	1.41	1.51

3. Results and discussion

3.1 XRD measurements

To check their crystal structures and phase purities, X-ray diffraction (XRD) measurements for all heated samples were carried out at room temperature. Fig. 1 represents the Rietveld refinement of powder X-ray diffraction patterns of $Pr_{0.67}Ba_{0.22}Sr_{0.11}Mn_{1-x}Fe_xO_3$ for $x = 0.00$ and $x = 0.05$ samples. A single phase was observed for all materials without any detectable impurities. The results reveal that prepared compounds crystallized in the orthorhombic structure with P_{nma} space group. The atomic positions are taken at 4c (x, 0.25, z) for (Pr, Ba and Sr), 4b (0.5, 0, 0) for (Mn and Fe), 4c (x, 0.25, z) for O_1 and 8d (x, y, z) for O_2.

The structural parameters, the goodness of the fit indicator χ^2 and the agreement factors (Braggs factor R_{Bragg} and structure factor R_F) are summarized in Table 1.

Lattice distortions may be caused by either Jan-Teller effect inherent to the Mn^{3+} high spin state, effect causing a distortion of MnO_6 octahedra. It is important to calculate the tolerance factor of Goldschmidt (t), given by:

$$t = \frac{r_A + r_O}{\sqrt{2}(r_B + r_O)} \tag{2}$$

Where $\langle r_A \rangle$; $\langle r_B \rangle$, and r_O are the average ionic radius of the A-site, B and O ions, respectively. The calculated tolerance factors for the samples were found to be t = 0.9168. The values of t (Table 1) have been estimated and they are within the range of stable perovskite structure.

3.2 DC conductivity study

Fig. 2 shows the conductance (G) as a function of temperature, respectively, for $Pr_{0.67}Ba_{0.22}Sr_{0.11}Mn_{1-x}Fe_xO_3$ ($x = 0.0$ and $x = 0.05$) between 80 and 340 K. The samples with $x = 0.0$ and 0.05 exhibit a metallic behavior at low temperatures and a semiconducting one at high temperatures. The temperatures T_{MS} of metal–semiconductor transition are 220 K and 90 K, respectively. This result can be explained by the effect of substitution of Mn^{3+} ions by Fe^{3+} ones. It is well-known that this substitution decreases the Mn^{3+}/Mn^{4+} ratio and increases the Fe^{3+}/Mn^{3+} one. In this case, the hopping electron from Mn^{3+} to Fe^{3+} is prohibited [23-26]. Moreover, the replacement of manganese by iron favors insulating character [23, 27]. Therefore, it can be seen that the conductivity and the temperature transition decrease with increasing iron content. Consequently, Fe doping leads to a depletion of the number of available hopping sites, which suppresses metallicity and pushes the system into the semiconductor side. For an undoped compound, the double substitution of Pr^{3+} by Ba^{2+} and Sr^{2+} can introduce a change in the conduction mechanism at a temperature T_{MS}.

In the hopping process, the carrier mobility is temperature dependent, which is usually characterized by the activation energy. The activation energies, in the high temperature's ranges, were calculated using the Arrhenius law and the expression could be described as, [28-30].

$$G_{DC}T = B_0 \exp\left(\frac{-E_a}{K_B T}\right) \tag{3}$$

Where B_0 is the pre-exponential factor, T is the absolute temperature, E_a is the activation energy and k_B is the Boltzmann constant.

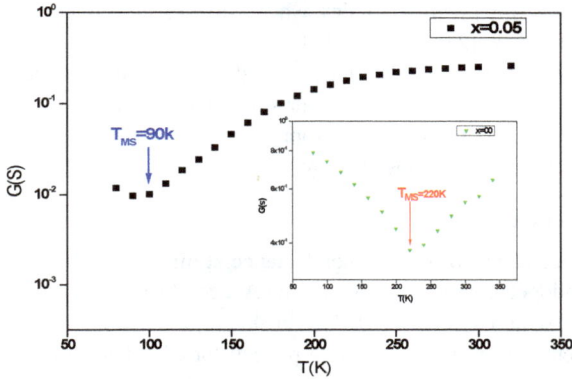

Fig. 2. Variation of the conductance (G) as a function of temperature for $Pr_{0.67}Ba_{0.22}Sr_{0.11}Mn_{1-x}Fe_xO_3$ (x = 0.00 and 0.05) samples.

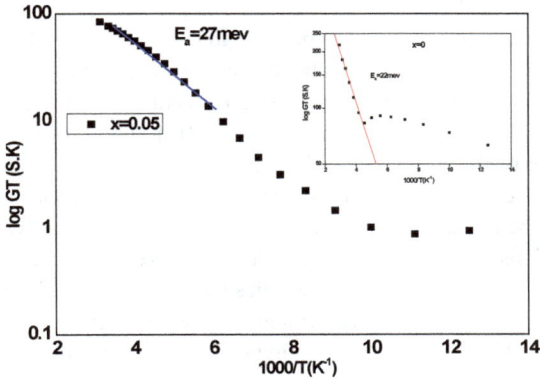

Fig. 3. Variation of the log (GD_{CT}) as a function of (1000/T) for $Pr_{0.67}Ba_{0.22}Sr_{0.11}Mn_{1-x}Fe_xO_3$ (x = 0.00 and 0.05) samples.

Fig. 3 shows the plots of log (GT) vs. 1000/T; we can observe that these curves are linear over the wide temperature range, confirming thermally activated polaron hopping. Fig. 3 shows that the values of activation energy increase with increasing Fe content. Similar behavior was observed in previous studies, where Mn was substituted by Cr, Ti, Ga or Fe [31-34]. Rahmouni *et al.* [23] have studied LaBaMnO-Fe doped system; they observed that the Fe content (0.00 ≤ x ≤ 0.20) affects strongly the activation energy (from 27 m eV for x = 0.00 to 90 m eV for x = 0.05). In our study, the activation energy increases with iron content but does not change much (from 22 m eV for x = 0.00 to 27 m eV for x = 0.05). Similar results have been observed by Sahastrabudhe *et al.* [35].

3.3   AC conductivity study

The frequency dependence of the AC conductance at different temperatures is shown in Fig. 4. For the undoped compound (Fig. 4a) the AC conductance spectra can be split into two parts. The first part is below 10^4 Hz, in this region we can observe that the AC conductance spectra are characterized by a plateau for each temperature. Above 10^4 Hz the conductance decreases with increasing frequency. The variation of conductance as a function of frequency shows a metallic behavior at low temperature and a semiconductor one at high temperature. For this compound, the temperature of transition T_{MS} is found to be 220 K as shown in the Fig. 4a. When the compound is doped (Fig. 4b), the same behavior was observed for x = 0.05 (Fig. 4b), the transition was observed at T_{MS} = 90 K.

3.4   Magnetic properties

Magnetization measurements versus temperature in the range 5–400 K and in a magnetic applied field of 0.05 T reveal that all elaborated compounds exhibit a paramagnetic (PM) to ferromagnetic (FM) transition with decreasing temperature (Fig. 5). With increasing Fe content, the Curie temperature T_C decreases from 216 K for x = 0.00 to 134 K for x = 0.05. The decrease of the Curie temperature T_C can be explained by the increase in the number of Fe^{3+}-O-Fe^{3+}, which induces a weakening of the ferromagnetism. In addition, this evolution can be explained by the fact that the presence of iron weakens the double exchange interactions to the profit of the super exchange ones in the complex Fe^{3+}-O-Mn^{3+}. In addition, the presence of iron modifies the Mn-O-Mn angles and the Mn/Fe-O distances thus decreasing the Curie temperature. Similar evolution was found by Nassri *et al.* [36] and Zouari *et al.* [37] after substitution by iron in B-site. The Curies temperatures deduced from the minimum of dM/dT versus temperature curves (the inset of Fig. 5).

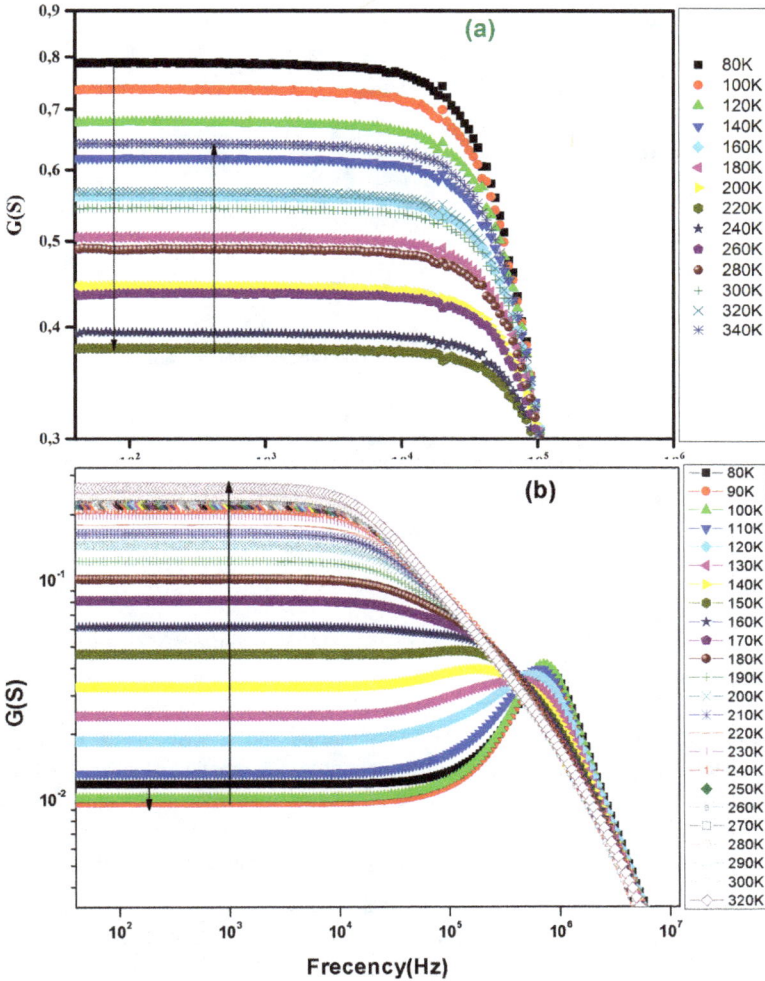

Fig. 4. Variation of the conductance (G) as a function of frequency at different temperatures for $Pr_{0.67}Ba_{0.22}Sr_{0.11}Mn_{1-x}Fe_xO_3$ (a). x = 0.00, (b). x = 0.05 samples.

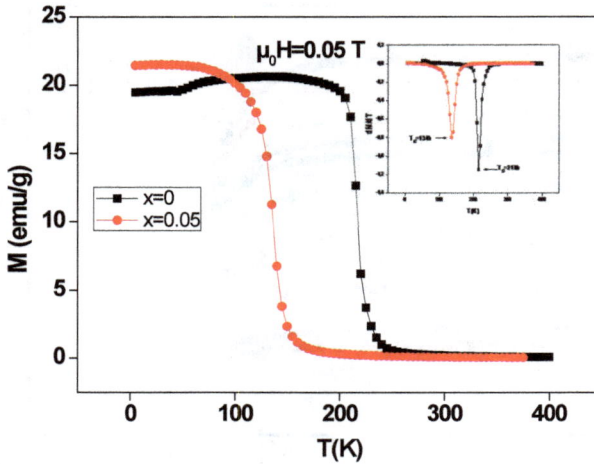

Fig. 5. M (T) curves for $Pr_{0.67}Ba_{0.22}Sr_{0.11}Mn_{1-x}Fe_xO_3$ at $\mu_0H = 0.05$ T magnetic field in FC regime. The inset is the dM/dT curves.

Our goal is to study the magnetic properties at low temperatures, so we have carried out the magnetization measurements as function of magnetic applied field up to 5T at several temperatures. The magnetization evolution vs. magnetic applied field at various temperatures for $x = 0.00$ and $x = 0.05$ was plotted and shown in Fig. 6. We can see clearly that the magnetization increased sharply at low magnetic fields ($\mu_0H < 0.5$ T) and then saturates for fields above 1 T, which confirms the existence of ferromagnetic state at low temperatures for all prepared compounds. On the other hand, we observe a linear behavior at high temperatures, presenting a sign of the paramagnetic state.

In order to study the nature of magnetic phase transition, the Arrott plots given M^2 vs. μ_0H/M are plotted (Fig. 7) [38]. The order of the FM – PM transition has been studied based on the Banerjee's criterion [39]. The Arrott curves have a positive slope which indicates that prepared materials undergo a second order FM–PM phase transition.

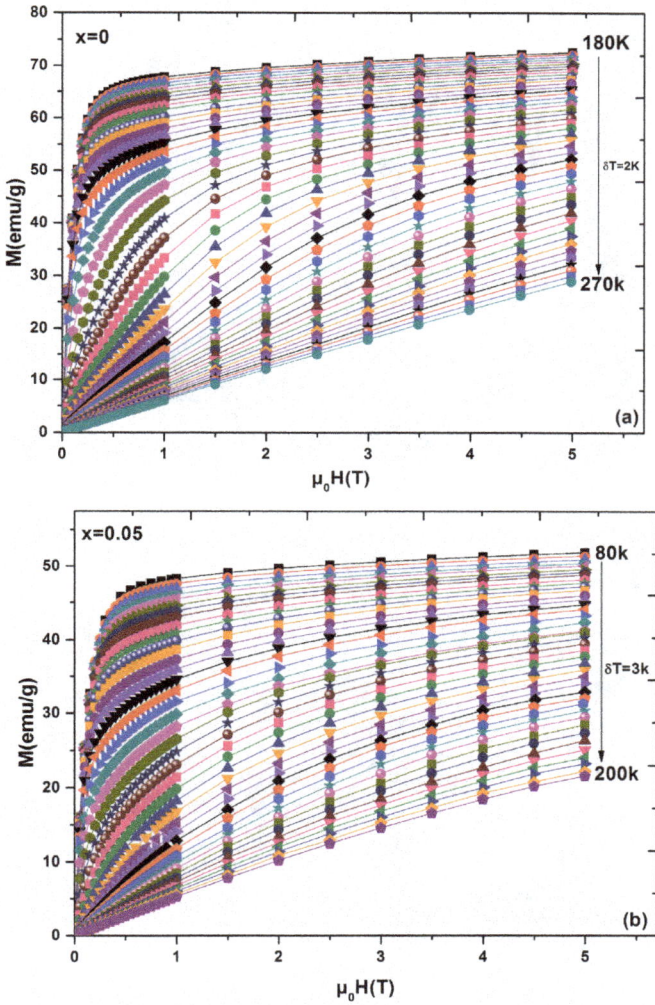

Fig. 6. Isothermal magnetization of $Pr_{0.67}Ba_{0.22}Sr_{0.11}Mn_{1-x}Fe_xO_3$ for (a). x = 0.00 and, (b). x = 0.05 measured at several temperatures.

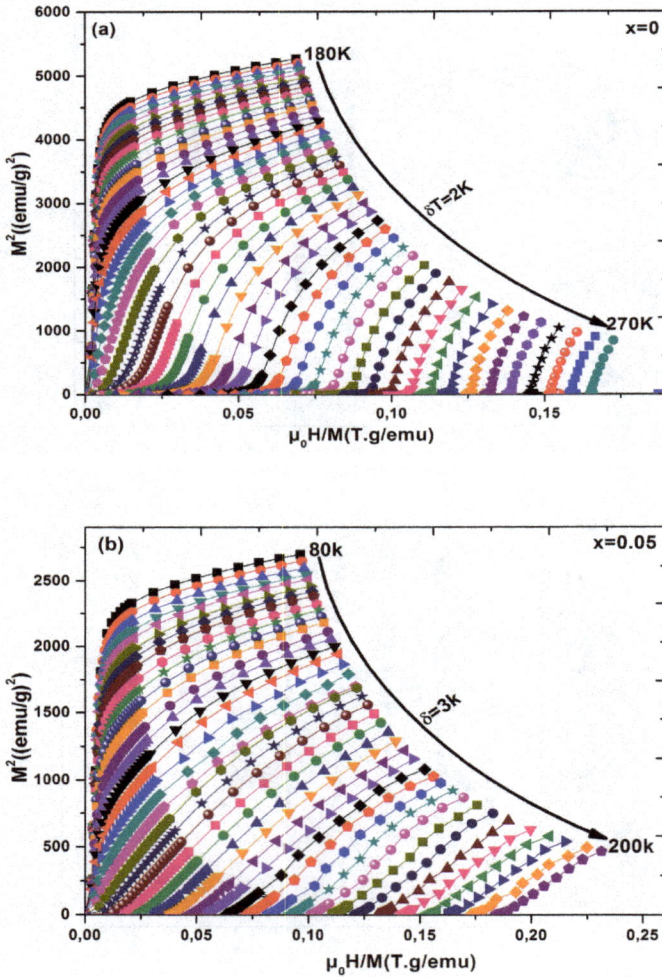

Fig. 7. μ_0H/M vs. M^2 curves of isotherms for $Pr_{0.67}Ba_{0.22}Sr_{0.11}Mn_{1-x}Fe_xO_3$ for (a). $x = 0.00$ and, (b). $x = 0.05$.

3.5   Magnetocaloric study

The magnetocaloric effect (MCE) is an intrinsic characteristic of most magnetic materials and the coupling of the magnetic sublattice with the magnetic applied field induces it. The MCE can be calculated indirectly using the magnetization measurements versus magnetic applied field at several temperatures. According to The Maxwell thermodynamic equation, the MCE is given by Eq. (4) [40].

$$\left(\frac{\partial S}{\partial H}\right)_T = \left(\frac{\partial M}{\partial T}\right)_H \tag{4}$$

Where S is the magnetic entropy, M is the magnetization and H the magnetic applied field.

After the integration of Eq. (4), the magnetic entropy change can be written as:

$$\Delta S_M(T, \mu_0 H_{max}) = \int_0^{\mu_0 H_{max}} \left[\frac{\partial M}{\partial T}\right]_H dH \tag{5}$$

If the magnetization measurements are performed at small discrete field and temperature intervals, ΔS_M can be approximated as:

$$\Delta S_M(T, \mu_0 H) = \sum_i \frac{M_{i+1}(T_{i+1}, \mu_0 H) - M_i(T_i, \mu_0 H)}{T_{i+1} - T_i} dH_i \tag{6}$$

Where M_i and M_{i+1} are the magnetization values measured at $\mu_0 H$, at T_i and T_{i+1} are the temperatures respectively.

Using the Maxwell thermodynamic equation, the magnetic entropy change was calculated for all prepared samples. The estimated $|\Delta S_M|$ versus temperature of $Pr_{0.67}Ba_{0.22}Sr_{0.11}Mn_{1-x}Fe_xO_3$ for $x = 0.00$ and $x = 0.05$ is given in Fig. 8 (a-b). From the obtained curves, we can observe that our materials have a large magnetocaloric effect near Curie temperatures and the maximum of the $|\Delta S_M|$ increases when $\mu_0 H$ increase. The maximum values of the magnetic entropy change ($|\Delta S_M^{max}|$) for $x = 0.00$ and $x = 0.05$ are found to be 4.72 and 1.90 $Jkg^{-1}K^{-1}$, respectively, under a magnetic field change of 5 T.

The relative cooling power (RCP) values are a useful way to evaluate the samples according to their efficiency. It represents the area under the $|\Delta S_M|$ versus temperature curves. It can be estimated by integrating the $|\Delta S_M(T)|$ curves over the full width at half-maximum [41-43]. The RCP is an important parameter to select magnetic refrigerants. It measures the heat transfer between hot and cold sources during a refrigeration cycle and it is calculated by the following expression:

$$RCP = -\Delta S_M^{max}(T, \mu_0 H_{max}) \times \delta T_{FWHM} \tag{7}$$

Where $-\Delta S_M^{max}$ the value of the maximum of the magnetic entropy is change and δT_{FWHM} is the full-width at half-maximum of $|\Delta S_M|$. The RCP values for $x = 0.00$ and $x = 0.05$ are found to be 218 and 175 Jkg^{-1}, respectively, under a magnetic field change of 5T. These results ensure that the currently investigated samples present good properties and parameters to be explored for magnetic refrigeration applications as reported by Masrour *et al.* [44-45].

Conclusions

To conclude our scientific studies, we investigated the effect of substitution of manganese by iron on the structural, electrical, magnetic and magnetocaloric properties of $Pr_{0.67}Ba_{0.22}Sr_{0.11}Mn_{1-x}Fe_xO_3$ ($x = 0.00$ and $x = 0.05$) manganite oxides. The desired samples were elaborated using the conventional solid-state reaction. XRD analysis reveals that prepared materials crystallized in the orthorhombic structure with the P_{nma} space group. The $G(T)$ curves are characterized by the appearance of metal–semiconductor transition at low temperature for $x = 0.00$ and 0.05. It is well known that this substitution decreases the Mn^{3+}/ Mn^{4+} ratio and increases the Fe^{3+}/ Mn^{3+} one. The deduced activation energy increases from 22 m eV for $x = 0.00$ to 27meV for $x = 0.05$.

Magnetic measurements show that all samples exhibit a paramagnetic (PM) to ferromagnetic (FM) transition with decreasing temperature. The Curie temperature T_C increased with increasing iron content. The Magnetic measurements show also that all materials present a second order PM to FM magnetic transition. The magnetocaloric effect study indicates that the investigated compounds have the appropriate properties to be suitable candidates to be used as refrigerants.

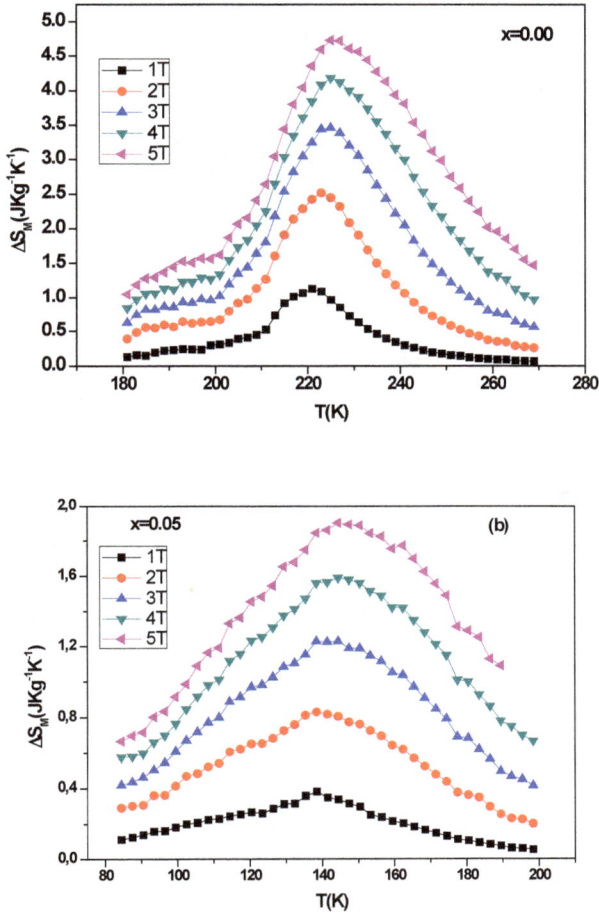

Fig. 8. Temperature dependence of the magnetic entropy change $|\Delta S_M|$ for (a). x = 0.00 and, (b). x = 0.05 at different applied magnetic field.

References

[1]   D. Turki, G. Remeny, S.H. Mahmood, E.K. Hlil, M. Ellouze, F. Halouani, Magnetic contributions to the specific heat of $La_{0.8}Ca_{0.2}Mn_{1-x}Co_xO_3$ perovskite, Materials Research Bulletin, 84 (2016) 245−253. https://doi.org/10.1016/j.materresbull.2016.08.018

[2]   F. Ben Jemaa, S. Mahmood, M. Ellouze, E.K. Hlil, E. Halouani, Critical behaviour and change in universality of $La_{0.67}Ba_{0.22}Sr_{0.11}Mn_{1-x}Co_xO_3$ manganites, Journal of Materials Science: Materials in Electronics, 26(7) (2015) 5381−5392. https://doi.org/10.1007/s10854-015-3085-1

[3]   F. Ben Jemaa, S.H. Mahmood, M. Ellouze, E.K. Hlil, F. Halouani, Structural, magnetic, magnetocaloric, and critical behaviour of selected Ti-doped manganites, Ceramics International, 41 (6) (2015) 8191 −8202. https://doi.org/10.1016/j.ceramint.2015.03.039

[4]   H. Omrani, M. Mansouri, W. Cheikhrouhou Koubaa, M. Koubaa, A. Cheikhrouhou, Structural, magnetic and magnetocaloric investigations in $Pr_{0.6-x}Er_xCa_{0.1}Sr_{0.3}MnO_3$ ($0 \leq x \leq 0.06$) manganites, Journal of Alloys and Compound, 688 (2016) 752 −761. https://doi.org/10.1016/j.jallcom.2016.07.082

[5]   J.M.D. Coey, M. Viret, S. von Molnar, Mixed-valence manganites, Advances in Physics, 48 (2) (1999) 167 −293. https://doi.org/10.1080/000187399243455

[6]   M. B. Salamon, M. Jaime, The physics of manganites: structure and transport, Review of Modern Physics, 73 (2001) 583-628. https://doi.org/10.1103/RevModPhys.73.583

[7]   C. N. R. Rao, A. Arulraj, A. K. Cheetham, B. Raveau, Charge ordering in the rare earth manganates: the experimental situation, Journal of Physics: Condensed Matter, 12(7) (2000) R83-R106. https://doi.org/10.1088/0953-8984/12/7/201

[8]   C. Zener, Interaction between the d-shells in the transition metals. II. Ferromagnetic compounds of manganese with perovskite structure, Physical Review, 82 (3) (1951) 403-405. https://doi.org/10.1103/PhysRev.82.403

[9]   A. J. Millis, P. B. little Wood, and B. I. Shraiman, Double exchange alone does not explain the resistivity of $La_{1-x}Sr_xMnO_3$, Physical Review Letters, 74, (1995) 5144– 5147. https://doi.org/10.1103/PhysRevLett.74.5144

[10] L. M. Rodriguez-Martinez and J. P. Atfield, Cation disorder and size effects in magnetoresistive manganese oxide perovskites, Physical Review B, 54(1996) R15622 −R15625. https://doi.org/10.1103/PhysRevB.54.R15622

[11]   M. Mansouri, H. Omrani, W. Cheikhrouhou-Koubaa, M. Koubaa, A. Madouri, A. Cheikhrouhou, Effect of vanadium doping on structural, magnetic and magnetocaloric properties of $La_{0.5}Ca_{0.5}MnO_3$, Journal of Magnetism and Magnetic Materials, 401, (2016) 593–599. https://doi.org/10.1016/j.jmmm.2015.10.066

[12]   G. M. Keith, C. A. Kirk, K. Sarma, N. M. Alford, E. J. Cussen, M. J. Rosseinsky, D. C. Sinclair, Synthesis, crystal structure, and characterization of $Ba(Ti_{1/2}Mn_{1/2})O_3$: a high permittivity 12R-type hexagonal perovskite, Chemistry of Materials, 16(10) (2004) 2007–2015. https://doi.org/10.1021/cm035317n

[13]   A. M. Smith, H. Duan, A. M. Mohs, S. Nie, Synthesis, bioconjugated quantum dots for in vivo molecular and cellular imaging, Advanced Drug Delivery Reviews, 60 (11) (2008) 1226-1240. https://doi.org/10.1016/j.addr.2008.03.015

[14]   A. Hädicke, W. Krech, Frequency-dependent Cooper pair tunneling in ultra-small superconductor-insulator-superconductor junctions, Physical Review B, 52 (1995) 13526–13531. https://doi.org/10.1103/PhysRevB.52.13526

[15]   R. Waser, Electronic properties of grain boundaries in $SrTiO_3$ and $BaTiO_3$ ceramics, Solid State Ionics, 75 (1995) 89–99. https://doi.org/10.1016/0167-2738(94)00152-I

[16]   S. K. Roy, M. E. Orazem, Analysis of flooding as a stochastic process in polymer electrolyte membrane (PEM) fuel cells by impedance techniques, Journal of Power Sources, 184 (1) (2008) 212–219. https://doi.org/10.1016/j.jpowsour.2008.06.014

[17]   S. Lanfredi, A. C. M. Rodrigues, Impedance spectroscopy study of the electrical conductivity and dielectric constant of polycrystalline $LiNbO_3$, Journal of Applied Physics, 86 (4) (1999) 2215–2219. https://doi.org/10.1063/1.371033

[18]   H. M. Rietveld, A profile refinement method for nuclear and magnetic structures, Journal of Applied Crystallography, 2 (1969) 65–71. https://doi.org/10.1107/S0021889869006558

[19]   T. Roisnel, J. Rodriguez-Carvajal, Computer Program FULLPROF, LLB-LCSIM, May, (2003).

[20]   R. D. Shannon, Revised effective ionic radii and systematic studies of interatomic distances in halides and chalcogenides, Acta Crystallographica A, 32(1976) 751–767. https://doi.org/10.1107/S0567739476001551

[21]   A. G. Souza Filho, J. L. B. Faria, I. Guedes, J. M. Sasaki, P. T. C. Freire, V. N. Freire, J. Mendes Filho, M. Xavier Jr, F. A. O. Cabral, J. H. de Araujo, J. A. P. da

Costa, Evidence of magnetic polaronic states in $La_{0.70}Sr_{0.30}Mn_{1-x}Fe_xO_3$ manganites, Physical Review B, 67 (2003) 052405–9. https://doi.org/10.1103/PhysRevB.67.052405

[22] J. Gutierrez, A. Pena, J. M. Barandiaran, T. Hernandez, J. L. Pizarro, L. Lezama, M. Insausti, Rojo, Structural and magnetic properties of $La_{0.7}Pb_{0.3}(Mn_{1-x}Fe_x)O_3$ (0<~x<~0.3) giant magnetoresistance perovskites, Physical Review B, 61 (2000) 9028–9035. https://doi.org/10.1103/PhysRevB.61.9028

[23] H. Rahmouni, B. Cherif, M. Baazaoui, K. Khirouni, Effects of iron concentrations on the electrical properties of $La_{0.67}Ba_{0.33}Mn_{1-x}Fe_xO_3$, Journal of Alloys and Compound, 575 (2013) 5– 9. https://doi.org/10.1016/j.jallcom.2013.04.077

[24] Jian-Wang Cai, Cong Wang, Bao-Gen Shen, Jian-Gao Zhao, Wen-Shan Zhan, Colossal magnetoresistance of spin-glass perovskite $La_{0.67}Ca_{0.33}Mn_{0.9}Fe_{0.1}O_3$, Applied Physics Letters, 71 (1997) 1727. https://doi.org/10.1063/1.120017

[25] M. Nadeem, M. J Akhtar, A.Y. Khan, R. Shaheen, M. N. Haque, Ac study of 10% Fe-doped $La_{0.65}Ca_{0.35}MnO_3$ material by impedance spectroscopy, Chemical Physics Letters, 366 (2002) 433– 439. https://doi.org/10.1016/S0009-2614(02)01662-7

[26] Xian-yu Wen-xu, Li Bao-he, Qian Zheng-nan, Jin Han-min, Effect of Fe doping in $La_{1-x}Sr_xMnO_3$, Journal of Applied Physics, 86 (1999) 5164–5168. https://doi.org/10.1063/1.371494

[27] H. Rahmouni, A. Selmi, K. Khirouni, N. Kallel, Chromium effects on the transport properties in $La_{0.7}Sr_{0.3}Mn_{1-x}Cr_xO_3$, Journal of Alloys and Compound, 533, (2013) 93–96. https://doi.org/10.1016/j.jallcom.2012.02.123

[28] K. P. Padmasree, D. K. Kanchan, A. R. Kulkani, Impedance and modulus studies of the solid electrolyte system $20CdI2–80[xAg2O–y(0.7V_2O_5–0.3B_2O_3)]$, where 1 $\leq$ x/y $\leq$ 3 Solid State Ionics,177(2006) 475–482. https://doi.org/10.1016/j.ssi.2005.12.019

[29] R. Brahem, H. Rahmouni, N. Farhat, J. Dhahri, K. Khirouni, L. C. Costa, Electrical properties of Sn-doped $Ba_{0.75}Sr_{0.25}Ti_{0.95}O_3$ perovskite, Ceramics International, 40 (2014) 9355–9360. https://doi.org/10.1016/j.ceramint.2014.02.002

[30] N. F. Mott, E. A. Davis, Electronic Process in Non-Crystalline Materials Oxford, 1979.

[31]   H. Rahmouni, R. Jemai, N.Kallel, A. Selmi, K. Khirouni, Titanium effects on the transport properties in $La_{0.7}Sr_{0.3}Mn_{1-x}Ti_xO_3$, Journal of Alloys and Compound, 497 (2010) 1–5. https://doi.org/10.1016/j.jallcom.2010.02.156

[32]   H. Rahmouni, A. Selmi, K. Khirouni, N. Kallel, Chromium effects on the transport properties in $La_{0.7}Sr_{0.3}Mn_{1-x}Cr_xO_3$, Journal of Alloys and Compound, 533 (2012) 93–96. https://doi.org/10.1016/j.jallcom.2012.02.123

[33]   R. Tlili, M. Khelil, M. Bejar, M. Bekri, E. Dhahri, K. Khirouni, Role of gallium ion on the conducting properties of $La_{0.7}(Ba, Sr)_{0.3}Mn_{1-x}Ga_xO_3$ ($x = 0.0, 0.1$ and 0.2) perovskite, Ceramics International, 42(9) (2016) 11256–11258. https://doi.org/10.1016/j.ceramint.2016.04.039

[34]   S. B. Ogale, R. Shreekala, R. Bathe, S. K. Date, S. I. Patil, B. Hannoyer, F. Petit, G. Marest, Transport properties, magnetic ordering, and hyperfine interactions in Fe-doped $La_{0.75}Ca_{0.25}MnO_3$: localization-delocalization transition, Physical Review B, 57 (1998) 7841–7145. https://doi.org/10.1103/PhysRevB.57.7841

[35]   M. S. Sahastrabudhe, S. I. Patil, S. K. Date, D. P. Adhi, S. D. Kulkarni, P.A. Joy, R.N. Bathe, Influence of magnetic (Fe^{+3}) and non-magnetic (Ga^{+3}) ion doping at Mn- site on the transport and magnetic properties of $La_{0.7}Ca_{0.3}MnO_3$, Solid State Communication, 137 (2006) 595–600. https://doi.org/10.1016/j.ssc.2006.01.011

[36]   A. Nasri, E. K. Hlil, A. F. Lehlooh, M. Ellouze, F. Elhalouani, Study of magnetic transition and magnetic entropy changes of $Pr_{0.6}Sr_{0.4}MnO_3$ and $Pr_{0.6}Sr_{0.4}Mn_{0.9}Fe_{0.1}O_3$ compounds, The European Physical Journal Plus,131(2016)110. https://doi.org/10.1140/epjp/i2016-16110-y

[37]   S. Zouari, M. L. Kahn, M. Ellouze, F. Elhalouani, Effect of iron substitution on the physico-chemical properties of $Pr_{0.6}La_{0.1}Ba_{0.3}Mn_{1-x}Fe_xO_3$ manganites (with $0 \leq x \leq 0.3$), The European Physical Journal Plus, 130 (2015) 177. https://doi.org/10.1140/epjp/i2015-15177-2

[38]   A. Arrott, Criterion for Ferromagnetism from Observations of Magnetic Isotherms, Physical Review, 108 (1957) 1394. https://doi.org/10.1103/PhysRev.108.1394

[39]   S. K. Banerjee, On a generalized approach to first and second order magnetic transitions Physical Letters, 12 (1964) 16-17. https://doi.org/10.1016/0031-9163(64)91158-8

[40] R. D. Michael, J. J. Ritter, R. D. Shull, Enhanced magnetocaloric effect in $Gd_3Ga_{5-x}Fe_xO_{12}$, Journal of Applied Physics, 73 (1993) 6946. https://doi.org/10.1063/1.352443

[41] K. A. Gschneidner, V. K. Pecharsky, A. O. Tosko, Recent developments in magnetocaloric materials, Reports on Progress in Physics, 68 (2005) 1479–1539. https://doi.org/10.1088/0034-4885/68/6/R04

[42] M-H. Phan, S-C.Yu, Review of the magnetocaloric effect in manganite materials, Journal of Magnetism and Magnetic Materials, 308 (2007) 325–340. https://doi.org/10.1016/j.jmmm.2006.07.025

[43] A. S. Erchidi Elyacoubi, R. Masrour, A. Jabar, Magnetocaloric effect and magnetic properties in $SmFe_{1-x}Mn_xO_3$ perovskite: Monte Carlo simulations, Solid State Communications, 271 (2018) 39–43. https://doi.org/10.1016/j.ssc.2017.12.015

[44] R. Masrour, A. Jabar, A. Benyoussef, M. Hamedoun, E.K. Hlil, Monte Carlo simulation study of magnetocaloric effect in $NdMnO_3$ perovskite, Journal of Magnetism and Magnetic Materials, 401 (2016) 91–95. https://doi.org/10.1016/j.jmmm.2015.10.019

[45] R. Masrour, A. Jabar, H. Khlif, F. Ben Jemaa, M. Ellouze, E.K. Hlil, Experiment, mean field theory and Monte Carlo simulations of the magnetocaloric effect in $La_{0.67}Ba_{0.22}Sr_{0.11}MnO_3$ compound, Solid State Communications, 268, (2017) 64–69. https://doi.org/10.1016/j.ssc.2017.10.003

Chapter 2

Effect of Al-substitution on Structural and Magnetic Properties of $Er_3Fe_{5-x}Al_xO_{12}$ Garnets

Ibrahim Bsoul[1,a], Khaled Hawamdeh[1,b], Sami H. Mahmood[2,c]

[1]Physics Department, Al al-Bayt University, Mafraq 13040, Jordan

[2]Physics Department, The University of Jordan, Amman 11942, Jordan

[a]ibrahimbsoul@yahoo.com, [b]khaled_hawamda@yahoo.com, [c]s.mahmood@ju.edu.jo

Abstract

$Er_3Fe_{5-x}Al_xO_{12}$ ($0.0 \leq x \leq 0.8$) garnets were prepared at 1300 °C by ball milling method. Rietveld refinement of the samples revealed a garnet structure with *Ia3d* symmetry. The lattice parameter, cell volume, X-ray density and magnetization of the prepared garnets decreased with the increase of Al content (x). The coercivity of the garnets increased with x, but remained generally low, being below 20 Oe. Low temperature magnetic measurements versus temperature indicated that the magnetization of $x = 0.0$ exhibited a compensation temperature at -186° C, however, $x = 0.8$ exhibited a minimum at a higher temperature of -134° C.

Keywords

Rare Earth Iron Garnet, Structural Characteristics, Magnetic Properties, Compensation Temperature, Thermomagnetic Curves

Contents

1.   Introduction

Ferrimagnetic garnets exemplified by yttrium iron garnet (YIG) were discovered in 1956, and received considerable interest due to their low dielectric losses and remarkable performance in microwave devices and magnetic bubbles for digital memories [1, 2], in addition to their importance in the field of fundamental magnetism. The garnets have a cubic crystal structure with space group *Ia3d* and chemical formula $\{R_3^{3+}\}_c[Fe_2^{3+}]_a(Fe_3^{3+})_d O_{12}^{2-}$, where R^{3+} is a trivalent rare-earth ion occupying dodecahedral (c) sites, and Fe^{3+} ions occupy octahedral [a-sublattice] and tetrahedral (d-sublattice) sites in the garnet lattice. The magnetic properties of the garnet are determined by the strength of the superexchange interactions between magnetic ions in the various sublattices [3-6]. Since Y^{3+} ion does not have magnetic moment, the magnetic properties of YIG are completely determined by a–d superexchange interactions between Fe^{3+} ions at tetrahedral and octahedral sites, resulting in a net moment of 5 μ_B per molecule at 0 K. If R^{3+} is a rare earth ion with net magnetic moment, however, the a–d superexchange interaction is much stronger than a–c and d–c interactions between R^{3+} and Fe^{3+} ions, and the rare earth [RE] sublattice couples antiferromagnetically with the net moment of the Fe^{3+} sublattices. When R^{3+} is a rare earth ion such as Gd^{3+} through Yb^{3+}, the rare earth sublattice magnetization is higher than the net magnetization of the Fe^{3+} sublattices, and the magnetization of the rare earth iron garnet is given by:

$$M(T) = M_c(T) - [M_d(T) - M_a(T)] \qquad (1)$$

Here the effect of M_c on M_a and M_d through a–c and d–c interactions was neglected [1]. Since the magnetization of the rare earth sublattice decreases faster than that of the iron sublattices, the magnetization of the rare earth iron garnet as a function of temperature exhibits a compensation temperature at which the magnetization vanishes. Generally, the compensation temperature is below room temperature, and above the compensation temperature, the net magnetization of the iron sublattices *[M_d (T) — M_a(T)]* exceeds that of the RE sublattice, resulting in a rise of the magnetization [1].

The partial substitution of Fe^{3+} ions by non-magnetic ions such as Al^{3+} ions was investigated by different researchers. The substitution of Al^{3+} ions in YIG was first reported to occur preferentially at tetrahedral (d) sites [7, 8], with a decreasing tendency as the concentration of Al^{3+} ions increases. The results of later detailed studies indicated that the fraction of Al^{3+} ions at the tetrahedral sites decreased smoothly from 1.0 at $x = 0$ to 0.6 at $x = 5$ [9, 10]. In addition, the variations of the magnetization of $Y_3Al_{1.25}Fe_{3.75}O_{12}$ with heat treatment indicated that the Al^{3+} ions must have redistributed themselves among tetrahedral and octahedral sites as a result of the heat treatment [11]. More recently, room temperature (RT) Mössbauer spectra of $Y_3Fe_{5-x}Al_xO_{12}$ ($x = 0.0, 0.25, 0.5, 0.75,$ and 1.0) garnets prepared by sol-gel method revealed Fe^{3+} and Al^{3+} cationic distribution at tetrahedral sites [12]. On the other hand, evidence of the distribution of substituted ions at octahedral (a) sites was provided based on the results of Mössbauer spectroscopy [13, 14]. On the other hand, Mössbauer spectroscopy indicated that the tetrahedral sites were the preferred sites of Al^{3+} ions in substituted holmium iron garnet [15]. The reported variations of site selectivity of Al^{3+} ions could be an indication of the sensitivity to the synthesis route, heat treatment, and the concentration of Al^{3+} ions in the compound.

Different synthesis routes and heat treatments were adopted for the preparation of iron garnets, including sol-gel method [12, 16, 17], co-precipitation method [18], and ball milling [13, 14, 19-21]. Even though YIG garnets with different scenarios of substitution for Fe^{3+}, and/ or Y^{3+} ions by other rare earth ions were synthesized and carefully characterized, to our best knowledge, no reports were provided on the synthesis and characterization of erbium iron garnet with partial substitution of Fe^{3+} ions by Al^{3+} ions. The present study is concerned with the preparation and investigation of the structural and magnetic properties of $Er_3Fe_{5-x}Al_xO_{12}$ ($0.0 \leq x \leq 0.8$) synthesized by solid state reaction. The structural refinement was used to investigate the site selectivity of Al^{3+} ions, and the magnetization measurements using vibrating sample magnetometry (VSM) were used to study the effect of Al-substitution on the magnetic properties of the prepared garnets.

2.   Experimental procedure

$Er_3Fe_{5-x}Al_xO_{12}$ garnets with $x = 0.0, 0.2, 0.4, 0.6$ and 0.8 were synthesized by solid state reaction method. Stoichiometric amounts of high purity Er_2O_3, Fe_2O_3 and Al_2O_3 starting powders were loaded into a hardened stainless-steel cup with ball to powder ratio of 8:1, and milled at 250 rpm for 4 h. The resulting precursor was pelletized and sintered at 1300 °C for 2 h. X-ray diffraction (XRD) patterns were collected using Philips PW 1720 X-ray diffractometer operating at (40 kV, 40 mA), with Cu-K$_\alpha$ radiation ($\lambda = 1.5405$ Å). The samples were scanned over the angular range $15° < 2\theta < 75°$ with 0.02° scanning step and speed of 1°/min. The XRD patterns were analyzed using X'pert High Score 2.0.1 software for phase identification, and Rietveld refinement of the crystal structure was performed using FullProf suite 2000 software. The magnetic measurements at room temperature or above were performed using Vibrating Sample Magnetometer (VSM MicroMag 3900, Princeton Measurements Cooperation), operating at a maximum applied magnetic field of $\pm$ 10 kOe, whereas, the magnetization measurements below room temperature were performed using a Quantum Design 9T- PPMS Ever Cool-II magnetometer.

3.   Results and discussion

3.1   XRD measurements

The structural phases of the investigated samples were identified by X'Pert HighScore software. XRD pattern of $Er_3Fe_5O_{12}$ sample is shown in Fig. 1, and the peak positions and intensities determined by the software are shown in Fig. 2, along with the standard pattern (ICDD: 01-081-0131) for $Er_3Fe_5O_{12}$ (ErIG). It is clear from these figures that the sample with $x = 0.0$ consists of a pure EIG phase with space group $Ia3d$; no secondary phases were observed. Similarly, preliminary analyses of the XRD patterns of all Al-substituted samples (not shown for brevity) indicated the presence of a major garnet phase consistent with the standard pattern for ErIG.

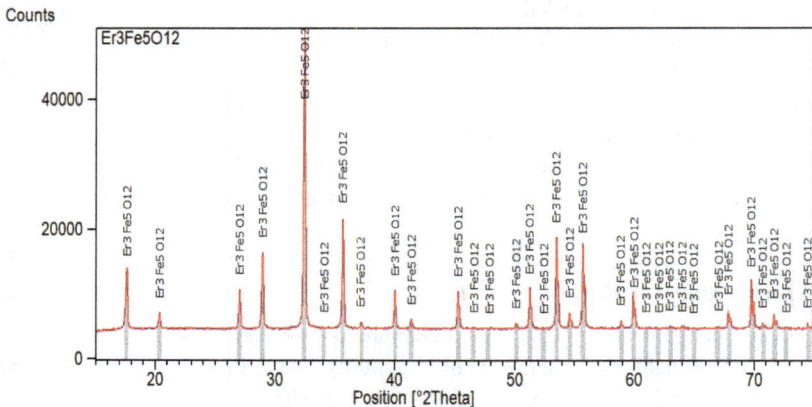

Fig. 1. XRD pattern for $Er_3Fe_5O_{12}$ (x = 0.0) sample.

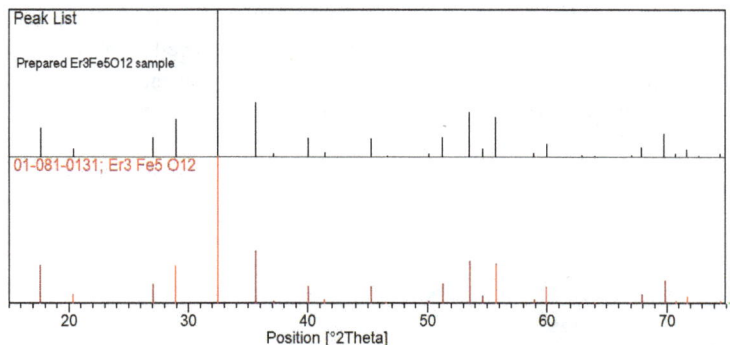

Fig. 2. XRD intensity profile of $Er_3Fe_5O_{12}$ (x = 0.0) sample (upper panel), and that of the standard (ICDD: 01-081-0131) pattern (lower panel).

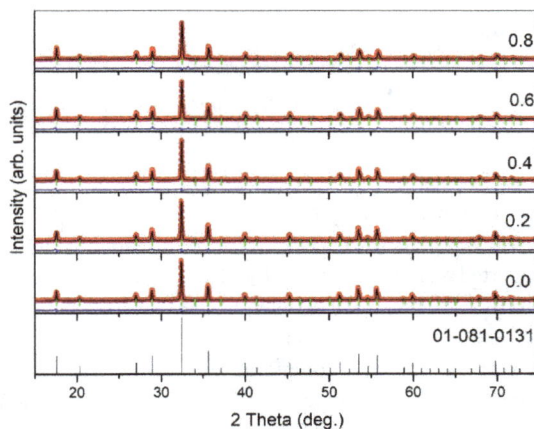

Fig.3. Rietveld refinement of the XRD patterns of all $Er_3Fe_{5-x}Al_xO_{12}$ samples.

Fig. 3 shows Rietveld refinement of the XRD patterns for all $Er_3Fe_{5-x}Al_xO_{12}$ samples, where the experimental data are represented by red circles, the theoretical pattern by black line, and the residual difference curve by blue line. The vertical (green) bars shown below the XRD patterns represent Bragg peak positions. From this figure, we can see that the theoretical pattern is in good agreement with the experimental data, where the residual difference curve is a straight horizontal line with small ripples only around the main structural peak positions. The lattice constant (a), the unit cell volume and the density of all examined samples were obtained from the output file of the refined XRD patterns.

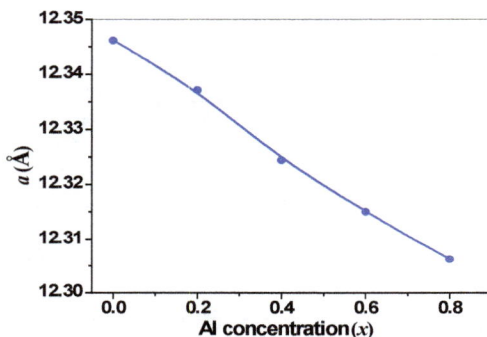

Fig.4. Lattice parameter (a) for $Er_3Fe_{5-x}Al_xO_{12}$ samples as a function of Al concentration (x).

Fig. 4 shows a monotonic decrease of the lattice constant (a) for $Er_3Fe_{5-x}Al_xO_{12}$ as x increases. The lattice constant of 12.346 Å for the sample with $x = 0.0$ is in good agreement with reported value for $Er_3Fe_5O_{12}$ [1], and the observed decrease with the increase of Al^{3+} content is consistent with the reported decrease of lattice constant of $Y_3Fe_{5-x}Al_xO_{12}$ with the increase of x [19]. This decrease, which also results in a monotonic decrease of the unit cell volume as shown in Fig. 5, is associated with the smaller ionic radius of Al^{3+} ion compared to Fe^{3+} ion. Also, a monotonic decrease in X-ray density was observed as x increased (Fig. 6), contrary to the expected increase due to the reduction of cell volume. This is opposite to what was observed with Dy-substituted Y in YIG [20], and can be associated with the higher rate of decrease of the molecular weight in comparison with the rate of decrease of the cell volume as x increased.

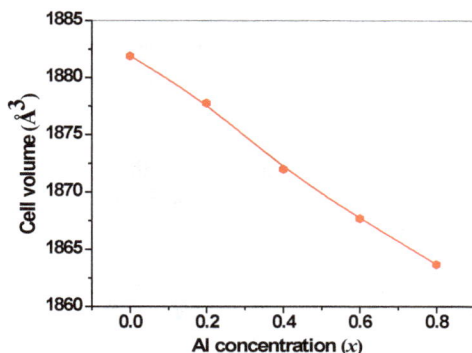

Fig. 5. Unit cell volume (V) for $Er_3Fe_{5-x}Al_xO_{12}$ samples as a function of Al concentration (x).

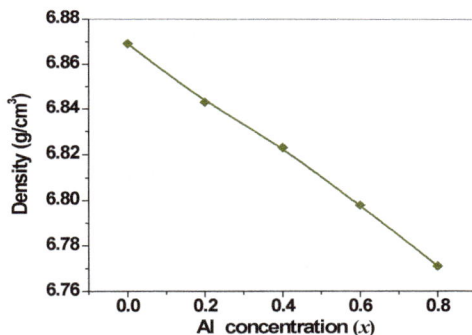

Fig. 6. Density of $Er_3Fe_{5-x}Al_xO_{12}$ samples as a function of Al concentration (x).

Further microstructural information was obtained from the output results (CIF file) of the Rietveld refinement of the XRD patterns. Diamond software was used to construct the crystal structure and oxygen polyhedra surrounding the metal ions, and the results for the sample with $x = 0.0$ are shown in Fig. 7. Three magnetic sublattices were observed in ErIG corresponding to magnetic ions occupying different crystallographic sites in the garnet lattice. Two of these arise from Fe^{3+} ions occupying 16a (octahedral) sites and 24d (tetrahedral) sites, whereas the third corresponds to Er^{3+} ions occupying 24c (dodecahedral) sites. For comparison purposes, we also constructed the unit cell (Fig. 8) for the sample $Er_3Fe_{5-x}Al_xO_{12}$ with $x = 0.8$. Similar constructions were obtained for the remaining Al-substituted samples (results are not shown to avoid redundancy). The fractions of Fe^{3+} ions and Al^{3+} ions at tetrahedral and octahedral sites in the investigated samples were also determined using the refinement output CIF file. The results indicated that the replacement of Fe^{3+} ions by Al^{3+} ions takes place only at tetrahedral (d) sites, and that the experimental values of Al^{3+} concentration in the investigated samples are in excellent agreement (within 0.04 % variation/error) with the corresponding stoichiometric values in the starting powders.

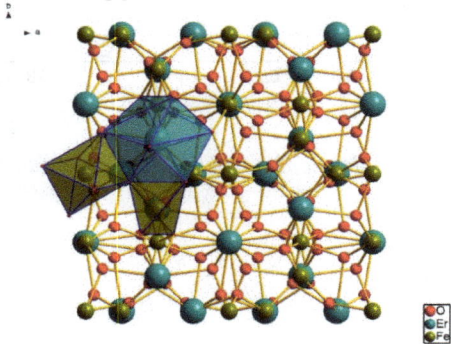

Fig.7. Crystal structure of $Er_3Fe_5O_{12}$ sample showing the polyhedra corresponding to the three magnetic sublattices in the garnet structure.

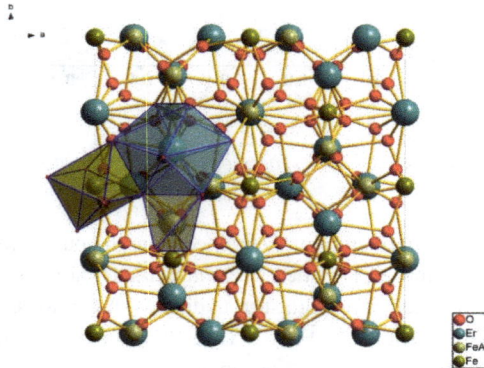

Fig.8. Crystal structure of $Er_3Fe_{4.2}Al_{0.8}O_{12}$ sample showing the polyhedra corresponding to the three magnetic sublattices in the garnet structure.

The substitution of Fe^{3+} ions by Al^{3+} ions resulted in small variations of the bond lengths and bond angles in the garnet structure. The trends of such variations became clear by comparing the parameters of the two extreme samples (with $x = 0.0$ and 0.8) as shown in Table 1. The observed structural parameters of the unsubstituted ErIG sample are in good agreement with the parameters of YIG [1], even though the ionic radius of Er^{3+} at dodecahedral sites (1.004 Å) is somewhat smaller than that of Y^{3+} (1.019 Å) [22]. An obvious reduction of the $Fe^{3+}(d)$–O^{2-} bond length from 1.851 Å for the sample with $x = 0.0$, to 1.809 Å for the sample with $x = 0.8$ was observed, whereas the $Fe^{3+}(a)$–O^{2-} bond length demonstrated smaller, unsystematic variations with Al substitution. The $Fe^{3+}(a)$–O^{2-}–$Fe^{3+}(d)$ bond angle of $125.88°$ for the sample with $x = 0.0$ is in very good agreement with that reported for YIG [1, 23], and this angle increased only slightly with Al substitution. Consequently, the reduction of the tetrahedral bond length is most probably responsible for the observed monotonic decrease of the lattice constant with increasing Al concentration.

Table 1. Refined bond lengths and bond angles $Er_3Fe_5O_{12}$ and $Er_3Fe_{4.2}Al_{0.8}O_{12}$.

	$Er_3Fe_5O_{12}$	$Er_3Fe_{4.2}Al_{0.8}O_{12}$
$Fe^{3+}(a)$–O^{2-} (Å)	2.023	2.045
$Fe^{3+}(d)$–O^{2-} (Å)	1.851	1.809
$Er^{3+}(c)$–O^{2-} (Å)	2.422	2.438
$Fe^{3+}(a)$–O^{2-}–$Fe^{3+}(d)$	125.88°	126.25°
$Er^{3+}(c)$–O^{2-}–$Fe^{3+}(a)$	101.49°	99.85°
$Er^{3+}(c)$–O^{2-}–$Fe^{3+}(d)$	123.89°	124.38°

The crystallite size was determined from the reflection at $2\theta = 55.8°$ using Scherrer's relation. No systematic variations were observed with the increase of Al concentration, indicating that the Al^{3+} substitution for Fe^{3+} may not have an appreciable effect on the crystallinity of $Er_3Fe_{5-x}Al_xO_{12}$ garnets. The crystallite size was found to be 70 ± 8 nm for all samples, and the fluctuations around an average value, with no systematic variation, may be an indication that the microstructural characteristics did not change appreciably with Al concentration. This is consistent with the observed small, insignificant changes in bond lengths and bond angles with Al substitution.

3.2   Magnetic measurements

Fig. 9 shows the hysteresis loops for $Er_3Fe_{5-x}Al_xO_{12}$ samples measured with a 100 Oe field step. The curves revealed a soft magnetic nature of erbium iron garnets with saturation magnetization of the unsubstituted sample of 15.2 emu/g, which is in good agreement with that of 14.2 emu/g reported for ErIG [24]. The saturation magnetization decreased monotonically with the increase of Al concentration as revealed by Fig. 10, reaching a value of 5.6 emu/g at $x = 0.8$.

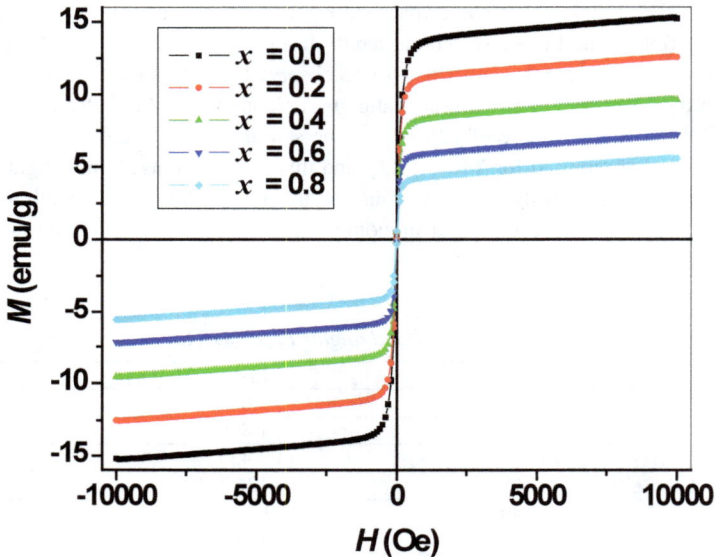

Fig.9. Hysteresis loops of $Er_3Fe_{5-x}Al_xO_{12}$ (x = 0.0, 0.2, 0.4, 0.6, 0.8) samples.

Fig. 9 shows the hysteresis loops for $Er_3Fe_{5-x}Al_xO_{12}$ samples measured with a 100 Oe field step. The curves revealed a soft magnetic nature of erbium iron garnets with saturation magnetization of the unsubstituted sample of 15.2 emu/g, which is in good agreement with that of 14.2 emu/g reported for ErIG [24]. The saturation magnetization decreased monotonically with the increase of Al concentration as revealed by Fig. 10, reaching a value of 5.6 emu/g at $x = 0.8$.

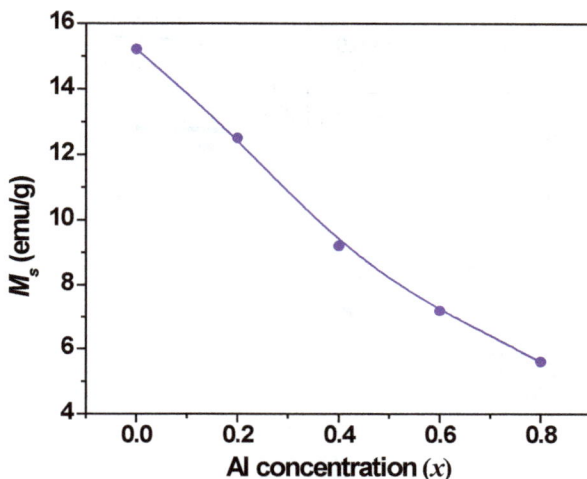

Fig. 10. Saturation magnetization (M_s) of $Er_3Fe_{5-x}Al_xO_{12}$ as a function of Al concentration (x).

The decrease of magnetization with the increase of Al content is understood if one recalls that the room temperature magnetization of ErIG is dominated by the net magnetization of the iron sublattices as mentioned in the introduction section. Accordingly, the substitution of Al at tetrahedral sites resulted in a decrease in saturation magnetization as a consequence of the decrease of the magnetization of the tetrahedral sublattice. Thus, the magnetic data supports the structural results concerning the site selectivity of Al^{3+} ions.

Since the coercivity (H_c) of all samples was rather low, hysteresis loops in a maximum applied field of 1 kOe were measured with a 2 Oe field step, and the results are shown in Fig. 11, along with an expanded view to investigate the effect of Al substitution on the width of the loop. The coercivity of all samples was obtained directly from the hysteresis loops, and the results in Fig. 12 indicated a low coercivity of 5.8 Oe for the unsubstituted sample, and monotonic increase with the increase of Al content. However, the coercivity remained below 20 Oe for all samples.

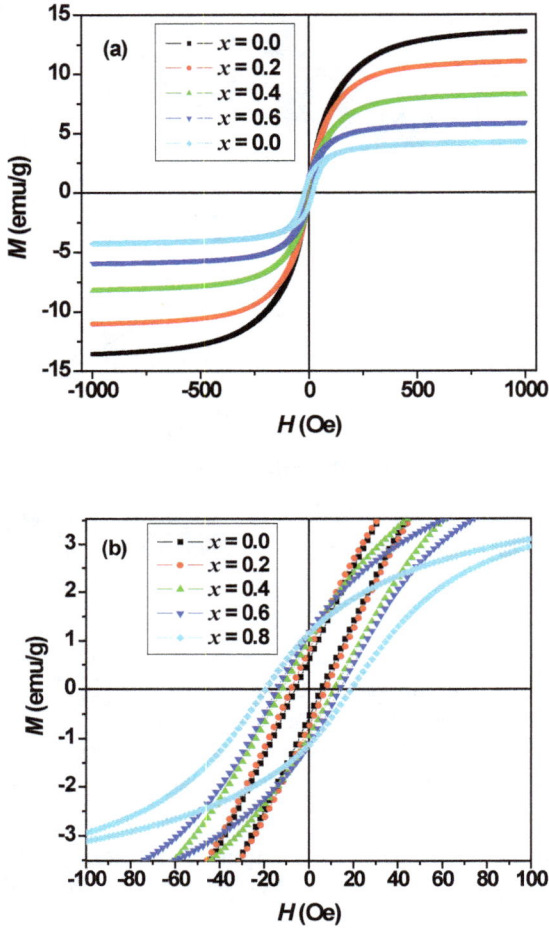

Fig. 11. Hysteresis loops for $Er_3Fe_{5-x}Al_xO_{12}$ samples measured with a field step of 2 Oe, under applied field of (a). ±1 kOe, (b). ±100 Oe.

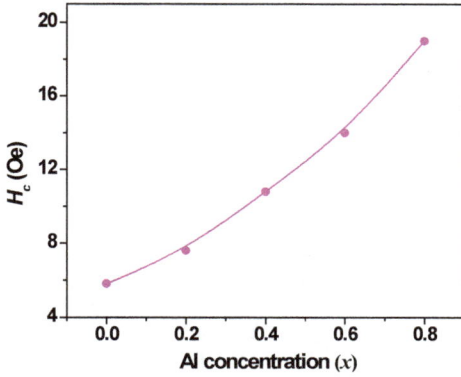

Fig. 12. Coercivity as a function of Al concentration for $Er_3Fe_{5-x}Al_xO_{12}$ samples.

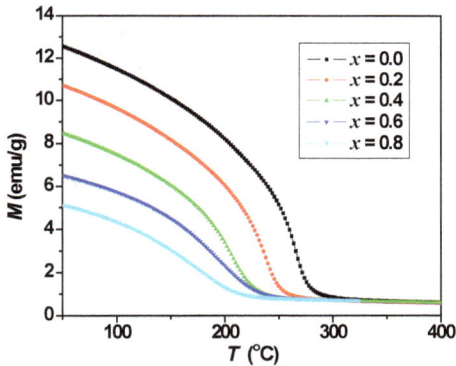

Fig. 13. Magnetization as a function of temperature for $Er_3Fe_{5-x}Al_xO_{12}$ samples at a constant applied field of 8 kOe.

3.3   Temperature dependence of the magnetic properties

The temperature dependence of the magnetization $M(T)$ was investigated by carrying out magnetization measurements as a function of temperature in a constant applied field of 8000 Oe. Fig. 13 shows magnetization curves as a function of temperature for $Er_3Fe_{5-x}Al_xO_{12}$ samples. The thermomagnetic curves demonstrated a non-linear decrease of magnetization with increasing temperature, and a sharper drop in magnetization was observed as the ferrimagnetic-paramagnetic critical transition temperature (Curie temperature; T_c) was approached. The critical transition temperature was obtained from the inflection point of the thermomagnetic curve revealed by the peak position in the derivative curve (dM/dT) as shown in Fig. 14.

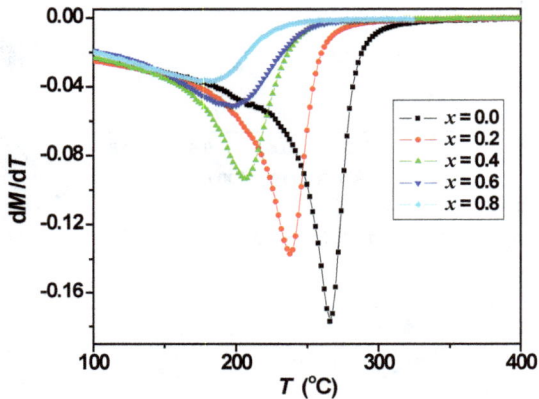

Fig.14. Derivative of the magnetization (dM/dT) of $Er_3Fe_{5-x}Al_xO_{12}$ samples as a function of temperature (T).

Curie temperature (T_c) as a function of Al^{3+} concentration in $Er_3Fe_{5-x}Al_xO_{12}$ is shown in Fig. 15, revealing a sharp decrease in T_c as result of Al^{3+} substitution. The structural analysis revealed a decrease of the Fe^{3+}(d)–O^{2-} bond length, and a small increase of the Fe^{3+}(a)–O^{2-}–Fe^{3+}(d) bond angle as a result of Al^{3+} substitution for Fe^{3+}, which, contrary to observed results, should lead to an increase in the strength of the a–d superexchange interaction, and a consequent increase in T_c. The observed decrease of T_c cannot therefore be attributed to variations of the interionic distances and bond angles, and can be associated with the weakening of the a–d superexchange interactions as a result of the decrease of the magnetization of the d-sublattice due to the replacement of magnetic Fe^{3+} ions by non-magnetic Al^{3+} ions in this sublattice.

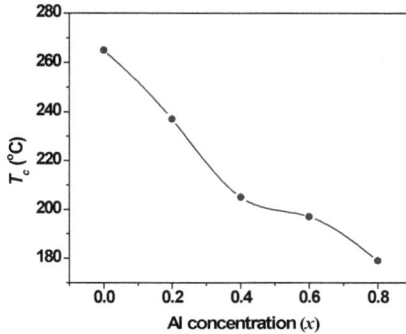

Fig. 15. Curie temperature (T_c) as a function of Al concentration (x) in $Er_3Fe_{5-x}Al_xO_{12}$ samples.

3.4 Low temperature measurements

The temperature dependence of the magnetization $M(T)$ below room temperature was investigated by measuring the magnetization of the samples as a function of temperature in a constant applied field of 100 Oe. Fig. 16 shows the magnetization curve as a function of temperature for $Er_3Fe_5O_{12}$ sample. The magnetization decreased sharply to zero with the increase of temperature, and then increased, exhibiting a compensation temperature $T_{comp} = -186$ °C, which is in good agreement with previously reported results [25]. At temperatures below T_{comp}, the magnetization of the Er^{3+} sublattice is larger than the net magnetization of the Fe^{3+} sublattices, resulting in a net magnetization of $Er_3Fe_5O_{12}$ parallel to the magnetization of the Er^{3+} sublattice. As the temperature increases, the magnetization of the Er^{3+} sublattice decreases faster than that of Fe^{3+} sublattices, so that the magnetization of the c-sublattice becomes equal (and antiparallel) to the net magnetization of the Fe^{3+} sublattices, and the magnetization of the sample vanishes at the compensation temperature. Above T_{comp}, the net magnetization of the Fe^{3+} sublattices continues to decrease at a slower rate compared to the magnetization of the Er^{3+} sublattice, resulting in a rise in net magnetization of the sample, which becomes parallel to the net magnetization of the Fe^{3+} sublattices in this temperature regime. The increase in magnetization continues with the increase of temperature, reaching a maximum value due to competition with the thermal (decreasing) effects, and then starts decreasing to zero at the Curie temperature [25].

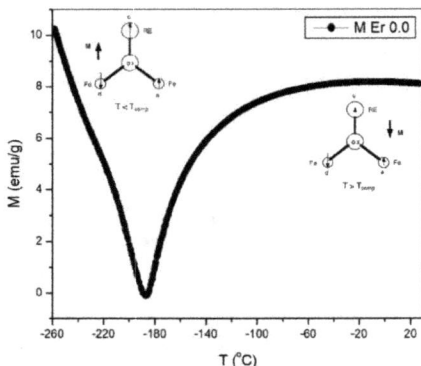

Fig. 16. Magnetization (M) as a function of temperature (T) for $Er_3Fe_5O_{12}$ sample.

The magnetization as a function of temperature for Al-substituted ErIG was also investigated in the temperature range below room temperature. In order to investigate the maximum effect of Al-substitution on the temperature dependence of the magnetic behavior of EIG, the thermomagnetic curves were measured for the two end samples with $x = 0.0$ and 0.8, and the results are shown in Fig. 17. The magnetization curve of the sample with $x = 0.8$ exhibited a minimum at a higher temperature of $-134°$ C compared to the unsubstituted sample, with obvious flattening and broadening of the minimum peak structure. This behavior may be explained in terms of significant changes in the magnetic structure and the temperature dependence of the magnetization of the magnetic sublattices as a result of the relatively high Al substitution in the tetrahedral sublattice. At this substitution level, the net magnetization of Fe^{3+} is low, which leads to the following effects: (1) The reduction of the tetrahedral sublattice magnetization reduces the strength of the a–d superexchange interaction, which may result in spin canting and a faster drop of the magnetization of the Fe^{3+} sublattices with the increase in temperature. (2) Since the c-sublattice is magnetized by coupling of the rare earth magnetization with the net magnetization of Fe^{3+} sublattices, the net magnetization of the c-sublattice is also expected to be lowered relative to the unsubstituted sample, and exhibit a higher degree of spin canting of the rare earth sublattice [26]. (3) This significant change in magnetic structure should, in principle, influence the temperature dependence of the magnetization of the magnetic sublattices. The higher magnetization of the Al-substituted sample relative to the unsubstituted sample, even up to temperatures exceeding the compensation temperature of the unsubstituted sample, is an indication of the dominance of the rare earth magnetic sublattice over a wider temperature range in the Al-substituted sample. This increase in magnetization in the low temperature range is a result of the significant reduction of the Fe^{3+} sublattice magnetization due to the depletion of Fe^{3+} ions at the tetrahedral site. Also, it was observed that the net magnetization of the Al-substituted sample did not vanish completely at the point of minimum magnetization, which requires further investigation.

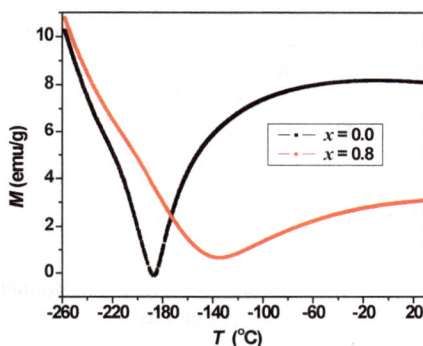

Fig.17. Magnetization (M) as a function of temperature (T) for $Er_3Fe_{5-x}Al_xO_{12}$ (x = 0.0 and 0.8) samples.

Conclusions

X-ray diffraction of the Al-substituted $Er_3Fe_{5-x}Al_xO_{12}$ (x = 0.0 to 0.8) samples were prepared by ball milling and sintering at 1300 °C indicated crystallization of the garnet phase with decreasing lattice parameter, cell volume, and X-ray density as a result of the increase of Al-substitution. The magnetization decreased significantly with the increase of Al content, and the coercivity increased, while remaining below 20 Oe for all samples. The Curie temperature also decreased appreciably with the increase of Al content, which is an indication of the reduction of the strength of superexchange interaction resulting from the reduction of the tetrahedral moment as a consequence of Al substitution. The unsubstituted sample exhibited the usual compensation point of erbium iron garnet, while the thermomagnetic curve of the sample with x = 0.8 exhibited a minimum at a higher temperature, with significant broadening. This could be an indication that Al substitution results in a significant change of the temperature dependence of the magnetic sublattices, in addition to the reduction of the overall magnetization above the compensation point of EIG.

References

[1]   M.A. Gilleo, Ferromagnetic insulators: Garnets, in: P. E, Wohlfarth (Ed.) Ferromagnetic materials, North-Holland, Amstrdam, 1980, pp. 1-53. https://doi.org/10.1016/S1574-9304(05)80102-6

[2]   S.H. Mahmood, Permanent Magnet Applications, in: S.H. Mahmood, I. Abu-Aljarayesh(Eds.) Hexaferrite Permanent Magnetic Materials, Materials Research Forum LLC, Millersville, PA, 2016, pp. 153-165. https://doi.org/10.21741/9781945291074

[3]   A. Paoletti, Physics of Magnetic Garnet, North Holland, Amstrdam, 1978.

[4]   A. R. Jha, Rare earth materials: properties and applications, CRC Press, Florida, 2014. https://doi.org/10.1201/b17045

[5]   M.S. Lataifeh, Room temperature magnetization measurements of some substituted rare earth iron garnets, Applied Physics A, 92 (2008) 681–685. https://doi.org/10.1007/s00339-008-4616-x

[6]   H. Xu, H. Yang, W. Xu, L. Yu, Magnetic properties of Bi-doped $Y_3Fe_5O_{12}$ nanoparticles, Current Applied Physics, 8 (2008) 1–5. https://doi.org/10.1016/j.cap.2007.04.002

[7]   M. Gilleo, S. Geller, Substitution for Iron in Ferrimagnetic Yttrium-Iron Garnet, Journal of Applied Physics, 29 (1958) 380–381. https://doi.org/10.1063/1.1723143

[8]   M. Gilleo, S. Geller, Magnetic and Crystallo graphic Properties of Substituted Yttrium-Iron Garnet, $3Y_2O_3 \cdot x\ M_2O_3 \cdot (5-x)Fe_2O_3$, Physical Review, 110 (1958) 73–78. https://doi.org/10.1103/PhysRev.110.73

[9]   S. Geller, H. Williams, G. Espinosa, R. Sherwood, Importance of intrasublattice magnetic interactions and of substitutional ion type in the behavior of substituted yttrium iron garnets, The Bell System Technical Journal, 43 (1964) 565–623. https://doi.org/10.1002/j.1538-7305.1964.tb00998.x

[10]   S. Geller, Crystal chemistry of the garnets, Zeitschrift für Kristallographie-Crystalline Materials, 125 (1967) 1–47. https://doi.org/10.1524/zkri.1967.125.125.1

[11]   H. Cohen, R. Chegwidden, Control of the Electromagnetic Properties of the
        Polycrystalline Garnet $Y_3Fe_{3.75}Al_{1.25}O_{12}$, Journal of Applied Physics, 37 (1966)
        1081–1082. https://doi.org/10.1063/1.1708343

[12]   C. S. Kim, B.K. Min, S.J. Kim, S.R. Yoon, Y.R. Uhm, Crystallographic and
        magnetic properties of $Y_3Fe_{5-x}Al_xO_{12}$, Journal of Magnetism and Magnetic
        Materials, 254 (2003) 553–555. https://doi.org/10.1016/S0304-8853(02)00864-8

[13]   Z. A. Motlagh, M. Mozaffari, J. Amighian, A.F. Lehlooh, M. Awawdeh, S.
        Mahmood, Mössbauer studies of $Y_3Fe_{5-x}Al_xO_{12}$ nanopowders prepared by
        mechanochemical method, Hyperfine Interactions, 198 (2010) 295–302.
        https://doi.org/10.1007/s10751-010-0234-z

[14]   A.F. Lehlooh, S. Mahmood, M. Mozaffari, J. Amighian, Mössbauer spectroscopy
        study on the effect of Al-Cr Co-substitution in yttrium and yttrium-gadolinium
        iron garnets, Hyperfine Interactions, 156-157 (2004) 181–185.
        https://doi.org/10.1023/B:HYPE.0000043224.54154.06

[15]   M.S. Lataifeh, A.F.D. Lehlooh, S. Mahmood, Mössbauer spectroscopy of Al
        substituted Fe in holmium iron garnet, Hyperfine Interactions, 122 (1999) 253–
        258. https://doi.org/10.1023/A:1012630730577

[16]   C. S. Kim, B. K. Min, S.Y. An, Y.R. Uhm, Mössbauer studies of $Y_3Fe_{4.75}Al_{0.25}O_{12}$,
        Journal of Magnetism and Magnetic Materials, 239 (2002) 54–56.
        https://doi.org/10.1016/S0304-8853(01)00625-4

[17]   E. Garskaite, K. Gibson, A. Leleckaite, J. Glaser, D. Niznansky, A. Kareiva, H.-J.
        Meyer, On the synthesis and characterization of iron-containing garnets
        ($Y_3Fe_5O_{12}$, YIG and $Fe_3Al_5O_{12}$, IAG), Chemical Physics, 323 (2006) 204-210.
        https://doi.org/10.1016/j.chemphys.2005.08.055

[18]   D. Rodic, M. Mitric, R. Tellgren, H. Rundlof, The cation distribution and
        magnetic structure of $Y_3Fe_{(5-x)}Al_xO_{12}$, Journal of Magnetism and Magnetic
        Materials, 232 (2001) 1–8. https://doi.org/10.1016/S0304-8853(01)00211-6

[19]   Q. I. Mohaidat, M. Lataifeh, Kh. Hamasha, S.H. Mahmood, I. Bsoul, M.
        Awawdeh, The structural and magnetic properties of aluminum substituted
        yttrium iron garnet, Materials Research, 21 (2018) 20170808 (pp. 1-7).
        https://doi.org/10.1590/1980-5373-mr-2017-0808

[20]   M. Lataifeh, Q.I. Mohaidat, S.H. Mahmood, I. Bsoul, M. Awawdeh, I. Abu-
        Aljarayesh, M. Altheeba, Structural, Mössbauer spectroscopy, magnetic

properties, and thermal measurements of $Y_{3-x}Dy_xFe5O_{12}$, Chinese Physics B, 27 (2018) 107501 (pp. 1-7). https://doi.org/10.1088/1674-1056/27/10/107501

[21] Q. I. Mohaidat, M. Lataifeh, S.H. Mahmood, I. Bsoul, M. Awawdeh, Structural, Mössbauer Effect, Magnetic, and Thermal Properties of Gadolinium Erbium Iron Garnet System $Gd_{3-x}Er_xFe_5O_{12}$, Journal of Superconductivity and Novel Magnetism, 30 (2017) 2135–2141. https://doi.org/10.1007/s10948-017-4003-y

[22] R. D. Shannon, Revised effective ionic radii and systematic studies of interatomic distances in halides and chalcogenides, Acta Crystallographica A, 32 (1976) 751–767. https://doi.org/10.1107/S0567739476001551

[23] Sandeep Kumar Singh Patel, Jae-Hyeok Lee, Biswanath Bhoi, Jung Tae Lim, Chul Sung Kim, S.-K. Kim, Effects of isovalent substitution on structural and magnetic properties of nanocrystalline Y_3-$_x$Gd$_x$Fe$_5$O$_{12}$ ($0 \leq x \leq 3$) garnets, Journal of Magnetism and Magnetic Materials, 452 (2018) 48–54. https://doi.org/10.1016/j.jmmm.2017.12.013

[24] Olga Opuchovic, Aivaras Kareiva, Kestutis Mazeika, D. Baltrunas, Magnetic nanosized rare earth iron garnets $R_3Fe_5O_{12}$: Sol–gel fabrication, characterization and reinspection, Journal of Magnetism and Magnetic Materials, 422 (2017) 425–433. https://doi.org/10.1016/j.jmmm.2016.09.041

[25] F. Bertaut, R. Pauthenet, Crystal structure and magnetic properties of ferrites having the general formula $5Fe_2O_3.3M_2O_3$, Proceedings of the IEE-Part B, 104 (1957) 261–264. https://doi.org/10.1049/pi-b-1.1957.0043

[26] M.S. Lataifeh, S. Mahmood, M.F. Thomas, Mössbauer spectroscopy study of rare-earth iron garnets at low temperature, Physica B: Condensed Matter, 321 (2002) 143–148. https://doi.org/10.1016/S0921-4526(02)00840-2

Magnetic Oxides and Composites II
Materials Research Foundations **83** (2020) 41-78

Materials Research Forum LLC
https://doi.org/10.21741/9781644900970-3

Chapter 3

A Brief Review on Magnetic and Magnetocaloric Properties of La-Type Manganites

Mohamed Ellouze[1a], F. Ben Jemaa[1b], S.H. Mahmood[2c], E.K. Hlil[3d]

[1]University of Sfax, Faculty of Sciences of Sfax, LAMMA, B. P. 1171 - 3000, Tunisia

[2]The University of Jordan, Amman, Jordan

[3]Institut Néel, CNRS et Université´ Joseph Fourier, BP 166, 38042 Grenoble, France

[a]Mohamed.Ellouze@fss.rnu.tn, [b]feresbenjemaa@yahoo.com [c]s.mahmood@ju.edu.jo, [d]el-kebir.hlil@grenoble.cnrs.fr

Abstract

In present chapter, experimental techniques and preparation conditions adopted for the synthesis of La-type manganites and their influence on the structural, magnetic and magnetocaloric properties are briefly reviewed. The effects of various strategies of substitutions on magnetic and magnetocaloric properties are addressed. Further, our synthesis and findings on Mn-Fe substituted manganites are presented. It was found that partial substitution of Mn by Fe results in decreasing the magnetization and the Curie temperature, but the magnetic entropy change values remained in the range suitable for magnetic refrigeration.

Keywords

Synthesis of Manganites, Structural and Magnetic Properties, Magnetic entropy, RCP

Contents

1.   Introduction

Magnetic materials have acquired great scientific, as well as industrial and technological interest due to their crucial role in the fabrication of essential components for a multitude of devices and machines in active use nowadays. The search for new magnetic materials for high performance magnetic entropy for applications has led to an exponential growth in both the scientific research in this field, and the investment in the development of such materials. This development in materials research was driven by the evolution of new technologies, and the ever-increasing demand for the improvement in efficiency, better machine designs, and device miniaturization.

A great deal of work had been carried out concerning the substitution effects of Mn cations in B-sites by trivalent and tetravalent ions such as Fe, Co, Mo, and Cr [1-5]. In fact, colossal magnetoresistance (CMR) and magnetocaloric effect (MCE) in various perovskite oxide manganites has become a topic of considerable research interest due to their potential applications for devices, and the challenge to fully understand the basic nature of the strong interplay between magnetic order, electronic transport, structural distortions, and elastic properties in these materials. Generally, the CMR and MCE effects are explained by double-exchange (DE) mechanism, which considers the transfer of one e_g electron between neighboring ions through the Mn^{3+}- O^{2-}- Mn^{4+} path [6, 7], and the dynamic Jahn-Teller (JT) distortions caused by a strong electron–phonon coupling in manganites [8]. To better understand the metal-insulator transition, the CMR and the MCE, it is essential to understand the nature of the paramagnetic (PM)-ferromagnetic (FM) transition. Moreover, many studies on the critical behaviors as well as the universality class around the Curie temperature (T_C) have indicated that the critical exponents play important roles in elucidating interaction mechanisms near T_C [9, 10]. The discontinuous FM-PM transition at T_C is accompanied by structural changes, and is known as a first order magnetic transition (FOMT) [11-13]. This discontinuous phase transition can be rounded to a continuous one of a second order magnetic phase transition (SOMT) upon the doping, reduced dimensionality and external fields [14]. To make these issues clear, it is necessary to study in detail the critical exponents associated with transition.

Also, perovskites attracted considerable interest due to their potential applications in solid oxide fuel cell (SOFC) cathodes, oxygen separating membranes, read heads in modern hard disk devices, and chemical sensors for gases [15–18]. In addition to their CMR and MCE, these oxides exhibit several important characteristics such as charge ordering, orbital ordering, isotope effect, etc. [19–21]. These effects were attributed to Mn^{3+}/Mn^{4+} mixed-valence state of the oxides, and as a consequence, a large number of studies were devoted to La-manganites with La ions substituted by rare earth or alkaline earth ions and Mn ions substituted by 3d-transition metal ions [22, 23]. Due to the similarity of the ionic radii of Mn^{3+} (0.65 Å) and Fe^{3+} ions (0.645 Å), Fe^{3+} ions substitute Mn^{3+} ions in six-fold octahedrally coordinated sites [24]. Accordingly, the substitution of Fe^{3+} for Mn^{3+} is not expected to modify the tolerance factor and, therefore, the Jahn-Teller distortion effect can be neglected. However, the deformation of the local structure in manganites which exhibit such distortions can be investigated by Mössbauer spectroscopy, which is an effective technique to examine deviations from octahedral symmetry at the Fe sites [25]. On the other hand, MCE in these oxides is defined by the adiabatic temperature change (ΔT_{ad}) or the isothermal magnetic entropy change (ΔS_M), which is a function of both magnetic field and temperature. Generally speaking, there are three principal methods to evaluate MCE [26]. First, by direct measurement of the adiabatic temperature change (ΔT_{ad}), which is carried out by exposing the thermally insulated sample to a magnetic field. The second method is based on Maxwell's relation, and ΔS_M is calculated from magnetization measurements, and the third method is based on heat capacity measurements in zero-field cooled (ZFC) and field cooled (FC) processes [27–29].

The scenarios for tailoring of manganites with modified magnetic properties for specific applications included the adoption of different synthesis techniques to control the grain size and morphology of the produced ceramic, the variations of the stoichiometry of the starting powders and the experimental conditions, and the substitutions for the metal ions in the standard compound [30–33].

These issues are addressed briefly in the following sections. Further, we dedicated forthcoming sections to the presentation and discussion of our findings concerning the structural, magnetic and magnetocaloric properties of La-based manganites with La partially substituted by Ba and Sr, and Mn by Fe.

2.   Sample preparation

Samples of $La_{0.67}Ba_{0.22}Sr_{0.11}Mn_{1-x}Fe_xO_3$ (x = 0.0, 0.1, 0.2, 0.3 and 1.0) were prepared by the conventional ceramic method. This method is widely used for commercial production of powder. This technique, however, could be ineffective in controlling the grain size and morphology of the powder product. The synthesis of manganites powders by the various techniques often involved variations of the experimental conditions to determine the optimal conditions for the production of highly pure, high quality powder. Among others, the main factors taken into consideration in synthesizing the powders included: heat treatment, chemicals used in the starting powders, and stoichiometry of the precursor powder.

Powder samples were prepared by mixing metal oxides and carbonates (like La_2O_3, BaO, $SrCO_3$, $MnCO_3$ and ($Fe_2O_3 \cdot H_2O$) up to 99.9% purity) in proportions consistent with the stoichiometry of the samples. The starting materials were intimately mixed in an agate mortar, and heated in air at 900 °C for 10 h and slowly cooled to room temperature, and then sintered at 1000°C for 12 h. The obtained powder mixture was pressed in the form of pellets and sintered at 1200 °C for 50 h. Finally, these pellets were sintered at 1400 °C for 4 h.

3.   Characterization techniques for the samples

The phase purity of the final product was examined by X-ray diffraction technique using Siemens D5000 X-ray diffractometer, with Cu-Kα radiation (λ = 1.5405 Å). The XRD data were also used for refining the lattice parameters by means of Rietveld analysis [34] using FULLPROF program software. The surface morphology and elemental analysis of our samples were evaluated respectively by scanning electron microscopy (SEM) imaging and the energy dispersive X-ray spectroscopy (EDX) facility available in (JEOL 840 A) electron microscope. The magnetization measurements versus temperature in the range 5 –500 K, and the isothermal magnetization curves in an applied magnetic field of up to 6 T were obtained using the SQUID (BS2 Quantum Design) facility at Louis Néel Laboratory of Grenoble. Isothermal curves around the ferromagnetic ordering transition temperature (T_C) of the samples were obtained in different steps. The low-temperature magnetic hysteresis loops of the samples were recorded under an applied magnetic field ranging from 0 to 6 T. The MCE was calculated from the magnetization measurements versus applied magnetic field up to 5 T at various temperatures.

4.   Effects of substitution for La and Mn

4.1   Crystal structure

X-ray powder diffraction (XRD) measurements for $La_{0.67}Ba_{0.22}Sr_{0.11}Mn_{1-x}Fe_xO_3$ (x = 0.0, 0.1, 0.2, 0.3 and 1.0) samples were carried out at room temperature and Rietveld refinement of the XRD patterns are shown in Fig. 1 (a-e). XRD analysis indicates that all samples, except that with x = 0.2 and 1.0 consist of an orthorhombic single-phase with P_{nma} space group. The weak diffraction peak at 2θ = 31.53° for the sample with x = 0.2 is due to the precipitation of a very small amount of Mn_3O_4 (see Fig. 1c) which is consistent with previously reported results [35, 36]. However, the sample with x = 1.0 contained small amounts of additional phases. The weak diffraction peaks at 2θ = 28.0° and 34.3° correspond to the precipitation of a very small amount of $(Ba, Sr)Fe_2O_4$ spinel phase, while the weak diffraction peaks at 2θ = 25.4°, 33.0° and 53.6° are consistent with the precipitation of a very small amount of α-Fe_2O_3 (see Fig. 1e). Rietveld refinement of the X-ray powder data confirmed the presence of these minor phases, and the good agreement between the observed and calculated inter-planer spacing for the perovskite major phase is coherent with the value of the Goldschmidt tolerance factor t_G:

$$t_G = \frac{r_A + r_O}{\sqrt{2}\,(r_B + r_O)} \qquad (1)$$

Here r_A, r_B and r_O are respectively the average ionic radii of the A and B perovskite sites and oxygen anion [37].

Rietveld refinement of the X-ray powder data was carried out to determine the lattice parameters, unit cell volume, and atomic positions of all samples using the FULLPROF code [34].

Table 1 shows the refined unit cell parameters for all samples and unconstrained refinement of the data gave good values of R_f and χ^2 parameters. The results of the refined structural parameter for $La_{0.67}Ba_{0.22}Sr_{0.11}Mn_{1-x}Fe_xO_3$ (x = 0.0, 0.1, 0.2, 0.3 and 1.0) are summarized in Table 2. The structural analysis indicated that La/Ba/Sr ions were located at 4c (x, 0.25, z) position, Mn/Fe ions at 4b (0, 0, 0.5), and oxygen ions occupied two different sites, namely O_1 at 4c (x, 0.25, z) and O_2 at 8d (x, y, z) positions. It is noticed that the lattice parameters and the unit cell volume increased significantly only upon complete substituting manganese by iron. However, the volume and lattice parameters should be essentially independent of the Fe content due to the similarity of the radii of Mn^{3+} and Fe^{3+} ions in the octahedral sites as indicated by the constancy of the average tolerance factor t_G (see Table 2). We can therefore associate this increase of the

Materials Research Forum LLC

https://doi.org/10.21741/9781644900970-3

lattice parameters and cell volume with the distortion of the original crystal lattice structure, and the adjustment of the Mn-O bond length and Mn-O-Mn bond angles. Our results are nearly comparable to those obtained by Jonker *et al*. [24] and Kallel *et al*. [38].

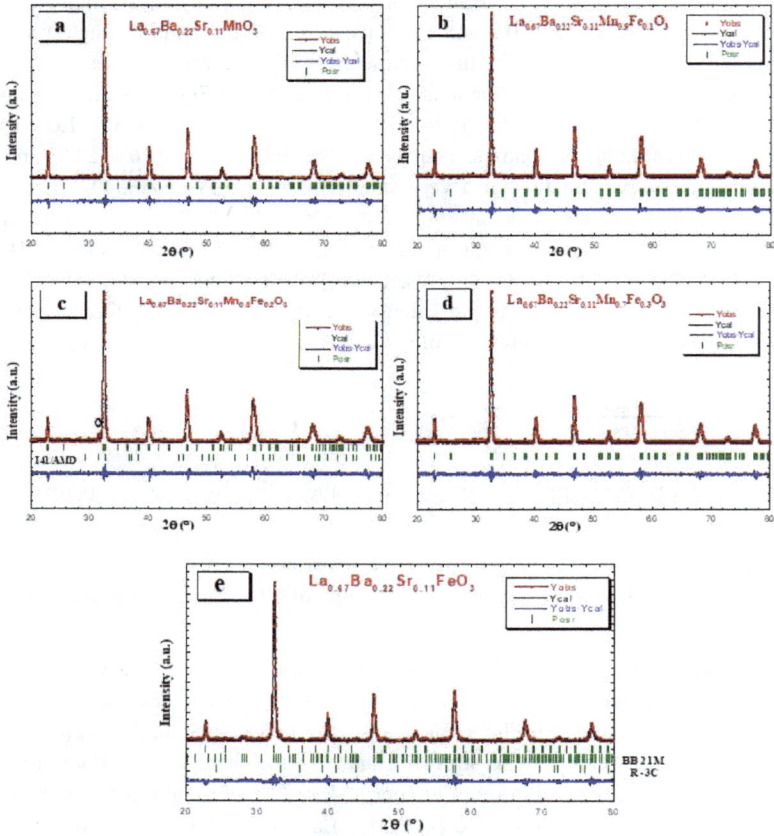

Fig. 1. Rietveld refinement of the XRD patterns of all elaborated samples.

Table 1. Cell parameters, unit cell volume and Curie temperature of $La_{0.67}Ba_{0.22}Sr_{0.11}$ $Mn_{1-x}Fe_xO_3$ (x= 0.0, 0.1, 0.2, 0.3 and 1.0) samples.

Samples	a (Å)	b (Å)	c (Å)	V (Å^3)	T_C (K)
$La_{0.67}Ba_{0.22}Sr_{0.11}MnO_3$	5.4929_9	7.7967_1	5.5269_7	236.70	345
$La_{0.67}Ba_{0.22}Sr_{0.11}Mn_{0.9}Fe_{0.1}O_3$	5.4982_4	7.7927_5	5.5366_1	237.22	190
$La_{0.67}Ba_{0.22}Sr_{0.11}Mn_{0.8}Fe_{0.2}O_3$	5.4843_2	7.8104_7	5.5349_5	237.09	130
$La_{0.67}Ba_{0.22}Sr_{0.11}Mn_{0.7}Fe_{0.3}O_3$	5.4987_8	7.8017_5	5.5398_8	237.66	80
$La_{0.67}Ba_{0.22}Sr_{0.11}FeO_3$	5.5527_1	7.8594_3	5.5378_9	241.68	-

The average crystallite size (D_{xrd}) was estimated from the XRD data using the Scherer relation [40] as shown in Eq.2.

$$D_{xrd} = \frac{K\lambda}{\beta \cos \theta} \qquad (2)$$

Where K, λ, β, and θ are the grain shape factor, the X-ray wavelength, the full width at half maximum (FWHM) of the diffraction peak, and the Bragg diffraction angle, respectively. The values of the effective crystallite sizes are summarized in Table 2. It is evident that the crystallite sizes of the samples were similar, with the exception of that with $x = 0.2$ which exhibited a better degree of crystallinity.

"Diamond" program was used to draw the unit cell (Fig. 2) and determine the distances between Mn ions and the nearest neighbor oxygen ions.

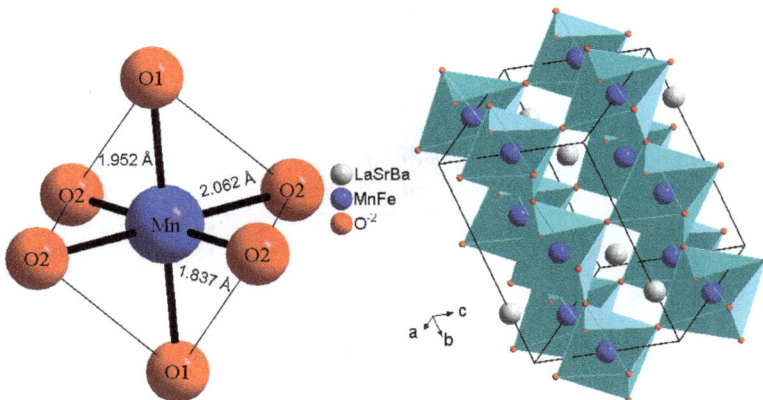

Fig. 2. Crystal structure of $La_{0.67}Ba_{0.22}Sr_{0.11}MnO_3$ and the coordination polyhedron on Mn.

The P_{nma} structure of $La_{0.67}Ba_{0.22}Sr_{0.11}Mn_{1-x}Fe_xO_3$ at room temperature is evident, and although the MnO_6 octahedral are almost regular, they are slightly distorted as observed in a previous work [40]. The interatomic distances and Mn–O–Mn angles for the $La_{0.67}Ba_{0.22}Sr_{0.11}Mn_{1-x}Fe_xO_3$ samples, in the orthorhombic structure, are listed in Table 3.

Table 2. Refined structural parameter for $La_{0.67}Ba_{0.22}Sr_{0.11}Mn_{1-x}Fe_xO_3$ (x= 0.0, 0.1, 0.2, 0.3 and 1.0), and the numbers in subscript represent the error bars.

Samples	$x = 0.0$	$x = 0.1$	$x = 0.2$	$x = 0.3$	$x = 1.0$
La/Ba					
x	0.0050_1	-0.0003_8	0.0081_5	0.0059_4	0.0177_1
y	0.25	0.25	0.25	0.25	0.25
z	0.0052_1	0.0029_5	0.0044_7	-0.0056_5	-0.0048_4
Mn/Fe					
x	0	0	0	0	0
y	0	0	0	0	0
z	0.5	0.5	0.5	0.5	0.5
O_1					
x	0.4828_3	0.4777_9	0.4897_5	0.4845_6	0.4898_9
y	0.25	0.25	0.25	0.25	0.25
z	-0.0052_1	0.0266_3	-0.0211_1	-0.0025_2	0.0225_9
O_2					
x	0.2555_7	0.2631_8	0.2679_5	0.2593_4	0.2729_2
y	-0.0024_5	0.0142_3	-0.0299_8	-0.0209_8	0.0293_5
z	0.7732_3	0.7629_2	0.7610_4	0.7737_7	0.7416_5
χ^2	1.43	1.52	1.57	1.38	1.33
t_G	0.9253	0.9255	0.9257	0.9259	0.9268
D_{hkl} (nn)	28.11	29.02	69.15	31.45	35.12

Table 3. Interatomic distance and angles for $La_{0.67}Ba_{0.22}Sr_{0.11}Mn_{1-x}Fe_xO_3$ (x = 0.0, 0.1, 0.2, 0.3 and 1.0) samples.

Distances	$x = 0.0$	$x = 0.1$	$x = 0.2$	$x = 0.3$	$x = 1.0$
Mn–O_1 (Å)	1.95	1.96	1.96	1.95	1.97
Mn–O_{21} (Å)	1.84	1.85	1.85	1.83	1.92
Mn–O_{22} (Å)	2.06	2.05	2.07	2.09	2.03
Angles					
Mn-O_1-Mn (°)	174.2	168.8	172.4	174.9	172.1
Mn-O_2-Mn (°)	175.8	173.5	166.1	169.8_2	164.8

4.2   Scanning electron microscopy investigation

Scanning Electron Microscopy (SEM) was employed to examine surface morphology of the $La_{0.67}Ba_{0.22}Sr_{0.11}Mn_{1-x}Fe_xO_3$ samples (x = 0, 0.1, 0.2, 0.3 and 1.0) and SEM images of all samples are shown in Fig. 3 (a-e). Micrographs of all samples show that agglomerated particles are generally parallelepipedal with a homogeneous microstructure, fine granulation and porous structure. We estimated the average grain size by SEM based on the average grain size of ten different grains and then compared with those calculated by the Deby-Scherrer formula using data of X-ray diffraction. The size of the grains observed by SEM are greater than those calculated by the Scherrer formula; this can be explained by the fact that each particle observed by SEM is composed of several crystallites [42].

The chemical composition of the samples was estimated from the elemental analysis using energy dispersive X-ray spectra (EDX). Fig. 4 shows XRD spectra for all samples. The observed energy peaks correspond to the various elements in the samples. The obtained chemical composition of the samples was found close to the nominal chemical stoichiometry of the starting powders.

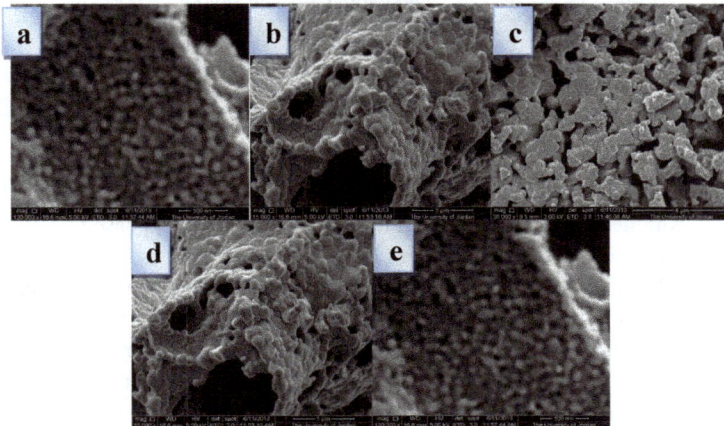

Fig. 3. Typical scanning electron micrographs (SEM images) of $La_{0.67}Ba_{0.22}Sr_{0.11}Mn_{1-x}Fe_xO_3$ samples; (a) x = 0.0, (b) x = 0.1, (c) x = 0.2, (d) x = 0.3 and, (e) x = 1.0.

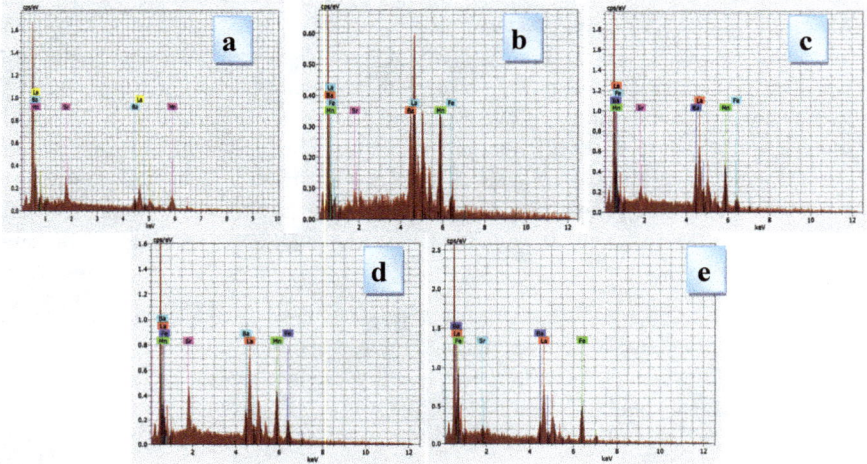

Fig. 4. EDX analysis spectral at room temperature of $La_{0.67}Ba_{0.22}Sr_{0.11}Mn_{1-x}Fe_xO_3$ samples; (a) x = 0.0, (b) x = 0.1, (c) x = 0.2, (d) x = 0.3 and, (e) x = 1.0.

4.3   Magnetism

Fig. 5 shows the magnetization of $La_{0.67}Ba_{0.22}Sr_{0.11}Mn_{1-x}Fe_xO$ (x = 0.0, 0.1, 0.2, 0.3 and 1.0) as a function of temperature for an applied field of 0.05 T. The samples with x = 0, 0.1 and 0.2 exhibited ferromagnetic (FM) to a paramagnetic (PM) phase transition as the temperature increased. Curie temperature (T_C) for each sample was determined from the minimum of $dM(T)/dT$ at the inflection point of the thermomagnetic curve, and the results are listed in Table 1. It is clear that T_C could be determined for x = 0.0, 0.1, 0.2 and 0.3, which indicates that the level of substitution of manganese by iron in these samples does not destroy the ferromagnetic behavior below a temperature which is sensitive to the value of x [43].

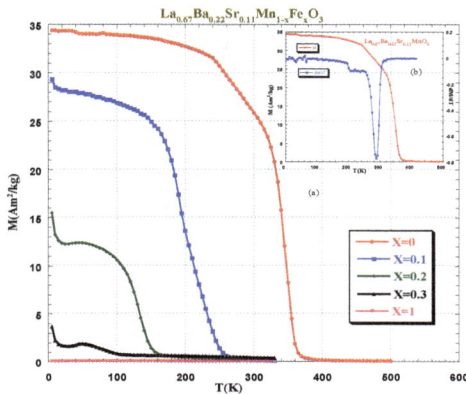

Fig. 5 (a). Temperature dependence of magnetization in a magnetic applied field of 0.05 T for all samples $La_{0.67}Ba_{0.22}Sr_{0.11}Mn_{1-x}Fe_xO$ (x = 0.0, 0.1, 0.2, 0.3 and 1.0). (b) The minimum values of dM/dT curves are defined as T_C of the samples.

Further, T_C for the un-substituted sample was found to be above room temperature, while even a low level of the substitution lowers T_C to below room temperature. The observed decrease in magnetization and T_C with increasing x could be associated with the fact that Fe^{3+} ions do not participate in the ferromagnetic double exchange (DE) interaction with Mn^{3+} ions, but encourage the antiferromagnetic Fe^{3+}–O–Mn^{3+}, Mn^{4+}–O–Fe^{3+} and Fe^{3+}–O–Fe^{3+} super-exchange (SE) interactions [32]. With regard to the sample with x =0.3, the magnetization exhibited a cusp at about 50 K. Such a behavior is typical of spin-glasses due to the competition between ferromagnetic (DE) and antiferromagnetic (super-exchange) interactions [44, 45]. Finally, the sample with x = 1.0 exhibited paramagnetic behavior down to the lowest temperature in this study, as expected for such iron oxides.

In order to better understand the magnetic properties of the samples, room temperature hysteresis loops were recorded in an applied magnetic field up to 1 T. Fig. 6 (a) shows that the magnetization for the un-substituted sample is almost saturated (52 A.m^2/kg) at 1 T applied magnetic field. This result is consistent with the thermomagnetic results which indicated that this sample is ferromagnetic at room temperature. However, the rest of the samples (x = 0.1, 0.2, 0.3 and 1.0) magnetization curves at room temperature did not saturate in magnetic fields up to 1T (Fig. 6(b)). Further, it was observed that the increased level of iron substitution resulted in a significant drop in the spontaneous magnetization, which is associated with the weakening of the Mn^{3+}–Mn^{4+} DE interactions. The magnetization curves for all samples demonstrated paramagnetic behavior. This is also consistent with the thermomagnetic results that T_C for all substituted samples, if any, is below room temperature. The variation in the magnetic behavior with increasing Fe substitution level is due to the progressive weakening of the net (DE) interaction due to the decrease in number of Mn^{3+}–O^{2-}–Mn^{4+} bonds [42-45]. The small hysteresis loop for the samples with $x = 0.2$ and 0.3 could be due to the presence of about 1 % of magnetically ordered phases, which reflects the presence of a small degree of in homogeneity in the magnetic structure in these samples.

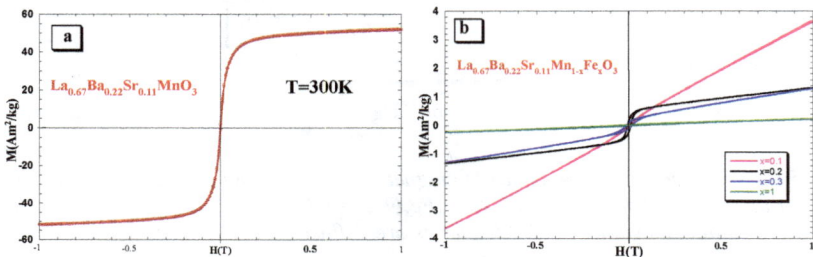

Fig. 6(a). Magnetization versus applied magnetic field at room temperature for $La_{0.67}Ba_{0.22}Sr_{0.11}MnO_3$, (b) Magnetization versus applied magnetic field at room temperature for $La_{0.67}Ba_{0.22}Sr_{0.11}Mn_{1-x}Fe_xO_3$ (x = 0.1, 0.2, 0.3 and 1).

The magnetization curves $M(H)$ at T = 5 K are shown in Fig. 7. The samples with $x = 0$ and 0.1 were obviously ferromagnetic at this temperature and almost saturated at an applied field of 6 T. The saturation magnetizations for these samples are about 90 and 70 A.m^2/kg, respectively. However, the curves of the samples with higher x values did not show saturation up to 6 T, which could be due to the increased SE antiferromagnetic coupling and the decreased DE ferromagnetic coupling with increasing Fe concentration, the competition between which could lead to frustration and the development of the spin-glass state [44-46]. The saturation magnetizations for these samples were determined from the low-of-approach to saturation in the high-field range, and found to be 54 and 29 A.m^2/kg for the samples with $x = 0.2$ and 0.3, respectively (Fig. 7–insets).

Magnetic Oxides and Composites II   Materials Research Forum LLC
Materials Research Foundations **83** (2020) 41-78   https://doi.org/10.21741/9781644900970-3

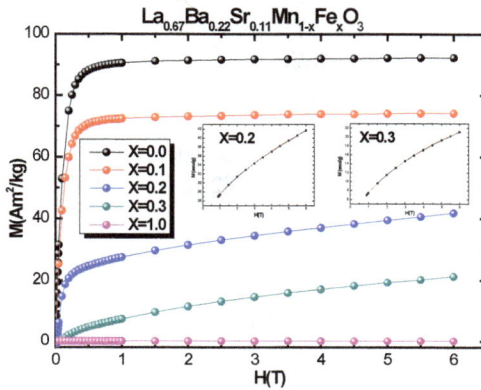

Fig. 7. Magnetization versus applied magnetic field at 5K for
La$_{0.67}$Ba$_{0.22}$Sr$_{0.11}$Mn$_{1-x}$Fe$_x$O$_3$ (x = 0.0, 0.1, 0.2, 0.3 and 1.0). The insets show the fitting of
isothermals curves corresponding respectively for each sample with x = 0.2 and x = 0.3.

4.4   Critical behavior

The experimental investigations of the critical behavior of manganites near the PM–FM phase transition using a variety of techniques provided a wide range of values for the magnetization critical exponent β which embraces the mean field, short-range isotropic 3D-Heisenberg, and 3D- Ising model estimates [49–51]. The critical behavior was first described within the long-range mean-field theory applied to the DE model [6, 52-54]. However, evidence that the FM–PM transition in CMR manganites should belong to the 3D-Heisenberg universality class was reported [59,50, 55, 56]. Further, the interpretations of a few relevant experimental results concerning the critical exponents and the universality class of manganites are still controversial [47, 49, 51, 56].

In this study, we conducted a study of the critical behavior of the La$_{0.67}$Ba$_{0.22}$Sr$_{0.11}$Mn$_{1-x}$Fe$_x$O$_3$ (x = 0.0, 0.1, 0.2) manganites by analyzing the critical exponents using different models, such as the modified Arrott plot (MAP) and the Kouvel–Fisher method, with the intention of understanding the magnetic interactions in these manganites. Based on the determined critical parameters β, γ, and δ we concluded that the critical phenomena near the PM– FM transition in these perovskites are best described by the 3D-Heisenberg model.

For this, we used vibrating sample magnetometer (BS2 magnetometer developed in Louis Neel Laboratory of Grenoble, France). The sensitivity of the magnetometer is better than $3 \cdot 10^{-7} Am^2$. The system is capable of providing measurements of magnetization in the temperature range between 1.5 and 300 K, where the temperature of the sample is regulated by a circulating helium gas temperature. The maximum applied magnetic field used in this study was 5 T. Isothermal magnetic curves were obtained in a temperature range around the critical temperature in steps of 2 K.

The scaling hypothesis provides a model for characterizing a second-order phase transition by the values of the critical exponent's β, γ, and δ associated with the spontaneous magnetization, the initial magnetic susceptibility, and the critical magnetization isotherm, respectively [48–50, 55-57]. In terms of these exponents, the magnetic parameters are given by the following relations:

- Below T_C, the spontaneous magnetization in the limit of zero-applied field is given by Eq. (3) as:

$$M_s(T) = M_0(-\varepsilon)^\beta \qquad T < T_C \tag{3}$$

- Above T_C, the initial susceptibility is given by Eq. (4) as:

$$x_0^{-1} = \frac{h_0}{M_0}\varepsilon^\gamma ; \qquad T > T_C \tag{4}$$

- At T_C, the dependence of the magnetization M on the magnetic field H is given by Eq. (5) as :

$$M = DH^{1/\delta} \qquad T = T_C \tag{5}$$

Where, $\varepsilon = (T - T_C)/T_C$ is the reduced temperature, and M_0, h_0, and D are the critical amplitudes.

The reliability of the critical exponents can be confirmed with the aid of the scaling theory. According to the scaling hypothesis, the magnetic equation of state in the critical regime obeys the scaling relation [48, 58]:

$$M(H,\varepsilon) = |\varepsilon|^\beta f_\pm(\frac{H}{|\varepsilon|^{\beta+\gamma}}) \tag{6}$$

In this last equation, the (+) and (−) signs refer to $T > T_C$ and $T < T_C$, respectively.

Accordingly, the scaled magnetization isotherms M/ε^β plotted against $H/\varepsilon^{\beta+\gamma}$ with the right choice of β, γ, and δ values will fall onto two universal curves: one above T_C and another below T_C. This is an important criterion for the determination of accurate and unambiguous values of the critical exponents.

Fig. 8 shows the isothermal magnetization curves (upper three panels) and Arrott plots (lower three panels) for the $La_{0.67}Ba_{0.22}Sr_{0.11}Mn_{1-x}Fe_xO_3$ samples in the temperature ranges 320–360 K for $x = 0.0$, 175–225 K for $x = 0.1$ and 100–150 K for $x = 0.2$, which are ranges around the Curie temperatures for the different samples. The Arrott plots M^2 vs. μ_0H/M in Fig. 8 are nonlinear and exhibit an upward curvature even at high fields, indicating that the phase transitions in these samples do not satisfy the mean-field theory. Moreover, the concave downward curvature clearly indicates a second-order phase transition according to the criterion suggested by Banerjee [59]. In general, in the high-field region, the effects of charge, lattice, and orbital degrees of freedom are suppressed in a FM phase and the order parameter can be identified with the macroscopic magnetization. Thus, the magnetic phase transition in $La_{0.67}Ba_{0.22}Sr_{0.11}Mn_{1-x}Fe_xO_3$ ($x = 0.0$, 0.1, 0.2) is a second-order transition. The nonlinearity of Arrott plots makes it difficult to evaluate the saturation magnetization, susceptibility, and critical temperature by extrapolation.

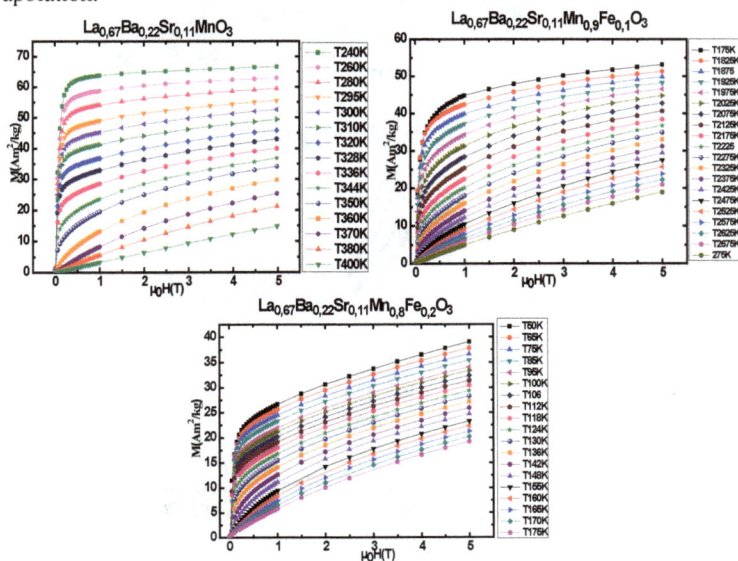

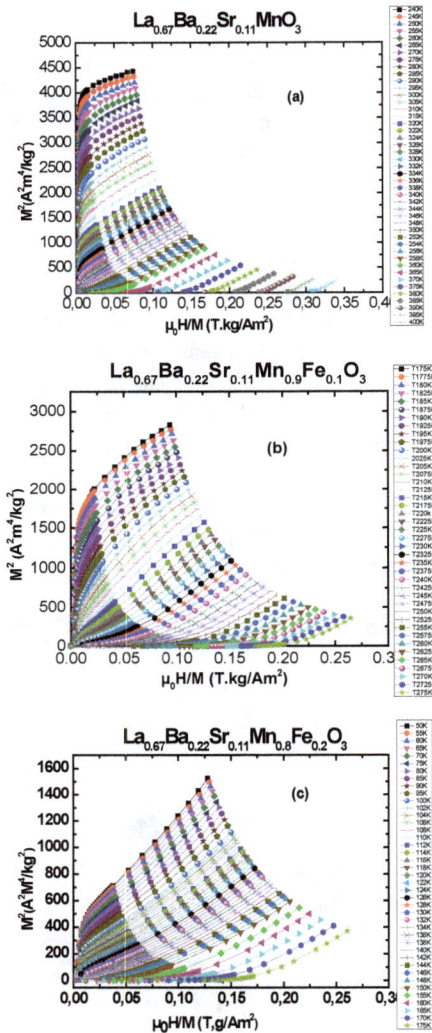

Fig. 8. Isothermal magnetization curves (upper three panels), and Arrott plots (lower three panels) at different temperatures for $La_{0.67}Ba_{0.22}Sr_{0.11}Mn_{1-x}Fe_xO_3$. (a). $x = 0$, (b). $x = 0.1$ and, (c). $x = 0.2$.

Materials Research Forum LLC
https://doi.org/10.21741/9781644900970-3

Arrott and Noakes suggested a powerful method, the MAPs, to analyze the magnetic data sing the following empirical relation [60]:

$$\left(\frac{H}{M}\right)^{\frac{1}{\gamma}} = \frac{T-T_C}{T_1} + \left(\frac{M}{M_1}\right)^{\frac{1}{\beta}}$$

(7)

Where T_1 and M_1 are material-dependent (the mean-field theory values of $\beta = 0.5$ and $\gamma = 1$ generate the regular Arrott plots). A linear relation between $M^{1/\beta}$ and $(H/M)^{1/\gamma}$ can then be sought by a proper choice of β and γ. Thus, by choosing the critical parameters of a given universality class, a linear relation means that the compound belongs to that class.

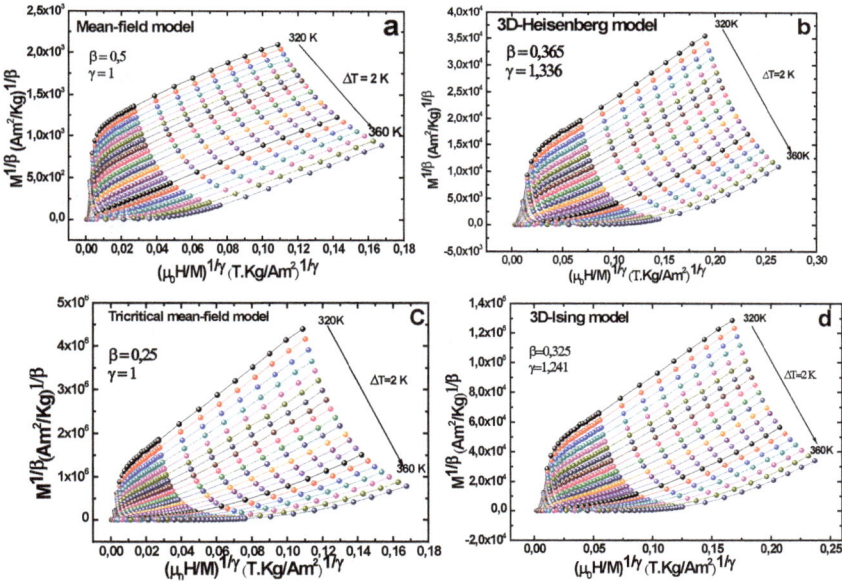

Fig. 9. Standard Arrot plot (a), and modified Arrott plots with 3D-Heisenberg model (b), tricritical mean-field model (c), and 3D-Ising model (d) for $La_{0.67}Ba_{0.22}Sr_{0.11}MnO_3$ compound.

Fig. 9 shows the MAPs derived from the sets of critical parameters corresponding to the four different models:

- panel (a) the mean-field model ($\beta = 0.5$, $\gamma = 1$);
- panel (b) the 3D-Heisenberg model ($\beta = 0.365$, $\gamma = 1.336$);
- panel (c) the 3D-Ising model ($\beta = 0.325$, $\gamma = 1.24$),
- panel (d) the tri critical mean-field model ($\beta = 0.25$, $\gamma = 1$).

The positive slope of the curves indicates second-order phase transition [59]. Further, with the exception of the tri critical model, all models gave quasi-straight lines which are nearly parallel in the high-field region. Thus, it is not obvious which model best represents the actual critical exponents for the samples.

Accordingly, in order to select the model which best describes this system we calculated their relative slopes (RS) defined as RS = $S(T)/S(T_C)$. Then, the most adequate model should be the one that possesses an RS value which is very close to unit. Fig. 10 shows the RS vs. T curves derived from the MAPs in Fig. 9 for the compound with $x = 0.0$. It is obvious that 3D-Heisenbergmodel is the best model which can describe this sample ($La_{0.67}Ba_{0.22}Sr_{0.11}MnO_3$). However, similar analysis (not shown) indicated that the behavior of the sample $La_{0.67}Ba_{0.22}Sr_{0.11}Mn_{0.9}Fe_{0.1}O_3$ is best described by 3D-Ising model.

The values of β and γ were then refined using Kaul's method [61]. In this method, the high-field quasi-straight lines obtained with the initial values of β and γ in the MAP (Fig. 9) were extrapolated to intercept the vertical and horizontal axes at $M_S^{1/\beta}$ in the FM regime and $\chi^{-1/\gamma}$ in the PM regime, respectively. The critical isotherm passing through the origin determines T_C. Consequently, $M_S(T)$ and $\chi_0^{-1}(T)$ were evaluated and plotted against temperature, and fitted with Eqs. (8) and (9) to obtain new values of β and γ which were used to construct new MAPs.

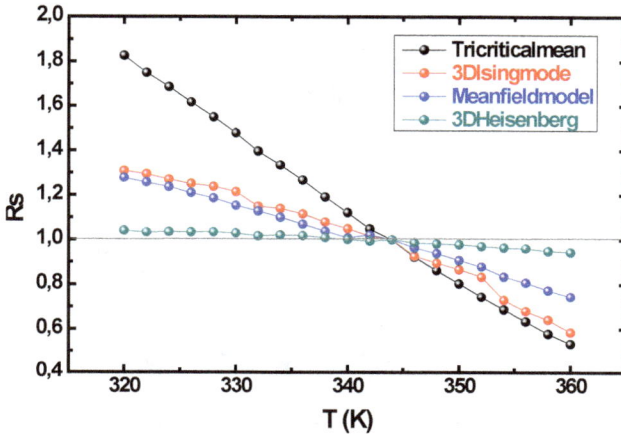

Fig. 10. Relative slope (RS) for La$_{0.67}$Ba$_{0.22}$Sr$_{0.11}$MnO$_3$ as a function of temperature using different models.

In Fig. 11, we plotted the temperature dependence of M_S(T) and $\chi_0^{-1}(T,0)$ for La$_{0.67}$Ba$_{0.22}$Sr$_{0.11}$Mn$_{1-x}$Fe$_x$O$_3$ (x = 0, 0.1, 0.2). Table 4 shows the results of the fits. The values of the critical parameters for the sample with x = 0.0 seem to agree with the 3D-Heisenberg model.

Upon doping with Fe^{3+}, the critical exponent β decreased and approached the value corresponding to the 3D-Isingmodel (see Table 4), while the critical exponent γ increased slightly. Accordingly, we may conclude that the samples with x = 0.1 and 0.2 seem to qualitatively agree with the 3D-Ising model. Other workers [55] observed similar results and argued that their critical exponents are between those of the 3D-Hesenberg model and those of the mean-field model, and concluded the possibility of short range to long-range order crossover in their samples.

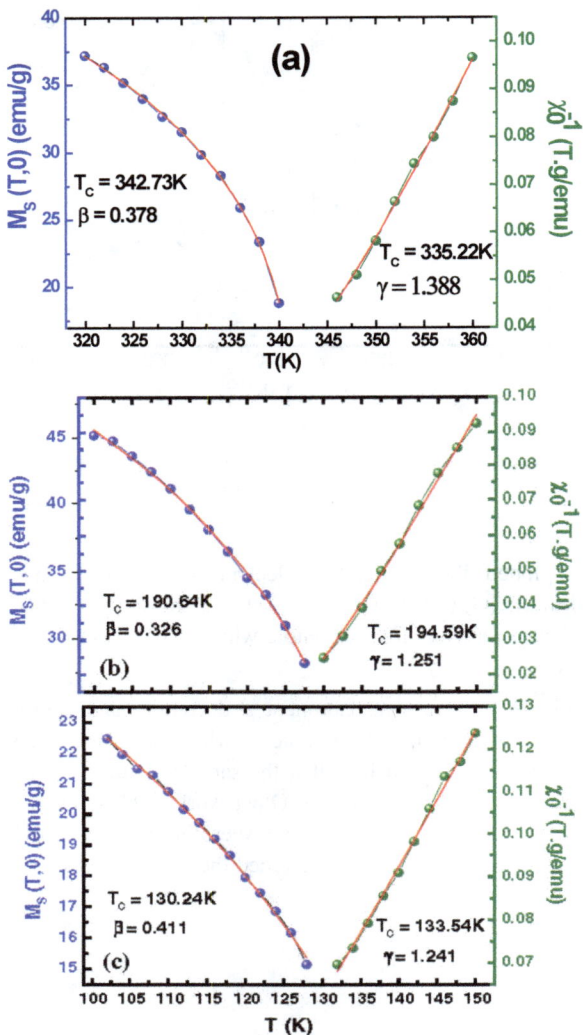

Fig. 11. Spontaneous magnetization M_S and the inverse of the initial susceptibility χ_0^{-1} vs. temperature for $La_{0.67}Ba_{0.22}Sr_{0.11}Mn_{1-x}Fe_xO_3$ with (a) x = 0, (b) x = 0.1, and (c) x = 0.2.

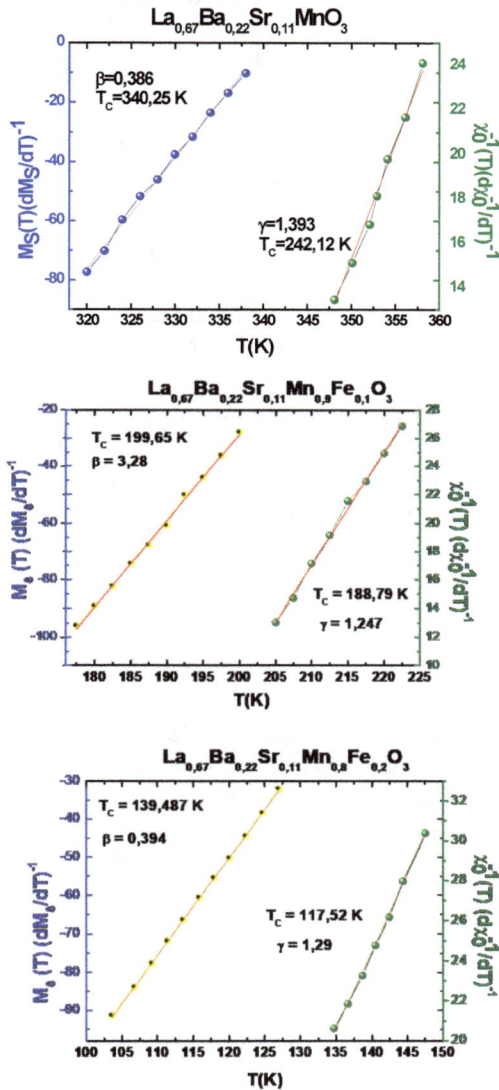

Fig. 12. Kouvel–Fisher plots of $M_S(T)[dM_S(T)/dT]^{-1}$ and $\chi_0^{-1}(T)[d\chi_0^{-1}(T)/dT]^{-1}$ vs. temperature for $La_{0.67}Ba_{0.22}Sr_{0.11}Mn_{1-x}Fe_xO_3$ with (a). x = 0, (b). x = 0.1 and, (c). x = 0.

A more accurate method to determine the critical exponents β and γ is by the Kouvel–Fisher (KF) method [62]:

$$\frac{M_S(T)}{\left(dM_S(T)/dT\right)} = \frac{T - T_C}{\beta} \tag{8}$$

$$\frac{\chi_0^{-1}(T)}{\left(d\chi_0^{-1}(T)/dT\right)} = \frac{T - T_C}{\gamma} \tag{9}$$

According to this method, plots of $M_S(T)[dM_S(T)/dT]^{-1}$ vs. T and $\chi_0^{-1}(T)[d\chi_0^{-1}(T)/dT]^{-1}$ vs. T should yield straight lines with slopes of $1/\beta$ and $1/\gamma$, respectively, and the intercepts on the T axes are equal to T_C. The linear fitting to the plots following the KF method for $La_{0.67}Ba_{0.22}Sr_{0.11}Mn_{1-x}Fe_xO_3$ give β, γ, and T_C for $T < T_C$ and $T > T_C$ (See Fig. 12, Table 4).

The exponent δ can be directly obtained by plotting the critical isothermal magnetization curve on a log–log scale according to Eq. 10 (Fig. 13). The high-field region of the data is fitted by a straight line with a slope of $1/\delta$. The results of the analysis are shown in Table 1. Further, the exponents derived from static scaling analysis are related by the Widom scaling relation [63]:

$$\delta = 1 + \frac{\gamma}{\beta} \tag{10}$$

Using this relation and the estimated values of β and γ, we obtained δ values (see Table 4), which are very close to the estimates of δ from the critical isotherms at T_C. Thus, the estimates of the critical exponents are consistent.

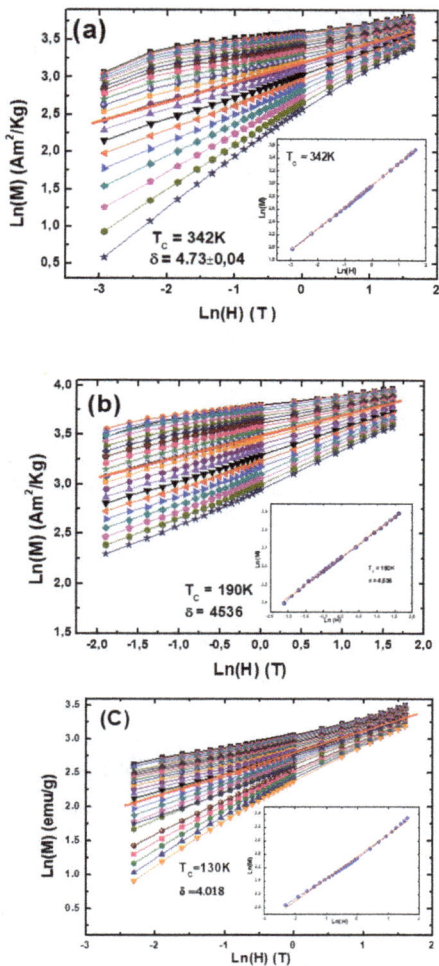

Fig. 13. M vs. H on a log–log scale at several temperatures close to T_C for the samples with (a). x = 0, (b). x = 0.1, and (c). x = 0.2. The straight line is the linear fit following Eq. 3 at $T \sim T_C$.

The obtained values of the critical exponents can be verified with the prediction of the scaling theory in the critical region using Eq. 4. Fig. 14 shows plots of $M|\varepsilon|^{-\beta}$ vs. $H|\varepsilon|^{-(\beta+\gamma)}$ for the three samples. The Fig. 14 demonstrates two different branches, one for temperatures below T_C and the other for temperatures above T_C. The inset shows the same data on a log–log scale. From this figure, it can be clearly seen that the scaling is well obeyed; all the points fall on the two branches. This indicates that the determined critical parameters are in good accordance with the scaling hypothesis.

The values of critical exponents of $La_{0.67}Ba_{0.22}Sr_{0.11}Mn_{1-x}Fe_xO_3$ and the theoretical values based on various models are summarized in Table 4 for comparison. From Table 4, we can clearly see that the values of the critical exponents for the sample with $x = 0.0$ agree with those of the 3D-Heisenberg-like ferromagnet. This behavior is similar to that of other CMR manganites belonging to the universality of 3D isotropic ferromagnets with short-range exchange couplings [51]. However, with the substitution of Mn^{3+} by Fe^{3+}, the critical exponent γ decreased and approached the 3D-Ising $\gamma = 1.24$. On the other hand, the critical exponent β increased to $\beta = 0.398$. This may indicate that the magnetic transition could still be described by short-range interactions (3D-Heisenberg/3D- Ising). However, the substitution of Fe^{3+} results in breaking some of the $Mn^{4+}–O^{2-}–Mn^{3+}$ bonds in a random fashion, and a consequent suppression of the DE interaction and long-range FM order. This was demonstrated by the significant decrease of T_C and saturation magnetization for Fe-substituted samples (Fig. 8, Table 4). Our results are nearly comparable to those obtained by Omri *et al.* [64]. In fact, we have found that the parent sample $La_{0.67}Ba_{0.22}Sr_{0.11}Mn_1O_3$ exhibits a sharp FM–PM transition. As Fe-substitution increases it causes a decrease in the magnetization and the magnetic transition become sincreasingly broader. These results have been explained by the fact that, the substitution of the Mn^{3+} by Fe^{3+} Ions reduces the Mn^{3+}/Mn^{4+} ratio, leading to a suppression of the DE interaction and enhancement of the super exchange interaction.

A similar behavior for manganites was reported by others authors [65, 66]. The broadening of the PM–FM transition was ascribed to magnetic inhomogeneity and the development of an AFM phase with increasing x. A significant degree of Fe^{3+} substitution reduces the Mn^{3+}/Mn^{4+} ratio significantly and weakens the DE interaction on the one hand, and increases the possibility of occurrence of $Fe^{3+}–O^{2-}–Fe^{3+}$ bonds and the subsequent AFM superexchange interaction. This may result in competition between ferromagnetism and antiferromagnetism, and lead to spin glass-like behavior at high levels of substitution as reported [65].

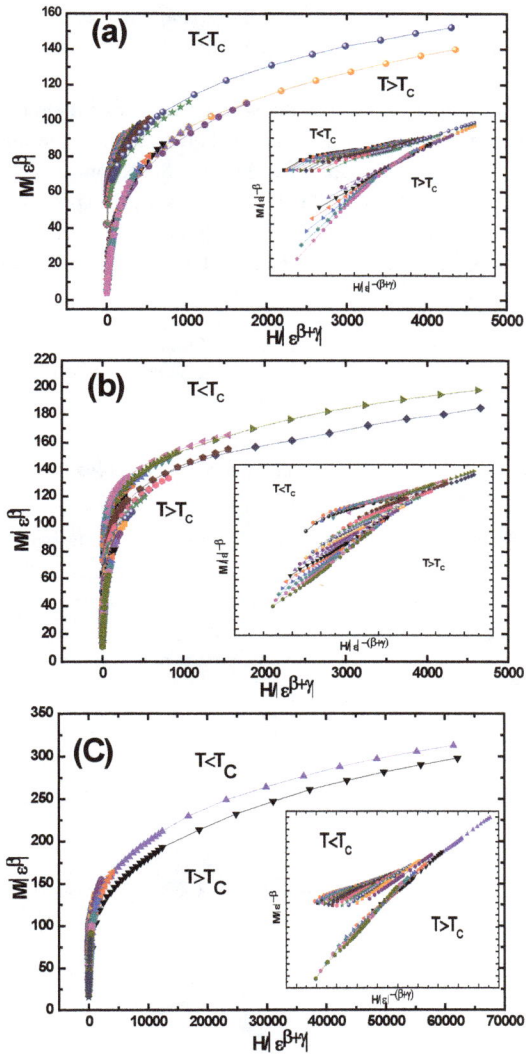

Fig. 14. Scaling plot $M|\varepsilon|^{-\beta}$ vs. $H|\varepsilon|^{-(\beta+\gamma)}$, indicating two universal curves below and above T_C for $La_{0.67}Ba_{0.22}Sr_{0.11}Mn_{1-x}Fe_xO_3$ with (a) x = 0, (b) x = 0.1, and (c) x = 0.2. Insets we show the same plots on log–log scales.

The critical exponents of a homogeneous ferromagnet are governed by the lattice dimension (3D), dimension of the order parameter (n = 3, magnetization), and range of interaction (short range, long range, or infinite). Fisher *et al.* [67] performed a renormalization group analysis of systems with an exchange interaction of the form J(r) = $1/r^{d+\sigma}$ (d is the dimension of the system and σ is the range of the interaction). If σ < 3/2, the mean-field exponents are valid. The Heisenberg exponents are valid for σ >2. For intermediate range, i.e., for J(r) ≈ $r^{-3-\sigma}$ with 3/2 < σ < 2 the exponents belong to a different universality class which depends upon σ. In a general way, the evolution of the critical exponents tends to move toward the values of the mean-field theory with long-range interaction.

Table 4. Comparison of the critical exponents of $La_{0.67}Ba_{0.22}Sr_{0.11}Mn_{1-x}Fe_xO_3$ (x = 0, 0.1, 0.2) with earlier reports, and with the various theoretical models.

Composition	Method	T_C(K)	β	γ	δ	Ref.
$La_{0.67}Ba_{0.22}Sr_{0.11}MnO_3$	M	342.73±0.17	0.378±0.003	1.388±0.001		[70]
	A	342.12±0.14	0.386±0.006	1.393±0.004		
	CI(exp)				4.73±0.003	
	CI(cal)				4.67±0.004	
$La_{0.67}Ba_{0.22}Sr_{0.11}Mn_{0.9}Fe_{0.1}O_3$	M	190.64±0.11	0.398±0.002	1.251±0.005		[70]
	A	188.79±0.18	0.395±0.003	1.247±0.003		
	CI(exp)				4.536±0.002	
	CI(cal)				4.14±0.007	
$La_{0.67}Ba_{0.22}Sr_{0.11}Mn_{0.8}Fe_{0.2}O_3$	M	130.24±0.12	0.411±0.001	1.241±0.004		[70]
	A	139.48±0.15	0.394±0.003	1.292±0.003		
	CI(exp)				4.018±0.006	
	CI(cal)				4.019±0.002	
Mean-field model	Theory	–	0.5	1	3	
3D-Heisenberg	Theory	–	0.365	1.336	4.80	
3D-Ising model	Theory	–	0.325	1.241	4.82	
Tricritical mean-field theory	Theory	–	0.25	1	5	
$La_{0.75}(Sr,Ca)_{0.25}Mn_{0.9}Ga_{0.1}O_3$	M	232.40±0.08	0.420±0.005	1.221±0.002	4.22±0.04	[64]
	A	232.36±0.06	0.428±0.005	1.286±0.004	3.87	
$La_{0.8}Ba_{0.2}Mn_{0.85}Fe_{0.15}O_3$	KF	158.24±0.01	0.370±0.02	1.359±0.02	4.67±0.01	[68]
$La_{0.8}Ba_{0.2}Mn_{0.8}Fe_{0.2}O_3$	MAP	125±0.08	0.365±0.01	1.227±0.03	4.52±0.03	[68]
$La_{0.8}Ba_{0.2}Mn_{0.8}Fe_{0.2}O_3$	KF	125.61±0.05	0.318±0.02	1.159±0.02	4.36±0.01	[68]
$La_{0.7}Sr_{0.3}MnO_3$		354.0	0.37	1.22	4.25	[49]
$La_{0.7}Sr_{0.3}Mn_{0.8}Ti_{0.2}O_3$		150.1	0.518	1.002	2.95	[69]

4.5 Magnetocaloric study

Magnetization measurements versus applied magnetic field up to 5 T were obtained at several temperatures below and above T_C for the samples with $x = 0$, 0.1, and 0.2, and in the temperature range between 100 K and 300 K for the sample with $x = 1.0$. The isothermal curves (Fig. 8) indicated that the magnetization for the un-substituted sample in the ferromagnetic regime increased sharply and saturated at an applied field below 0.5 T. This behavior is characteristic for a soft magnetic material, and is consistent with the room temperature hysteresis behavior.

However, the curves for the substituted samples continued to rise and did not saturate even at 5 T. Fig. 8 shows M^2 vs. ($\mu_0 H/M$) Arrott plots derived from the magnetization curves. The plots exhibited a positive slope around T_C which confirms that our samples (with the exception of that with $x = 1.0$) exhibit second order ferromagnetic to paramagnetic phase transition [59, 71]. The plots for the sample with $x = 1$, demonstrated the absence of such transition.

In order to study the magnetocaloric properties of our samples, the temperature dependence of magnetic entropy change was evaluated (Fig. 15). The magnetic entropy change, $-\Delta S_M$, induced by the magnetic field change was calculated according to the classical thermodynamic theory based on Maxwell's relations using the following equation according to [72]:

$$\Delta S_M (T)_{\Delta H} = \int_{H_1}^{H_2} \left(\frac{\partial M (T, H)}{\partial T} \right)_H dH \tag{11}$$

The numerical integration of the latter formula gives the values of ΔS_M at different values of fields and temperatures:

$$\Delta S_M = \sum_i \frac{M_{i+1} - M_i}{T_i - T_{i+1}} \Delta H_i \tag{12}$$

Here $M_i(T_i, H)$ and $M_{i+1}(T_{i+1}, H)$ are the experimental values of magnetization measured at temperatures T_i and T_{i+1}, respectively, under applied magnetic field H_i. Fig. 15(a) shows the temperature dependence of the magnetic entropy change ($-\Delta S_M$) vs. temperature for the un-substituted sample at different external applied magnetic fields of 1 to 5 T.

As expected, the magnetic entropy change enhanced with increasing applied magnetic field, and exhibited a maximum value around T_C. The magnitude of the maximum entropy change increased from 0.675 J.kg^{-1}.K^{-1} at an applied field of 1 T to 2.258 J kg^{-1}K^{-1} at an applied field of 5 T.

Figs. 15 (b) and (c) show $-\Delta S_M$ of La$_{0.67}$Ba$_{0.22}$Sr$_{0.11}$Mn$_{1-x}$Fe$_x$O$_3$ (x = 0.1 and x = 0.2) samples as a function of temperature at different external magnetic fields. The magnitude of the maximum change in magnetic entropy for the sample with x = 0.1 improved only slightly at high fields with respect to the un-substituted sample, and the peak became sharper around T_C. However, the maximum entropy changes for the sample with x = 0.2 dropped down to about half of the value for the previous two samples (1.03 J K^{-1}kg^{-1} at 139 K). This is due to the reduction in the rate of change in magnetization with increasing temperature as a result of the increase in the antiferromagnetically aligned moments of Mn and Fe ions. Our magnetic results are consistent with those obtained by P. Nisha *et al.* [73]. The small value of magnetic entropy change can be associated with the fact that the magnetic phase transition in these samples is of second order.

Generally, materials with first order transitions exhibit much larger magnetocaloric effect than materials with second order transitions. However, the importance of the studied system stems from the ability to tune their properties and control the structural and morphological factors, in addition to the ease of synthesizing such materials with good chemical stability. In Table5, we list the MCE values of the present compound in comparison with those reported in the literature having the same chemical elements.

The relative cooling power (RCP) was calculated using the formula:

$$RCP = |\Delta S_M^{MAX}(T, H)| \times \delta T_{FWHM} \tag{13}$$

Where δT_{FWHM} is the full-width at half-maximum of $|\Delta S_M^{Max}|$ versus temperature [74, 75]. Relatively high RCP values of 193.27 J/kg and 153.2 J/kg at ΔH = 5 T were obtained for the samples x = 0 and x = 0.1. Despite the fact that iron doping reduces T_C and $|\Delta S_M|$ at relatively high iron concentration, the observed properties of the investigated system are promising and open the way for investigations of materials useful for magnetic refrigeration.

Materials Research Forum LLC
https://doi.org/10.21741/9781644900970-3

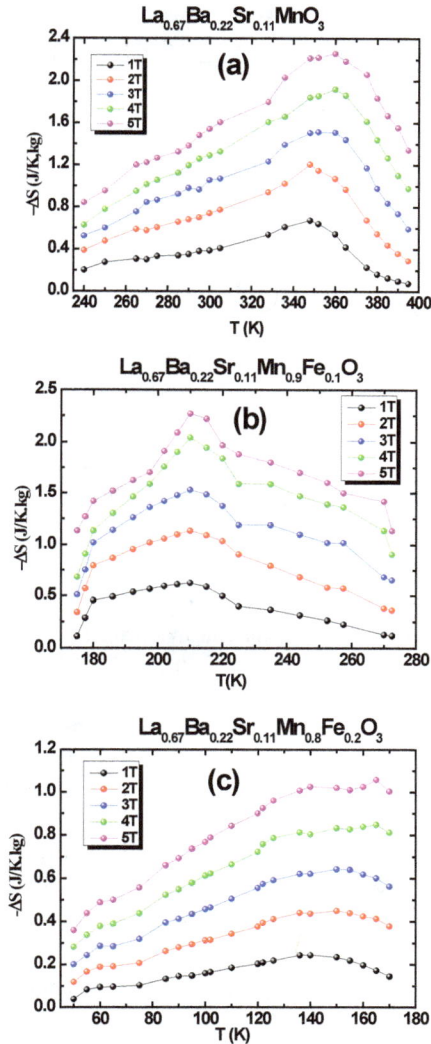

Fig. 15. Temperature dependence of magnetic entropy change for samples: (a) $La_{0.67}Ba_{0.22}Sr_{0.11}MnO_3$, (b). $La_{0.67}Ba_{0.22}Sr_{0.11}Mn_{0.9}Fe_{0.1}O_3$, (c). $La_{0.67}Ba_{0.22}Sr_{0.11}Mn_{0.8}Fe_{0.2}O_3$.

Table 5. MCE values of the compound in comparison with reported one in the literature having the same chemical elements.

Samples	$|\Delta S_M^{Max}|$ (J kg^{-1} K^{-1})	RCP (Jkg^{-1})	ΔH (T)	Ref.
Gd	9.5	410	5	[76]
Gd$_5$(Si$_2$Ge$_2$)	18.5	535	5	[76]
La$_{0.8}$Ba$_{0.2}$MnO$_3$	4.15	230	5	[46]
La$_{0.8}$Ba$_{0.2}$Mn$_{0.95}$Fe$_{0.05}$O$_3$	3	238	5	[46]
La$_{0.8}$Ba$_{0.2}$Mn$_{0.9}$Fe$_{0.1}$O$_3$	2.62	211	5	[46]
La$_{0.67}$Ba$_{0.22}$Sr$_{0.11}$MnO$_3$	2.258	193	5	[77]
La$_{0.67}$Ba$_{0.22}$Sr$_{0.11}$Mn$_{0.9}$Fe$_{0.1}$O$_3$	2.261	153	5	[77]
La$_{0.67}$Ba$_{0.22}$Sr$_{0.11}$Mn$_{0.8}$Fe$_{0.2}$O$_3$	1.03	91	5	[77]
La$_{0.67}$Ba$_{0.33}$MnO$_3$	1.48	161	5	[78]
La$_{0.67}$Ba$_{0.33}$Mn$_{0.98}$Ti$_{0.02}$O$_3$	3.19	307	5	[79]
La$_{0.7}$Sr$_{0.3}$Mn$_{0.093}$Fe$_{0.07}$O$_3$	4	25	2	[79]

Conclusions

We investigated the structural, magnetic and magnetocaloric properties of nanopowder samples La$_{0.67}$Ba$_{0.22}$Sr$_{0.11}$Mn$_{1-x}$Fe$_x$O$_3$ (x = 0, 0.1, 0.2, 0.3, and 1.0) prepared by the conventional ceramic method. By Rietveld refinement, we generally found that all samples were almost single-phase and crystallize in the orthorhombic structure with P_{nma} space group. However, a weak diffraction peak corresponding to a very small amount of Mn$_3$O$_4$ precipitates was identified in the pattern of the sample with x = 0.2. The substitution of Mn by Fe was found to influence the magnetic characteristics significantly as a result of the development of SE interactions at the expense of the ferromagnetic interactions. Relatively large values of MCE wereobtained at low Fe substitution levels, and $|\Delta S_M^{MAX}|$ reached the highest value for the sample with x = 0.1 at an applied magnetic field of 5 T.

The critical behavior of La$_{0.67}$Ba$_{0.22}$Sr$_{0.11}$MnO$_3$, La$_{0.67}$Ba$_{0.22}$Sr$_{0.11}$Mn$_{0.9}$Fe$_{0.1}$O$_3$ and La$_{0.67}$Ba$_{0.22}$Sr$_{0.11}$Mn$_{0.8}$Fe$_{0.2}$O$_3$ perovskites was comprehensively studied by the isothermaldc-magnetization around the Curie point T_C. Refined values of the parameters T_C, β, γ, and δ were obtained by means of the MAP and Kouvel–Fisher methods. The values of critical exponents for the parent sample with x = 0 agree with those of a 3D-Heisenberg-like ferromagnet with short-range exchange couplings. However, with the substitution of Fe^{3+} ion, the critical exponent β increased, and the critical exponent γ decreases approaching the 3D-Ising value of $\delta = 1.24$. These results show a change in the universality class of these manganites with iron doping, which can be explained by the suppression of the DE interaction caused by Fe^{3+} ions at the Mn^{3+}site, and the enhancement of the super exchange interaction. These findings indicated that the critical behavior of the magnetic transition in manganites is sensitive to Mn site doping.

References

[1]   S. Ben Abdelkhalek, N. Kallel, S. Kallel, O. Pena, M. Oumezzine, Critical behavior and magnetic entropy change in the $La_{0.6}Sr_{0.4}Mn_{0.8}Fe_{0.1}Cr_{0.1}O_3$ perovskite, Journal of Magnetism and Magnetic Materials, 324 (2012) 3615-3619. https://doi.org/10.1016/j.jmmm.2012.06.024

[2]   The-Long Phan, T. A. Ho, P. D. Thang, Q. T. Tran, T. D. Thanh, N. X. Phuc, M. H. Phan, B. T. Huy, S. C. Yua, Critical behavior of Y-doped $Nd_{0.7}Sr_{0.3}MnO_3$ manganites exhibiting the tricritical point and large magnetocaloric effect, Journal of Alloys and Compounds, 615 (2014), 937-945. https://doi.org/10.1016/j.jallcom.2014.06.107

[3]   N. Dhahri, J. Dhahri, E. K. Hlil, E. Dhahri, Critical behavior in Co-doped manganites $La_{0.67}Pb_{0.33}Mn_{1-x}Co_xO_3$ ($0 \leq x \leq 0.08$), Journal of Magnetism and Magnetic Materials, 324 (2012) 806-811. https://doi.org/10.1016/j.jmmm.2011.09.024

[4]   L. Chen, J. H. He,Y. Mei, Y. Z. Cao, W.W. Xia, H. F. Xu, Z.W. Zhu, Z.A. Xu, Critical behavior of Mo-doping $La_{0.67}Sr_{0.33}Mn_{1-x}Mo_xO_3$ perovskite system, Physica B, 404 (2009) 1879-1882. https://doi.org/10.1016/j.physb.2008.07.023

[5]   M. Oumezzine, O. Pena, S. Kallel, M. Oumezzine, Crossover of the magnetocaloric effect and its importance on the determination of the critical behaviour in the $La_{0.67}Ba_{0.33}Mn_{0.9}Cr_{0.1}O_3$ perovskite manganite, Journal of Alloys and Compounds, 539 (2012) 116-123. https://doi.org/10.1016/j.jallcom.2012.06.043

[6]   C. Zener, Interaction between the d-Shells in the Transition Metals. II. Ferromagnetic Compounds of Manganese with Perovskite Structure, Physical Review, 82 (1951) 403-405. https://doi.org/10.1103/PhysRev.82.403

[7]   P.W. Anderson and H. Hasegawa, Considerations on Double Exchange, Physical Review, 100 (1955) 675-681. https://doi.org/10.1103/PhysRev.100.675

[8]   A. J. Millis, B.I. Shraiman, R. Mueller, Dynamic Jahn-Teller Effect and Colossal Magnetoresistance in $La_{1-x}Sr_xMnO_3$, Physical Review Letters, 77 (1996) 175-178. https://doi.org/10.1103/PhysRevLett.77.175

[9]   M. Triki, E. Dhahri, E.K. Hlil, Unconventional critical magnetic behavior in the Griffiths ferromagnet $La_{0.4}Ca_{0.6}MnO_{2.8\square0.2}$ oxide, Journal of Solid State Chemistry, 201 (2013) 63-67. https://doi.org/10.1016/j.jssc.2013.02.019

[10]   S. Ghodhbane, E.Tka, J. Dhahri, E. K. Hlil, A large magnetic entropy change near room temperature in $La_{0.8}Ba_{0.1}Ca_{0.1}Mn_{0.97}Fe_{0.03}O_3$ perovskite, Journal of Alloys and Compounds, 600 (2014) 172-177. https://doi.org/10.1016/j.jallcom.2014.02.096

[11]   J. Mira, J. Rivsa, F. Rivadulla, C.V.Vazquez, M.A.L. Quintela, Change from first- to second-order magnetic phase transition in $La_{2/3}(Ca,Sr)_{1/3}MnO_3$ perovskites, Physical Review B, 60 (1999) 2998-3001. https://doi.org/10.1103/PhysRevB.60.2998

[12]   P. Zhang, P. Lampe, T.L. Phan, S. C. Yu, T. D. Thanh, N. H. Dan, V. D. Lam, H. Srikanth, M. H. Phan, Influence of magnetic field on critical behavior near a first order transition in optimally doped manganites: The case of $La_{1-x}Ca_xMnO_3$ ($0.2 \le x \le 0.4$), Journal of Magnetism and Magnetic Materials, 348 (2013) 146-153. https://doi.org/10.1016/j.jmmm.2013.08.025

[13]   M. H. Phan, V. Franco, N. S. Bingham, H. Srikanth, N. H. Hur, S. C. Yu, Tricritical point and critical exponents of $La_{0.7}Ca_{0.3-x}Sr_xMnO_3$ ($x = 0, 0.05, 0.1, 0.2, 0.25$) single crystals, Journal of Alloys and Compounds, 508 (2010) 238 - 244. https://doi.org/10.1016/j.jallcom.2010.07.223

[14]   T.-L. Phan, Q.T. Tran, P.Q. Thanh, P.D.H. Yen, T.D. Thanh, S.C. Yu, Critical behavior of $La_{0.7}Ca_{0.3}Mn_{1-x}Ni_xO_3$ manganites exhibiting the crossover of first- and second-order phase transitions, Solid State Communications, 184 (2014) 40-46. https://doi.org/10.1016/j.ssc.2013.12.032

[15]   E. Bucher and W. Sitte, Defect chemical analysis of the electronic conductivity of strontium-substituted lanthanum ferrite, Solid State Ionics, 173 (2004) 23-28. https://doi.org/10.1016/j.ssi.2004.07.047

[16]   S. P. Simner, J. F. Bonnett, N. L. Canfield, K. D. Meinhardt, J. P. Shelton, V. L. Sprenkle, and J. W. Stevenson, Development of lanthanum ferrite SOFC cathodes, Journal of Power Sources, 113 (2003) 1-10. https://doi.org/10.1016/S0378-7753(02)00455-X

[17]   S. Tanasescu, N. D. Totir, D. I. Marchidan, and A. Turcanu, The influence of compositional variables on the thermodynamic properties of lanthanum strontium ferrite manganites and lanthanum strontium manganites, Materials Research Bulletin, 32 (1997) 915-923. https://doi.org/10.1016/S0025-5408(97)00054-8

[18]   K. Yan, X. Fu, and Y. Cui, Influence of stoichiometric ratio and Mn-doping on the surface morphology and dielectric properties of perovskite $(La,Sr)FeO_3$ films,

Journal of Inorganic and Organometallic Polymers and Materials, 22 (2012) 59-63. https://doi.org/10.1007/s10904-011-9567-6

[19] M. Paraskevopoulos, F. Mayr, J. Hemberger, A. Loidl, R. Heichele, D. Maurer, V. Müller, A. Mukhin, A. Balbashov, Magnetic properties and the phase diagram of $La_{1-x}Sr_xMnO_3$ for $x \leq 0.2$, Journal of Physics: Condensed Matter, 12 (2000) 3993-4011. https://doi.org/10.1088/0953-8984/12/17/307

[20] E. Dixon, J. Hadermann, M.A. Hayward, Structures and Magnetism of $La_{1-x}Sr_xMnO_{3-(0.5+x)/2}$ ($0.67 \leq x \leq 1$) Phases, Chemistry of Materials, 24 (2012) 1486-1495. https://doi.org/10.1021/cm300199b

[21] A. Urushibara, Y. Moritomo, T. Arima, A. Asamitsu, G. Kido, Y. Tokura, Insulator-metal transition and giant magnetoresistance in $La_{1-x}Sr_xMnO_3$, Physical Review B, 51 (1995) 14103-14109. https://doi.org/10.1103/PhysRevB.51.14103

[22] N. Chau, H. N. Nhat, N. H. Luong, D. L. Minh, N. D. Tho, N. N. Chau, Structure, magnetic, magnetocaloric and magnetoresistance properties of $La_{1-x}Pb_xMnO_3$ perovskite, Physica B: Condensed Matter, 327 (2003) 270-278. https://doi.org/10.1016/S0921-4526(02)01759-3

[23] N. H. Luong, D. T. Hanh, N. Chau, N. D. Tho and T. D. Hiep, Properties of perovskites $La_{1-x}Cd_xMnO_3$, Journal of Magnetism and Magnetic Materials, 290-291 (1) (2005) 690-693. https://doi.org/10.1016/j.jmmm.2004.11.338

[24] G. H. Jonker, Magnetic compounds with perovskite structure IV Conducting and non-conducting compounds, Physica, 20 (1956) 707-722. https://doi.org/10.1016/S0031-8914(56)90023-4

[25] R. D. Shannon, Revised effective ionic radii and systematic studies of interatomic distances in halides and chalcogenides, Acta Crystallographica, A32 (1976) 751-767. https://doi.org/10.1107/S0567739476001551

[26] J. Mira, J. Rivas, L. E. Hueso, F. Rivadulla, M. A. Lopez Quintela, Drop of magnetocaloric effect related to the change from first- to second-order magnetic phase transition in $La_{2/3}(Ca_{1-x}Sr_x)_{1/3}MnO_3$, Journal of Applied Physics, 91 (2002) 8903-8907. https://doi.org/10.1063/1.1451892

[27] J.Z. Sun, W.J. Gallagher, P.R. Duncombe, L. Krusin-Elbaum, R.A. Altman, A. Gupta, Y. Lu, G.Q. Gong, G. Xiao, Observation of large low-field magnetoresistance in trilayer perpendicular transport devices made using doped manganate perovskites, Applied Physics Letter, 69 (1996) 3266-3271. https://doi.org/10.1063/1.118031

[28]   S. Kallel, N. Kallel, O. Peña, M. Oumezzine, Large magnetocaloric effect in Ti-modified $La_{0.7}Sr_{0.3}MnO_3$ perovskite, Materials Letters, 64 (2010) 1045-1048. https://doi.org/10.1016/j.matlet.2010.02.005

[29]   N. A. Viglin, S. V. Naumov, Y. M. Mukovskii, A Magnetic resonance study of $La_{1-x}Sr_xMnO_3$ manganites, Physics of the Solid State, 43 (2001) 1855–1863. https://doi.org/10.1134/1.1410634

[30]   A. S. Erchidi Elyacoubi, R. Masrour, A. Jabar, Magnetocaloric effect and magnetic properties in $SmFe_{1-x}Mn_xO_3$ perovskite: Monte Carlo simulations, Solid State Communications, 271 (2018) 39-43. https://doi.org/10.1016/j.ssc.2017.12.015

[31]   R. Masrour, A. Jabar, A. Benyoussef, M. Hamedoun, E. K. Hlil, Monte Carlo simulation study of magnetocaloric effect in $NdMnO_3$ perovskite, Journal of Magnetism and Magnetic Materials, 401 (2016) 91-95. https://doi.org/10.1016/j.jmmm.2015.10.019

[32]   R. Masrour, A. Jabar, H. Khlif, F. Ben Jemaa, M. Ellouze, E. K. Hlil, Experiment, mean field theory and Monte Carlo simulations of the magnetocaloric effect in $La_{0.67}Ba_{0.22}Sr_{0.11}MnO_3$ compound, Solid State Communications, 268 (2017) 64-69. https://doi.org/10.1016/j.ssc.2017.10.003

[33]   A. S. Erchidi Elyacoubi, R. Masrour, A. Jabar, M. Ellouze, E. K. Hill, Magnetic properties and magnetocaloric effect in double Sr_2FeMoO_6 perovskites, Materials Research Bulletin, 99 (2018) 132-135. https://doi.org/10.1016/j.materresbull.2017.10.037

[34]   H. M. Rietveld, A profile refinement method for nuclear and magnetic structures, Journal of Applied Crystallography, 2 (1969) 65-71. https://doi.org/10.1107/S0021889869006558

[35]   P. A. Joy, C. R. Sankar, S. K. Date, The limiting value of x in the ferromagnetic compositions $La_{1-x}MnO_3$, Journal of Physics: Condensed Matter, 14 (2002) L663-L669. https://doi.org/10.1088/0953-8984/14/39/104

[36]   A. G. Mostafa, E. K. Abdel-Khalek, W. M. Daoush, S. F. Moustfa, Study of some co-precipitated manganite perovskite samples doped iron, Journal of Magnetism and Magnetic Materials, 320 (2008) 3356-3360. https://doi.org/10.1016/j.jmmm.2008.07.025

[37]   P. Schiffer, A.P. Ramírez, W. Bao, S.-W. Cheong, Low Temperature Magnetoresistance and the Magnetic Phase Diagram of $La_{1-x}Ca_xMnO_3$, Physical Review Letters, 75 (1995) 3336-3339. https://doi.org/10.1103/PhysRevLett.75.3336

[38]   N. Kallel, M. Oumezzine, H. Vincent, Neutron powder diffraction study of
       structural and magnetic structure of $La_{0.7}Sr_{0.3}Mn_{1-x}TixO_3$ (x = 0, 0.10, 0.20, and
       0.30), Journal of Magnetism and Magnetic Materials, 320 (2008) 1810-1816.
       https://doi.org/10.1016/j.jmmm.2008.02.106

[39]   A. Guinier. in: X. Dunod (Ed.), Théorie et Technique de la Radiocristallographie,
       3rd ed., 482 (1964).

[40]   M. Ellouze, W. Boujelben, A. Cheikhrouhou, H. Fuess, R. Madar, Vacancy effects
       on the crystallographic and magnetic properties in lacunar $Pr_{0.7}Ba_{0.3-x}MnO_3$ oxides,
       Solid State Communications, 124 (2002) 125-130. https://doi.org/10.1016/S0038-
       1098(02)00482-9

[41]   J. Gutiérrez, A. Peña, J. M. Barandiarán, J. L. Pizarro, T. Hernández, L. Lezama,
       M. Insausti et T. Rojo, Structural and magnetic properties of
       $La_{0.7}Pb_{0.3}(Mn_{1-x}Fe_x)O_3$ (0<~x<~0.3) giant magnetoresistance perovskites, Physical
       Review B, 61 (2000) 9028-9035. https://doi.org/10.1103/PhysRevB.61.9028

[42]   K. H. Ahn, X. W. Wu, K. Liu, C. L. Chien, Magnetic properties and colossal
       magnetoresistance of La(Ca)MnO3 materials doped with Fe, Physical Review B,
       54 (1996) 15299- 15302. https://doi.org/10.1103/PhysRevB.54.15299

[43]   Y. L. Chang, Q. Huang, C. K. Ong, Effect of Fe doping on the magnetotransport
       properties in $Nd_{0.67}Sr_{0.33}MnO_3Nd_{0.67}Sr_{0.33}MnO_3$ manganese oxides, Journal of
       Applied Physics, 91 (2002) 789-793. https://doi.org/10.1063/1.1421044

[44]   M. Baazaoui, S. Zemni, M. Boudard, H. Rahmouni, A. Gasmi, A. Selmi, M.
       Oumezzine. Magnetic and electrical behaviour of $La_{0.67}Ba_{0.33}Mn_{1-x}Fe_xO_3$
       perovskites, Materials Letters, 63 (2009) 2167-2170.
       https://doi.org/10.1016/j.matlet.2009.07.019

[45]   N. Kallel, Ben S. Abdelkhalek, S. Kallel, O. Péna, M. Oumezzine, Structural and
       magnetic properties of $(La_{0.70-x}Y_x)Ba_{0.30}Mn_{1-x}Fe_xO_3$ perovskites simultaneously
       doped on A and B sites ($0.0 \leq x \leq 0.30$) Journal of Alloys and Compounds, 501
       (2010) 30-36. https://doi.org/10.1016/j.jallcom.2010.04.073

[46]   W. Chérif, M. Ellouze, A.-F. Lehlooh, F. Elhalouani, Structure, ferromagnetism
       and magnetotransport properties of nanopowders of $Pr_{0.67}Ca_{0.33}Fe_xMn_{1-x}O_3$
       manganites oxide prepared by sol–gel method, Journal of Alloys and Compounds,
       543 (2012) 152-158. https://doi.org/10.1016/j.jallcom.2012.06.014

[47]   J. M. D. Coey, M. Viret, S. Von Molnar, Mixed-valence manganites, Advances in
       Physics, 48 (1999)167-293. https://doi.org/10.1080/000187399243455

[48]   M. R. Said, Y. A Hamam, I. Abu-Aljarayesh, S. Mahmood, Critical exponents of (Fe,Mn)$_3$Si, Journal of Magnetism and Magnetic Materials, 195 (1999) 679-686. https://doi.org/10.1016/S0304-8853(99)00285-1

[49]   K. Ghosh, C. J. Lobb, R. L. Greene, S. G. Karabashev, D.A. Shulyatev, A. A. Arsenov, Y. Mukovskii, Critical phenomena in the double-Exchange ferromagnet La$_{0.7}$Sr$_{0.3}$MnO$_3$, Physical Review Letters, 81 (1998) 4740-4743. https://doi.org/10.1103/PhysRevLett.81.4740

[50]   Y. Motome, N. Furukawa, Critical phenomena of ferromagnetic transition in double-exchange systems, Journal of Physical Society of Japan, 70 (2001) 1487-1490. https://doi.org/10.1143/JPSJ.70.1487

[51]   P. Zhang, P. Lampen, T. L. Phan, S. C. Yu, T. D. Thanh, N. H. Dan, V. D. Lam, H. Srikanth, M. H. Phan, Influence of magnetic field on critical behavior near a first order transition in optimally doped manganites: The case of La$_{1-x}$Ca$_x$MnO$_3$ ($0.2 \leq x \leq 0.4$), Journal of Magnetism and Magnetic Materials, 348 (2013) 146-153. https://doi.org/10.1016/j.jmmm.2013.08.025

[52]   P.W. Anderson, H. Hasegawa, Considerations on double exchange, Physical Review, 100 (1955) 575-681. https://doi.org/10.1103/PhysRev.100.675

[53]   H E. Stanley, Introduction to Phase Transitions and Critical Phenomena, Oxford University Press (1971), London and New York.

[54]   C. Martin, A. Maignan, M. Hervieu, B. Raveau, Magnetic phase diagrams of L$_{1-x}$A$_x$MnO$_3$ manganites (L = Pr, Sm, A = Ca, Sr), Physical Review B, 60 (1999) 12191-12199. https://doi.org/10.1103/PhysRevB.60.12191

[55]   J. Fan, B. Hong, L. Zhang, Y. Shi, W. Tong, L. Ling, L. Pi, Y. Zhang, Heisenberg-like ferromagnetism and percolative conductivity in the half-doped manganite Nd$_{0.5}$Ca$_{0.25}$Sr$_{0.25}$MnO$_3$, Journal of Magnetism and Magnetic Materials, 322 (2010) 3692-3695. https://doi.org/10.1016/j.jmmm.2010.07.027

[56]   N. Moutis, I. Panagiotopoulos, M. Pissas, D. Niarxhos, Structural and magnetic properties of La$_{0.67}$(Ba$_x$Ca$_{1-x}$)$_{0.33}$MnO$_3$ perovskites ($0 < \sim x < \sim 1$), Physical Review B, 59 (1999) 1129-1133. https://doi.org/10.1103/PhysRevB.59.1129

[57]   T. l. Phan, P. Q. Thanh, N. H. Sinh, K. W. Lee, S.C. Yu, Critical behavior and magnetic entropy change in La$_{0.7}$Ca$_{0.3}$Mn$_{0.9}$Zn$_{0.1}$O$_3$ perovskite manganite, Current Applied Physics, 11 (2011) 830-833. https://doi.org/10.1016/j.cap.2010.12.002

[58]   J. C. Debnath, P. Shamba, A. M. Strydom, J. L. Wang and S. X. Dou, Investigation of the critical behavior in Mn$_{0.94}$Nb$_{0.06}$CoGe alloy by using the field dependence of magnetic entropy change, Journal of Applied Physics, 113 (2013) 0939021-0939025. https://doi.org/10.1063/1.4794100

[59]   B. K. Banerjee, On a generalised approach to first and second order magnetic transitions, Physics Letters, 12 (1964) 16-17. https://doi.org/10.1016/0031-9163(64)91158-8

[60]   A. Arrott, J. E. Noakes, Approximate equation of state for nickel near its critical temperature, Physical Review Letters, 19 (1967) 786-789. https://doi.org/10.1103/PhysRevLett.19.786

[64]   A.Omri, A. Tozri, M. Bejar, E. Dhahri, E. K. Hlil, Critical behavior in Ga-doped manganites $La_{0.75}(Sr,Ca)_{0.25}Mn_{1-x}Ga_xO_3$ ($0 \leq x \leq 0.1$), Journal of Magnetism and Magnetic Materials, 324 (2012) 3122-3128. https://doi.org/10.1016/j.jmmm.2012.05.013

[65]   M. Baazaoui, S. Zemni, M. Boudard, H. Rahmouni, A. Gasmi, A. Selmi, M. Oumezzin, Magnetic and electrical behaviour of $La_{0.67}Ba_{0.33}Mn_{1-x}Fe_xO_3$ perovskites, The International Journal of Nanoelectronics and Materials, 3(2010) 23-26. https://doi.org/10.1016/j.matlet.2009.07.019

[66]   S. M. Yusuf, J. M. De Teresa, P. A. Algarabel, J. Blasco, M. R. Ibarra, A. Kumar, C. Ritter, Nature of the magnetic ordering for small mean-size and large-size mismatch of A-site cations in CMR manganites, Physica B, 401 (2006) 385–386. https://doi.org/10.1016/j.physb.2006.05.083

[67]   M.E. Fisher, S-K. Ma, B. G. Nickel, Critical Exponents for Long-Range Interactions, Physical Review Letters, 29 (1972) 917-920. https://doi.org/10.1103/PhysRevLett.29.917

[68]   S. Ghodhbane, A. Dhahri, N. Dhahri, E. K. Hlil, J. Dhahri, M. Alhabradi, M. Zaidi, Critical behavior in Fe-doped manganites $La_{0.8}Ba_{0.2}Mn_{1-x}Fe_xO_3$ ($x = 0.15$ and $x = 0.2$), Journal of Alloys and Compounds, 580 (2013) 558-563. https://doi.org/10.1016/j.jallcom.2013.06.181

[69]   N.V. Khiem, L.V. Bau, Critical exponents for the ferromagnetic-paramagnetic transition in $La_{0.7}Sr_{0.3}Mn_{0.8}Ti_{0.2}O_3$, Journal of the Korean Physical Society, 52(2008) 1518-1521. https://doi.org/10.3938/jkps.52.1518

[70]   F. Ben Jemaa, S. H. Mahmood, M. Ellouze, E. K. Hlil, F. Halouani, Critical behavior in Fe-doped manganites $La_{0.67}Ba_{0.22}Sr_{0.11}Mn_{1-x}Fe_xO_3$ ($0 \leq x \leq 0.2$, Journal of Materials Science, 49 (2014) 6883-4887. https://doi.org/10.1007/s10853-014-8390-1

[71]   N. K. Singh, K. G. Suresh, A. K. Nigam, Itinerant electron metamagnetism and magnetocaloric effect in $Dy(Co,Si)_2$, Solid State Communications, 127 (2003) 373-377. https://doi.org/10.1016/S0038-1098(03)00441-1

Materials Research Forum LLC
https://doi.org/10.21741/9781644900970-3

[72] R. D. McMichael, J. J. Ritter, R. D. Shull, Enhanced magnetocaloric effect in $Gd_3Ga_{5-x}Fe_xO_{12}$, Journal of Applied Physics, 73 (1993) 6946-4650. https://doi.org/10.1063/1.352443

[73] P. Nisha, S. Savitha Pillai, A. Darbandi, M. Varma, K.G. Suresh, H. Hahn, Critical behaviour and magnetocaloric effect of nano crystalline $La_{0.67}Ca_{0.33}Mn_{1-x}Fe_xO_3$ (x = 0.05, 0.2) synthesized by nebulized spray pyrolysis, Journal of Materials Chemistry and Physics, 136 (2012) 74-74. https://doi.org/10.1016/j.matchemphys.2012.06.029

[74] S. Tapas, I. Das, S. Banerjee, Magnetocaloric effect in Ho_5Pd_2: Evidence of large cooling power, Applied Physics Letters, 91(2007) 082511-082515. https://doi.org/10.1063/1.2775050

[75] M.H. Phan, S. C. Yu, Review of the magnetocaloric effect in manganite materials, Journal of Magnetism and Magnetic Materials, 308 (2007) 325-340. https://doi.org/10.1016/j.jmmm.2006.07.025

[76] J. S. Lee, Evaluation of the magnetocaloric effect from magnetization and heat capacity data, Physica Status Solidi B, 7 (2004) 1765 – 1768. https://doi.org/10.1002/pssb.200304685

[77] F. Ben Jemaa, S. Mahmood, M. Ellouze, E. K. Hlil, F. Halouani, I. Bsoul, M. Awawdeh, Structural, magnetic and magnetocaloric properties of $La_{0.67}Ba_{0.22}Sr_{0.11}Mn_{1-x}Fe_xO_3$ nanopowders, Solid State Sciences, 37 (2014) 121-130. https://doi.org/10.1016/j.solidstatesciences.2014.09.004

[78] W. Zhong, W. Cheng, C. T. Au, Y. W. Du, Dependence of the magnetocaloric effect on oxygen stoichiometry in polycrystalline $La_{2/3}Ba_{1/3}MnO_{3-\delta}$, Journal of Magnetism and Magnetic Materials, 261 (2003) 238-243. https://doi.org/10.1016/S0304-8853(02)01479-8

[79] X. Bohigas, J. Tejada, E. D. Barco, X. X. Zhang, M. Sales, Tunable magnetocaloric effect in ceramic perovskites, Applied Physics Letters, 73 (1998) 390-394. https://doi.org/10.1063/1.121844

Chapter 4

Graphene-based Materials and their Nanocomposites with Metal Oxides: Biosynthesis, Electrochemical, Photocatalytic and Antimicrobial Applications

Ratiram Gomaji Chaudhary[1,a], Ajay K. Potbhare[1,b], Prashant B. Chouke[1,c], Alok R. Rai[2,d], Raghvendra Kumar Mishra[3,e], Martin F. Desimone[4,f] and Ahmed A. Abdala[5,g]

[1]Post Graduate Department of Chemistry, Seth Kesarimal Porwal College of Arts, Science and Commerce, Kamptee – 441001, Maharashtra, India

[2]Post Graduate Department of Microbiology, Seth Kesarimal Porwal College of Arts, Science and Commerce, Kamptee – 441001, Maharashtra, India

[3]IMEDA Materials,Technogetafe, Calle Eric Kandel, 2, 28906 Getafe, Madrid, Spain

[4]Universidad de Buenos Aires, Consejo Nacional de Investigaciones Científicasy Técnicas (CONICET). Instituto de la Químicay Metabolismodel Fármaco (IQUIMEFA), Facultad de Farmacia y Bioquimica, Junin 956 Piso 3°, (1113) Buenos Aires, Argentina

[5]Chemical Engineering Program, Texas A& M University at Qatar, POB 23874, Doha, Qatar

[a]chaudhary_rati@yahoo.com, [b]ajaypotbhare2@gmail.com, [c]prashant_bchouke@rediffmail.com, [d]alok.rrai@gmail.com, [e]raghvendramishra4489@gmail.com, [f]martinfdesimone@gmail.com, [g]ahmed.abdala@qatar.tamu.edu

Abstract

Metal oxides and their nanocomposites are used in various technological applications. Biofabrication of carbon-based metal oxide nanocomposites preparation using plants, microbes, cell cultures and enzymes are the most attractive technique because of non-toxic nature, and sustainable process. Phytochemicals play important role in size lessening of the particles by performing as structure-directing, capping and reducing agents. In this chapter, we shed light on eco-friendly, money-spinning, and phytosynthesis of carbon based nanomaterials (CNMs) like graphene oxide, reduced graphene oxide, and metal doped-rGO nanocomposites using green reducers. Moreover, electrochemical, photocatalytical and biological applications of CNMs and their nanocomposites with metal oxides are discussed.

Keywords

Carbon Nanomaterials, Biosynthesis, Electrochemical Performances, Photocatalytical Activity, Biological Assay, Metal Oxides, Graphene Oxide, Reduced Graphene Oxide

Contents

1. Introduction

Nowadays, the growing global populations invariably exert tremendous pressure on the scientific community to tackle several environmental issues and day to day challenges including water pollution, microbial diseases, increased carbon dioxide level, use of hazardous chemicals, proper drug delivery, electronic devices, sensors and energy storage etc. In this context, the development of multifunctional materials for environmental remediation and modernization is a key approach to tackle many of these challenges. Carbon-based nanomaterials including carbon nanotubes, nanodots, nanofibers, nanowires, fullerenes, and graphene could be multifunctional smart materials because of their stupendous mechanical, thermal, chemical, and electrical properties. Carbon is one of the most plentiful element in the world as it represents about 0.2 % of the earth's crust, and it is also the second common element in the human body. One of the outstanding characteristics of carbon is its high kinetic stability. In addition to the recognized types of carbon allotropy, e.g. graphite and diamond, several new carbon allotropies have been discovered in the recent few decades including buckminsterfullerene, carbon nanotubes, carbon nanobuds, nanoribbons, carbon balls, carbon dots, and graphene. The properties of each allotrope depends on the structural arrangement of the adjacent carbon atoms. For example, diamond has the highest hardness among all materials, while graphite is brittle [1, 2]. Carbon nanomaterials are the most promising materials in the field of smart nanotechnology. Carbon nanomaterials and their composites or metal-doped carbon nanomaterials have interesting optical and electrical properties that are important for smart devices [3-5,2]. In addition, carbon nanomaterials-based composites exhibit excellent mechanical, electrical and thermal properties making them potential for various purposes. Carbon-based nanomaterials can also be used in multifunctional biomedical devices and products such as sensors, pharmaceuticals and drugs [5,6]. Nevertheless, for such applications to be commercialized, major challenges remain to be overcome.

This chapter makes available a concise compilation of the synthesis, assets, and appliances of carbon-based nanomaterials. The research on application of carbon nanomaterials (CNMs) as multifunctional devices is eye-catching [7, 8]. Among these materials, we focus on graphene, carbon-nanotubes, carbon-nanofibers, and fullerenes, which have capable appliances in the field of photocatalytic activity, electrochemical performances, bio-sensing, antibacterial coating and drug delivery [9]. Moreover, the interesting photoluminescence properties of CNMs, which are reported by several researchers will be discussed [10, 11]. Bucky balls, nanoribbons, nanosphere, nanocages and nanotubes have been fascinating to researchers and the scientific community for decades.

Magnetic Oxides and Composites II   Materials Research Forum LLC
Materials Research Foundations 83 (2020) 79-116   https://doi.org/10.21741/9781644900970-4

Likewise, CNTs, which were first discovered in 1991 by Sumio Iijima *et al.* [13], are the most important nanostructured materials in the recent decades because of their unique mechanical and electrical characteristics contrast to additional carbon allotropy due to its one-dimensional structure and few nanometers diameter [12].

Further, a cylindrical and concentric carbon structure, well-developed materials of carbon called multi-walled carbon nanotubes (MWCNTs), was synthesized [14]. In 1993, Ijma and Donald Bethune have published separate studies on the synthesis of SWCNT [15,16]. After these independent reports, nanotubes grow to be the subject of research in various science fields [17]. CNTs implication in nanotechnology is evidenced by their potential facets of applications. In engineering, CNTs are use in developing polymer nanocomposites for the aeronautics industry [18]. They have also been used in nanodevices and electronic nanocircuits to fabricate new computer chips [19].

Recently, graphene, single sheet of sp^2 carbon, has attracted substantial attention amid chemists, physicists, and material scientists because of its 2-D structure and combination of extraordinary properties [20]. Graphene is highly ordered with many extraordinary properties with very high surface area (2630 m^2g^{-1}), extremely high Young's modulus (1000 GPa), high thermal conductivity (~5 $kWm^{-1}K^{-1}$), very fast electron mobility (2.5×105 $cm^2v^{-1}s^{-1}$), excellent optical transparency (~97.7%) [25] and, strong chemical stability [21]. It also provides detection of single gas molecule [22] and room temperature long-range ballistic transport [24]. These exceptional graphene characteristics have led to exploring graphene in broad variety of appliances in various technologies such as lithium-ion batteries [26], catalysis [27], chemical sensors [28], biosensors [29], anti-bacterial agents[30], flexible thin film transistors [31], drug delivery [32], solar cells [33], photovoltaic cells [34], imaging [35], p-n junction materials [36], super capacitors [37], touch panel screens [38], electromagnetic shielding applications [39], and water purification for removal of different noxious wastes such as oils, alkenes, aromatics, organic solvents, dyes, and ionic solutions [40-42].

Current developments of graphene-based semiconductors as 'Nano-photocatalyst have been considered as arising agent to address the budding environmental concerns. Scalable production of carbon based materials is necessary to meet up the large demand, but large-scale production of good worth graphene using rate effective routes remains a challenge to date [43]. Graphene can be synthesized from pristine graphite *via* several routes including vacuum thermal annealing, non-catalytic synthesis, micromechanical cleavage, thermal and chemical lessening of graphene oxide (GO) and liquid phase exfoliation [44-47]. Moreover, untraditional and harmful techniques for graphene fabrication such as the chemical explosion, arc discharge and unzipping of CNT are also investigated. Epitaxial growth from Silica Carbide produces good quality graphene but is limited by the inability to isolate single or bilayer graphene with high conductivity required for device applications [48]. Mechanical cleavage is also limited by its inability of scalable production. Regardless of the recent breakthrough in the CVD graphene synthesis, the production cost remains significantly high. On the other hand, graphene production via thermal or chemical reduction of GO is the least costly and the most appropriate technique for scalable fabrication. However, the quality of the produced graphene, termed as reduced graphene oxide (rGO) usually contains significant concentration of oxygen functional groups (C/O ratio of 5/1 to 10/1). Nonetheless, these oxygen functional groups allow further functionalization of rGO and are also very important for applications that require chemically reactive graphene materials such as in composites, sensors, coating and catalysis.

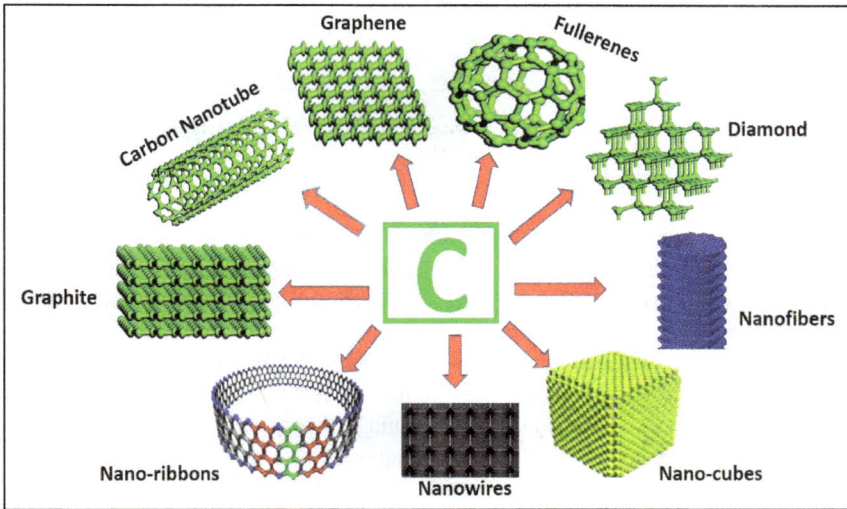

Fig. 1 Schematic diagram of carban nanomaterials.

2. Synthesis of graphene oxide

Production of GO from graphite is a two-steps process. In the first step, graphite is simultaneously intercalated using intercalant (H_2SO_4) and single or multiple oxidizing agents such as $KClO_3$, $KMNO_4$, and $NaNO_3$ to introduce oxygen-functional groups into the graphite basal plane and edges as shown in Fig. 2. These oxygen groups include epoxy (C–O–C), hydroxyl (OH), carbonyl (C=O) and carboxyl (R–COOH) groups [11, 16]. In the second step graphite oxide is exfoliated to GO by dispersion in water usually assisted by sonication [20].

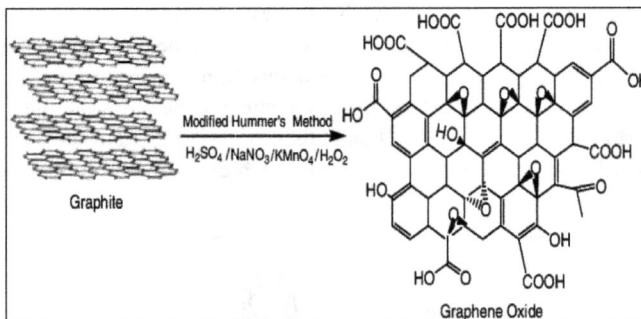

Fig. 2 Synthesis of graphene oxide by modified Hummer's method. Reproduced from [51].

There are few historical methods for synthesis of graphite oxide dated back to 1958. These methods vary in the type of oxidizing agents and intercalants employed, leading to production of GO with different assets. The Brodie technique was the first reported method, where graphite oxidation is accomplished by addition of $KClO_3$ (oxidant) to a mixture of graphite and fuming HNO_3 (intercalant) [21]. The Staudenmaier technique was later produced via modifying the Brodie technique by replacing HNO_3 with a mixture of H_2SO_4 and HNO_3 and thus fabricating GO with a superior C/O ratio [49]. Because these two methods required very lengthy oxidation process that extends to a week, a rapid process that can be completed in 120 min was developed by Hummers [22] to become currently the most widely used GO synthesis method [50]. Hummer's method uses a mixture of $NaNO_3/KMnO_4$ as the oxidizing agents and H_2SO_4 as the intercalant. The Hummer's method, however, suffers from the release of the hazardous NOx and ClO_2 gases. Newly, an improved Hummers' technique was investigated by the Tour group at Rice University by eliminating $NaNO_3$ form the oxidizing agents and adding H_3PO_4 to the H_2SO_4 intercalant. This improved routine eliminates the production of toxic gases and produces a more oxidized GO with a more regular carbon framework and larger sheet size [16].

Magnetic Oxides and Composites II   Materials Research Forum LLC
Materials Research Foundations **83** (2020) 79-116   https://doi.org/10.21741/9781644900970-4

3.   Reduction of graphene oxide

As discussed in the earlier section, GO is produced from oxidation of graphite leading to decorating the graphene sheets by various oxygen moieties with C/O ratio of ~2/1. The large concentrations of oxygen containing groups in the carbon basal plane makes GO not only electrically insulating, but also less thermally and chemically stabile compared to graphene [52]. Nevertheless, GO is considered as precursor for the synthesis of the electrically conducting and thermally stable reduced graphene oxide (rGO) *via* electrochemical [53], thermal [54,55], photocatalytic [56, 57], hydrothermal [58], or chemical reduction. GO should be reduced to a C/O ratio of ~10 or more to restore graphene inherent electrical conductivity and thermal stability.

Electrochemical reduction of graphene provides a simple, rapid, large-scale, economic, and benign process for manufacture of high quality rGO from GO *via* the removal of O_2 functionalities and recovery of the sp^2 carbon structure [52]. It can proceed via single or two-step approaches. However, an electrochemical process is limited because of its inability to provide rGO with high C/O ratio.

Solvothermal synthesis of rGO can be traced back to 1859 when Brodie heated what he then called graphitic acid in naphtha at 250 °C for three to four hours [59]. The C/O ratio was increased from 2/1 for graphic acid to 5/1 for rGO. Recently, many studies reported hydrothermal reduction of GO at temperatures between 80 and 250 °C for various durations [60, 58]. The degree of reduction can be tuned by the temperature and time of the hydrothermal step. Nonetheless, the degree of reduction obtained with hydrothermal treatment is lower than that achieved by thermal or chemical reduction.

One of the main advantages of the hydrothermal methods is its application for in-situ reduction of GO during synthesis of graphene based nanocomposites with metal and metal oxide [61-62]. Thermal exfoliation/reduction thermally reduces and exfoliates graphite oxide *via* very rapid heating. It is a dry process that produces thermally reduced graphene (TRG) with C/O ratio of 10/1 or larger. The mechanism of thermal reduction relies on the elimination of large fraction of the oxygen groups in the form of CO_2 creating high pressure in the gallery of graphite oxide. At high temperature, the rate of generation of CO_2 becomes higher than the rate of CO_2 diffusion out of the graphite oxide gallery leading to collapse of the stacked graphite oxide sheets and production of TRG with single or few layers [63, 54]. Thermal exfoliation process is excellent for large scale production and was the first graphene production processes to make it to commercialization.

Photocatalytic reduction takes place under continuous ultra-violet or laser radiation [64,56]. A rapid, dry, convenient, and environmentally attractive process for the reduction of GO by subjecting GO to laser pulses of different intensity and controlling the laser pulse energy, number of radiation cycles, the degree of reduction can be controlled.

Chemical reduction of colloidal GO with hydrazine, sodium boro-hydride, and hydroquinone is one of the most facile techniques for manufacture of ultrathin graphene sheets with large lateral dimension with restored conductivity and thermal stability [65-66] and is relatively rapid, economically viable, and suitable for subsequent functionalization of rGO. However, the toxicity of the used chemicals poses major environmental and safety concerns.

3.1   Green reduction of graphene oxide

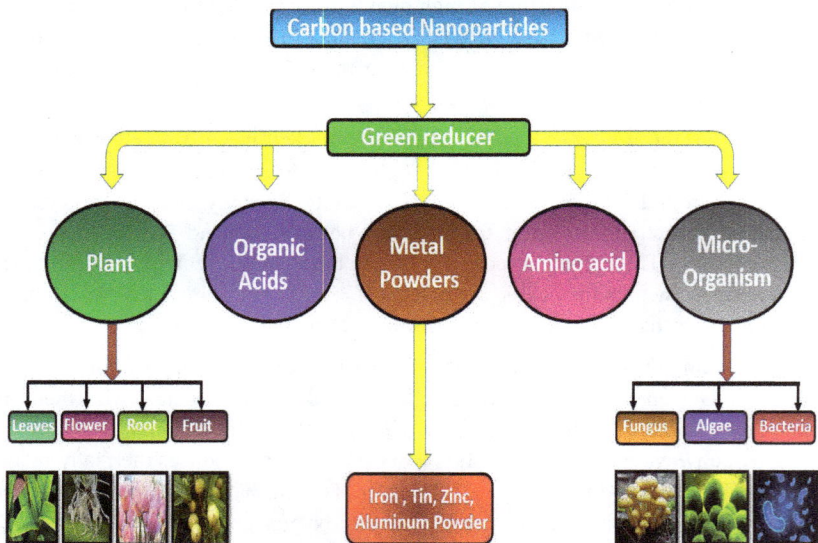

Fig. 3 A schematic for different routes for green reduction of graphene oxide.

As a green alternative, different plant extracts, fruit extracts, biochemical substances, organic acids, amino acids, fungus, bacteria, cellulosic compounds or metallic powders are successfully used for GO reduction and are considered as "Green reducers" because they are less corrosive, carcinogenic, and toxic compared to chemical reducing agents. Few reviews on GO reductions already exist [67]. However, no compilation of the green methods for reduction of GO to rGO is available. Therefore, this section summarizes the reduction of GO with environmentally friendly agents (Fig. 3).

A variety of phytochemicals derived from natural sources are utilized as green reducers in synthesis of metal, metal oxide nanoparticles (NPs), nanocomposites (NCs) etc. For example, leaves of *aspidopterys cordata* are highly efficient as a green reducing agent for the synthesis of copper oxide NPs [68]. In addition, *Colocasia esculenta* and *M. ferrea Linn* are abundant in some locations, whereas orange peel is a waste material. *C. esculenta leaves*, *M. ferrea Linn* leaves and other plants contain mainly pectins, flavonoids, ascorbic acid, apigenin, luteolin and various other flavones with high tendency to be oxidized [69].

In the presence of reactive oxygen, these phytochemicals are converted to quinone forms and, therefore, have strong possibility to reduce the GO oxygen containing groups. Among these green reducers are aqueous extracts acquired from *Citrus* peel, leaves of *C. esculenta* and *M. ferrea Linn*are readily obtainable, non-edible or residue parts of plant extracts [70].

3.1.1 Fruit extract

Fruit extract is affluent resource of phytochemicals e.g. polyphenols, flavonoids and catechins. These phytochemicals can reduce the salts of silver and gold to their metallic form [71]. In addition to preparation of metal NPs, grapes extract has outstanding presentation as selective reducer for ketone and nitro compounds and therefore should be potential reducing agent for GO [72]. Moreover, phytochemicals content in the fruit extracts are of immense therapeutic value and environmentally benevolent, making them very attractive substitute to the current toxic GO reducing agent.

3.1.2 Natural organic acid

Natural organic acids are some of the best green reducers for the synthesis of rGO and they are widely used as effective antioxidants in food stuff dispensation, wrapping and toiletries to avoid rotting [73]. An exceedingly rGO was also reported at room temperature or elevated temperatures using different organic acid as both reducing agent and stabilizing agent [74]. Interestingly, the purified rGO is re-dispersible in water, and organic solvents were prepared with high room temperature dispersibility of 1.2 mg/mL in water and 4 mg/mL in DMSO, possibly the highest dispersibility ever reported in both water and organic solvents for rGO [75].

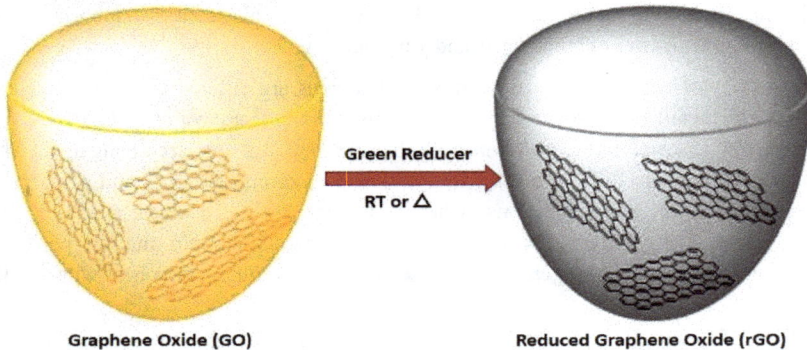

Green Reducer

RT or Δ

Graphene Oxide (GO)

Reduced Graphene Oxide (rGO)

Fig. 4 Schematic representation of rGO.

Amino acids are low price, non-hazardous and biocompatible materials. The amine groups present in amino acid can reduce GO under mild conditions. It has verified to be a proficient absorbent for metal salts with a low possibility of causing secondary contamination. An environmentally friendly and simple move toward for the reduction of GO via amino acid to produces stable rGO suspension was reported [78]. The schematic representation of rGO is shown in Fig.4.

3.1.3 Metal powder

Most metal powders are abundant and cheap, and many have mild reductive ability and non-toxic nature [76]. Therefore, they are good candidates for reduction of GO. A green process for synthesis of rGO by iron as reducer with several benefits in terms of yield, cost, and reduction time is reported [77].

3.1.4 Cellulose

Cellulose is the major rich natural polysaccharide available at low price with minimal ecological impact, and good sustainability [80]. Nonetheless, cellulose is difficult to dissolve and process in a liquid because of the strong hydrogen bonds formed due to the hydroxyl groups in glucose residues. The polar hydroxyl groups give cellulose mild reducing affinity [81]. Therefore, the cellulose functional groups can be used for deoxygenating graphene oxide and provide strong binding sites to reduced graphene oxide. A facile, efficient and environmentally benign process was developed for reduction of GO using cellulose to prepare biocompatible rGO. Moreover, huge area, free-standing rGO-CL composites were synthesized by vacuum-assisted self-assembly [82]. Table 1 represents green reducing agents capable of reducing GO.

Table 1. Green reducing agents capable of reducing GO.

Reducing agents	GO Conc. (mg/mL)	T (°C)	Time (h)	Sheet thickness (nm) GO	rGO	Ref.
Fruit juice	0.6	95	6-12	N/A	N/A	[71]
Gallic acid	4.0	RT & 95	24 & 7	1.0	1.4	[73]
Formic acid	2.5	100	18	N/A	N/A	[74]
Caffeic acid	0.1	95	12	1.2	0.85	[75]
Tin powder	1.0	RT	0.5-3	1.5	0.55	[76]
Aluminum Powder	1.0	RT	0.5	1.5	0.88	[77]
L-glutathione	0.1	50	6	1.0	8	[78]
L- lysine	1.0	90	9	1.0	0.8	[79]
Natural Cellulose	0.8	80	12	1.0	1.9	[81]
Coconut water	0.7	100	12, 24	N/A	N/A	[83]
Reducing sugar	0.1	95	1	0.97	1.1	[84]
E. coli	5.0	37	48	1	1.6	[85]
Green protein	1.0	90	1	1	1.4	[86]
G. biloba	0.5	30	12	N/A	N/A	[87]
Iron powder	0.5	RT	0.5	1	1-5	[88]
Glycine	0.3	95	24	N/A	N/A	[89]
H. sabdariffa	0.1	95	1	N/A	N/A	[90]
Ascorbic acid	0.1	23	12	1.2	2.0	[91]
Melatonin	0.1	40-80	3	0.9	1.8	[92]
Metal salts	0.1	RT	1	1.5	0.86	[93]
Manganese powder	0.5	RT	2	1.5	0.92	[94]
Metallic Zinc	0.5	RT	6	1.22	0.93	[95]
Glucosamine	0.5	90	7	N/A	N/A	[96]
Casin	1.0	90	7	N/A	N/A	[97]
Green tea	1.0	80	8	1.2	1.8	[98]
Humanin	1.0	40	1	1.0	1-3	[99]
Yeast	0.5	35-40	72	0.8	1.2	[100]
Benzyl alcohol	8.0	100	120	N/A	N/A	[101]
Benzylamine	0.5	90	1.5	N/A	N/A	[102]
Cysteine	0.5	28	12-72	1.0	0.8	[103]
L- valine	0.1	90	Few	1.0	1	[104]
vitamin C	0.1	95	1	1.0	1.0	[105]
Phyto extracts	0.5	RT	5–8	1.2	0.8	[106]
Beta carotene	2.0	95	24	N/A	N/A	[107]
Carrot root	0.5	RT	48	1.7	2.5	[108]
Cinnamon	1.6	RT	0.75	0.8	1.5	[109]
Clove extract	1.6	100	0.5	N/A	N/A	[110]

4.   Applications of graphene based nanocomposites

4.1   Electrochemical performance

Researchers have focused to study carbon based graphene due to its inimitable optical, electrical, electrochemical and mechanical assets. In addition, the very high surface area $(2630 \, m^2 \, g^{-1})$, structural stability, and electrical conductivity as well as the ability to use it as a support for metal dopants makes graphene substances very favorable for power storage appliances [111]. Moreover, graphene composites containing uniformly dispersed SiO_2, TiO_2 , Co_3O_4, Fe_3O_4, SnO_2 , Cu_2O and CuO synergitically prevent restacking of graphene sheets and reduce the metal/metal oxide volume expansions during electrochemical cycling [112]. Among these composites, graphene-MoO_2 nanocompsites have good performance due to the electrochemical deployment of the metal-oxide and the rapid ionic transport right through the electrode interior volume [113]. Existing supercapacitors are based on high surface area activated carbon. They provide the energy density of ~ 4 – 5 watt.hour per kilogram (Wh/kg) that is less than 20% of the density provided by lead acid batteries [114]. Therefore, research to enhance the energy density of supercapacitors while mainating their cylcying lifetime and the high power density is urgent [115]. Carbon based nanomaterials are excellent candidates for such targets becuase of their very high surface area and electronic conductivity.

SWNTs have already demonstrated good performance as organic electrolytes [116], but their high cost hinders the commercialization of SWNTs supercapacitors. On the other hand, to perform SWNTs in terms of surface area and electrical conductivity, additionally to its better mechanical strength and chemical stability [117] and tons of low price graphene can now be produced in low price through reduction of GO [118]. Supercapacitors based on rGO were developed with capacitance of ~130 and 100 F/g in aqueous and organic electrolytes [119]. The electrochemical study of Co_3O_4 NPs and Co_3O_4/rGO NCs was analyzed *via* cyclovoltametry in the range of 0 V and 3.0 V and scan rate of 50 mV/s; the results have been displayed in Fig. 5 [120]. The cyclic voltametry capacity of the NCs revealed the reduction of Co_3O_4 NPs to Co metal, clustering of Co and Li_2O, insertion of Li^+ into rGO and development of solid electrolyte interphase (SEI) layer on the active materialas (schematically described in Fig. 6). Moreover, no significant drops in peak intensity was examined during cycling for both either Co_3O_4NPs or Co_3O_4/rGO NCs indicating the excellent reversibility of Li^+ storage and the elevated structural stability [121]. The electrochemical conversion reaction in Co_3O_4 NPs-based anodes procceds according to the reaction [120]:

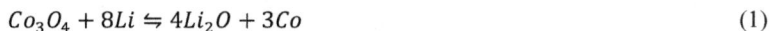

$$Co_3O_4 + 8Li \leftrightarrows 4Li_2O + 3Co \qquad (1)$$

Materials Research Forum LLC
https://doi.org/10.21741/9781644900970-4

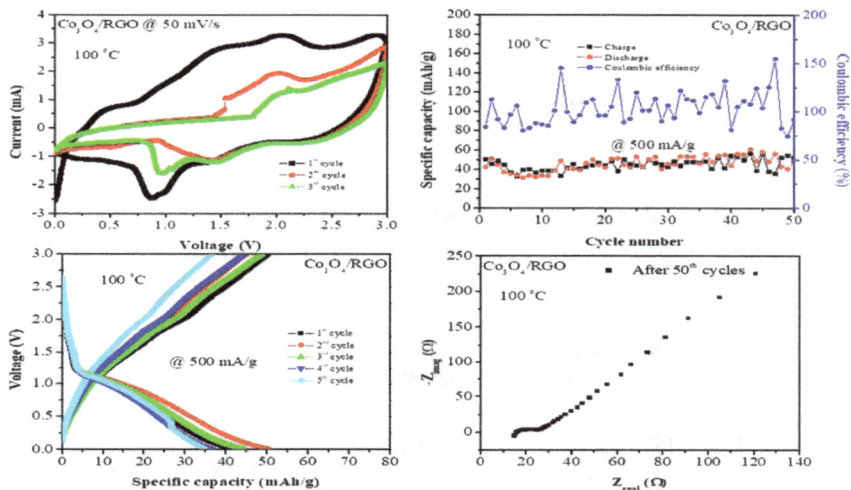

Fig. 5 Electrochemical study of Co₃O₄/Co₃O₄/rGO NCs. Cyclic voltammetric (CV), galvanostatic charge/discharge curves, cycling performance and EIS spectra of Co₃O₄ NPs and Co₃O₄/rGO nanocomposites performed at 100 °C. (a, b). Cyclic voltammetric (CV) and, (c, d). Galvanostatic charge-discharge curves. Reproduced from [121].

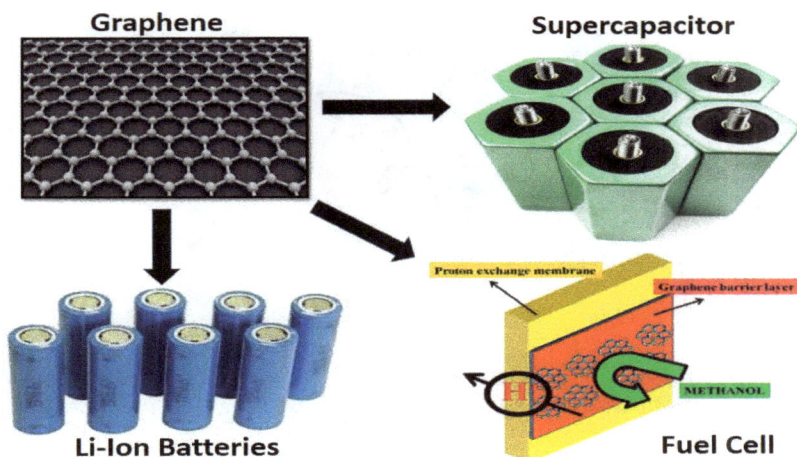

Fig. 6 Schematic diagram of grapene-based electrode materials [121].

4.2 Heterogeneous photocatalysis

Commercially available heterogeneous catalysts consist of highly porous support structure such as alumina and silica, with surface areas of the order of several hundred square meters per gram, and a catalytically active material impregnated into these porous support [122].

4.2.1 Principles of photocatalysis and reaction mechanism

The basic main beliefs of heterogeneous photocatalysis have been broadly reported [123]. The photocatalytic activity is based on large surface area, band gap energy of the semiconductor NPs. For example in TiO_2, electrons are promoted from the filled valence band (VB) to the empty conduction band (CB) as the absorbed photon energy exceeds the band gap of the semiconductor photocatalyst (Fig. 7). There are eight steps for the catalytic transfer of reactant (R) into product (S). These steps include adsorption of (R), absorption of light photons by the catalyst (C), formation of electron-hole (e^--h^+) pairs, and migration of e^- and h^+ to the active sites on the surface of C. It should be emphasized that these eight steps are considered as three distinct phenomena: surface chemical reactions, electronic reactions, and mass transfer.

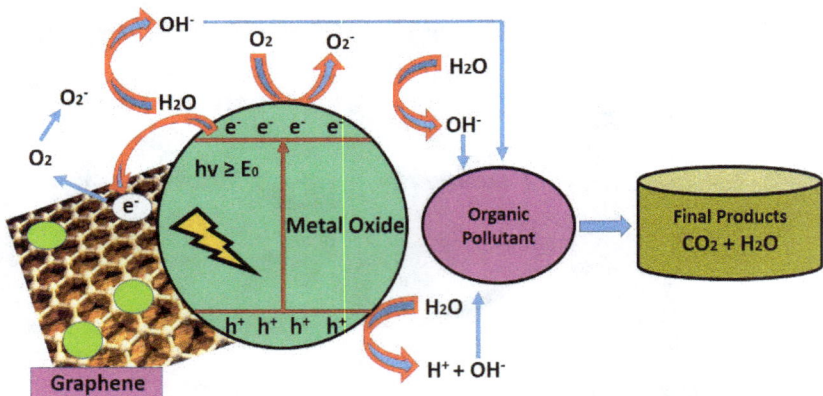

Fig. 7 Schematic diagram of working principle of heterogeneous photocatalysts.

When a photoactive material is exposed to photons with energy higher than the band gap energy of the material, an electron is excited from the valence band to the conduction band and a hole is formed simultaneously in the valence band as follows [124] (Equations 2-6).

Photon activation:

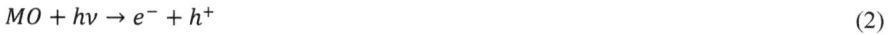

$$MO + hv \rightarrow e^- + h^+ \tag{2}$$

Oxygen adsorption

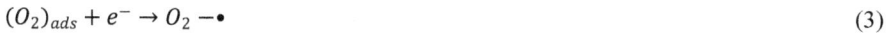

$$(O_2)_{ads} + e^- \rightarrow O_2 -\bullet \tag{3}$$

Water ionization

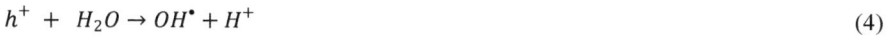

$$h^+ + H_2O \rightarrow OH^\bullet + H^+ \tag{4}$$

Hydroxyl Ionization

$$h^+ + OH^- \rightarrow OH^\bullet \tag{5}$$

Superoxide protonation:

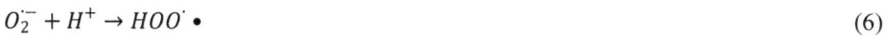

$$O_2^{\cdot-} + H^+ \rightarrow HOO^\cdot \bullet \tag{6}$$

The $HOO^\cdot$ radicals (Equations (2-6)) also have scavenging properties similar to oxygen, thus extending the photo hole lifetime as follows (Equations 6 and 7):

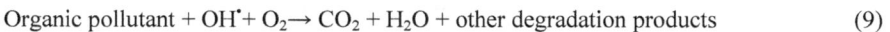

$$e^- + HOO^- + H^+ \rightarrow H_2O_2 \tag{7}$$

$$e^- + H_2O_2 \rightarrow OH^\bullet + OH^- \tag{8}$$

Organic pollutant $+ OH^\bullet + O_2 \rightarrow CO_2 + H_2O +$ other degradation products $\tag{9}$

Redox processes occur at the surface of the photo-excited catalyst, and rapid electron-hole recombination would happen unless oxygen (or other electron acceptors) is available to search electrons to form superoxides (O_2^-) hydroperoxyl radicals (HO_2) and subsequently H_2O_2 [125]. Instantaneous recombination of the e^--h^+ pair has the most negative impact on the catalytic efficiency, as high temporal resolution transient photoluminescence study shows that more than 90% of the e^--h^+ pairs recombined within 10 nanoseconds rendering their photocatalytic capability [126].

4.2.2 Improving photocatalytic efficiency

Improving the photocatalytic activity is the subject of significant number of research projects and publications. To achieve the enhancement of the oxidant, yield the reorganization of rate limiting properties is very essential. A thorough perceptive of the steps involved in heterogeneous photocatalysis is not yet ascertained. Important aspects such as the photocatalyst particle size, structure and composition dictate the photocatalytic activity. Therefore, two routes are usually followed to enhance the photocatalytic activity, *i.e.*, enhancement of surface area and exterior of electron acceptors [127].

4.2.3.1 Surface area

One of the unique properties of nanoparticles is the large surface area, that strongly influences number of key properties. For example, electrical resistivity of granular materials scale with the total are of grain boundaries, and the chemical activity of heterogeneous catalysts is proportional to their overall surface area per unit volume. In general, nanomaterials with large surface area act as a good catalyst or substrate. When the substrate is grafted with the active moieties such as acids sites or impregnated with components like platinum atoms that speed up the catalytic reactions. Examples of conventional substrates include silica (SiO_2), gamma alumina (γ-Al_2O_3), anatase Titania (a-TiO_2), and Zirconia (ZrO_2). Mixed oxides are also used as high-surface-area silica-alumina [128].

4.2.3.2 Exterior of electron acceptors

An alternate way to enhance photocatalytic character is *via* outside electron acceptors. Such external, dissolved electron acceptors are species that tend to adsorb and reduce by conduction band electrons on the surface of the photocatalyst and therefore they enhance the photocatalytic activities because exclusion of electron from the photocatalyst conduction band suppresses electron-hole recombination. Several reports are available with increase photocatalytic efficiency along with addition of dissolved electron acceptors [129-131]. Various electrophilic species act as electron acceptors in solution despite the simplest molecular oxygen and hydrogen peroxide. Both compounds adsorb to the MO surface enabling the detain of the photo-excited conduction band electrons. Molecular oxygen has two main functions in photocatalysis; (i) combine with organic radicals formed upon oxidation and (ii) hydrogen abstraction from organic solutes and forming peroxyl radicals. Based on organic substrate structure peroxyl radical could induce further oxidation. One-electron reduction of H_2O_2 yields hydroxide ion and additional hydroxyl radical. Therefore, the photocatalytic activity increases as H_2O_2 is also a direct source of hydroxyl radicals [132].

4.2.4 Photocatalytic degradation of dyes

Due to the rapid growth of water pollution, the need for clean water has become the order of the day. As a result, the new purification technology is eco-friendly and most sought process in the current situation. Our future mainly depends on identifying these requirements, development of proficient, sustainable and environment friendly water treatment technologies [133]. The harmful influence of wastewater treatment affects the whole ecosystem causing damage to the human health, hygiene and all other creatures necessitating the search of green, energy-saving and cost effective water purification methods. Textile wastewater streams are contaminated with aromatic compounds and organic dyes presenting hazard to the environment and therefore are required to be treated before being reused or discharged [134]. To overcome the challenges in wastewater treatment technology, several materials have been explored for application as photocatalyst for degradation of pollutants. In 1841, Robert Bunsen [135] proposed replacing the rare and costly platinum current collector of Grove cell with more cost-effective carbon material.

4.2.5 Photoatyltic applications of graphene/metal oxide nanocomposites

Hybrids of metal oxides and graphene were used to develop nanocomposites with tailored properties for variety of applications such as solar cells, photocatalysis, sensors, and energy storage. High separation cost and generation of secondary pollutants are some of the side effects of traditional water treatment methods based on adsorption, coagulation and membrane separation. Photocatalysts like ZnO, TiO_2, PbO, Fe_2O_3, Co_3O_4, Bi_2O_3, and ZnS [136-141] have exhibitd high effectiveness in degrading range of organic pollutants into biodegradable or less noxious organic compounds as well as inorganic radicals and halide ions. However, the difficult post-separation easy agglomerations of these inorganic catalysts limit their large-scale appliances. Related research communities have turned their attention towards the study of new catalysts. Moreover, wide range of metal/metal oxide/polymers nanoparticles are investigated for their catalytic performance especially in organic transformation [142-149]. Excellent review exist on the photocatalytic applications of graphene metal oxide [150-152].

4.3   Antibacterial activity of graphene-metal oxide nanocomposites

An array of antibacterial active materials are used to effectively protect the public health including graphene [153] and metal oxides/polymer materials [154, 155] as well as quaternary ammonium compounds that are identified to impede microbe attachment and propagation on surfaces. However, concerns about the antibiotic resistance, pollution, high cost, and complexity of the production process are major disadvantages of these materials [156]. Recently published data revealed that metal/metal oxide NPs could be used as an excellent antibacterial agents [157, 158]. Very recently, graphene [159] paper was used as biocompatible substrate for adhesion and proliferation of different cells, for example L-929 cells, neuroendocrine PC12 cells, oligodendroglia cells, and osteoblasts, where GO nanocarriage was used to deliver water-insoluble drugs into cells.

The cytotoxicity of graphene based nanocomposites was studied to evaluate whether GO is able to induce oxidative stress in bacterial cell due to the presence of structural and physiochemical properties of carbon nanomaterials (Fig. 8). Moreover, the different metal oxides NPs damaged the membrane by releasing the metal ion and producing reactive oxygen species and ROS dependent oxidative stress. For example, recent study confirmed the presence of ROS in TiO_2 suspension under dark circumstances [160], but other studies did not. Moroever, in another study, revealed that presence of TiO_2 in microbial growth medium induced the generation of $O_2\bullet$ [161].

Reduction of ROS by NPs is closely correlated with the antibacterial assay of the NPs and quantitative relationships between production of ROS and the NPs antibacterial activities is derived. For example, death rate of *E. coli* cells correlates linearly with the concentration of $\bullet OH$ radicals produced by TiO_2 [162]. Moreover, the antibacterial activity of ZnO suspension correlates linearly to H_2O_2 concentration. Yet, not all ROS (*i.e.* $^1O_2$ and $O_2\bullet$) are accounted for when the antibacterial activity of NPs is correlated.

Because there are thousands of novel nanoparticles that could be employed, it is very challenging to experimentally test the antibacterial activities of each NPs require establishing quantitative relationships between ROS formation and antimicrobial activities of the nanoparticles for fast selection. The dominant role of ROS on the mechanism of the antibacterial activities of CNMs is reported by several research groups [163]. However, other groups claim CNMs can be antibacterial active through through ROS-independent oxidative stress as there was no ROS formation or ROS-mediated injure after C_{60} treatment nor there effect on C_{60} antimicrobial activities upon exposure to light and oxygen [164]. Similar claims on the ROS-independent antibacterial mechanism in graphene materials were reported and a consequent examination has verified this assumption and revealed that C_{60} rather than ROS that directly interacts with the membrane of the cell and decouples the electron transport from energy transduction through respiration [165]. Results from Ellman's assay revealed diverse oxidation affinity of graphene-based materials toward GSH confirming their ability to mediate $O_2^{\cdot-}$-independent oxidative stress analogous to metal oxides [166]. Moreover, the antibacterial assay of graphene is suggested to be due to the interruption of electron transport within the membrane respiratory chain mediated by ROS independent oxidative stress. Other results suggested graphene-Cu and graphene-Ge surfaces have elevated antibacterial activities due to the electron transfer from the microbial membrane to the graphene surface and further transfer to the conducting Cu or semiconducting Ge substrate [167]. In contrary, the antibacterial assay of graphene-SiO_2 surface was negligible due to the inhibted electrons transfer from the microbial membrane to the underlying SiO_2 substrate through graphene is not possible [168]. Moreover, stimulating the formation of electron–hole pairs on the GO surface by light irradiation accelerates electron transfer from bacterial antioxidant biomolecules to GO resulting in GO reduction [169]. In summary, these results conclude that graphene-based samples bactericidal effects can be induced through ROS-independent oxidative stress. Moreover, GO was reported to have antimicrobial activities evaluated to rGO because of the high concentration of surface oxygen functionalities that facilitate oxidation reactions [170].

In comparison, metal oxide nanocomposite shows better antibacterial activities [171]. An interesting finding related to the size of GO sheets is the higher activities for the smaller size as fourfold increase in the antimicrobial effect was observed when GO sheet size was decreased from 0.65 to 0.01 μm^2. This was attributed to the higher edge-defects density with smaller sheet area [172, 173]. The possible free radical generation mechanism is shown in following way (Equations 10 to 14):

$$MO + hv \rightarrow MO(e^- + h^+) \tag{10}$$

$$H_2O + MO \rightarrow MO + \dot{O}H + H^+ \tag{11}$$

$$O_2 + MO(e^-) \rightarrow MO + \dot{O}^- \tag{12}$$

$$O_2 + H^+ \rightarrow HO_2 \tag{13}$$

$$HO_2^- + H^+ \rightarrow H_2O_2 \qquad\qquad (14)$$

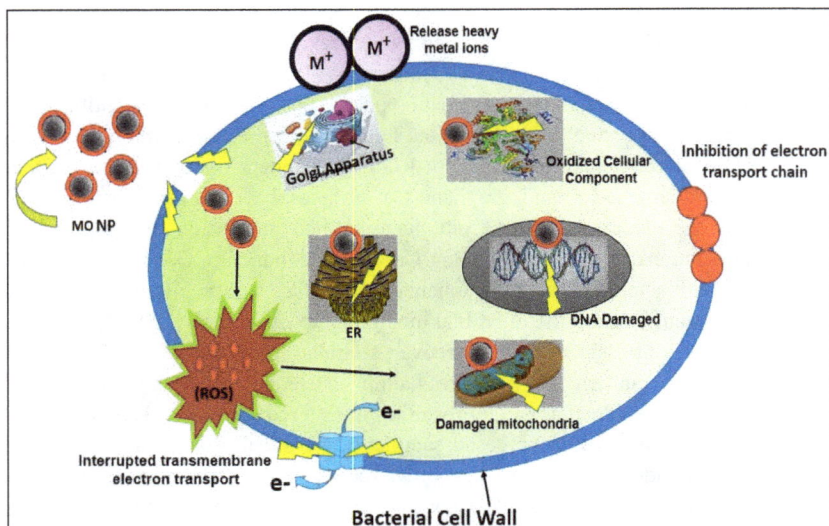

Fig. 8 The mechanism of carbon-based NPs induced toxicity in the bacterial microorganism.

Conclusions

We have discussed the reduction of GO using various green reducing agents for electrochemical, photocatalytic and antimicrobial applications. The use of eco-friendly green reducers to prepare graphene/rGO attracts significant interest in the systematic society. Numerous effective green reducing agents for GO reduction are available such as plant extract, fruit extract, root extract, bark extract, bacteria, algae, fungus, metal powder, organic and amino acids. Such reagents are excellent alternatives to hydrazine, borohydride or other noxious agents as they are easy to handle and produces biocompatible byproducts. There are no limitations for the green reducing agents because of the presence of some different types of function group and structural directing agents. Apart from reducing agents, green reducers can be employed as surfactant or capping agents for size reduction. In this chapter, we have discussed few fascinating appliances of graphene-based nanoparticles. Green reduction of graphene-based nanomaterials shows great microbial activity towards different pathogenic bacteria, efficient photocatalytic degradation towards organic pollutant and heavy metals ion as it has large surface area and large band gap energy.

References

[1]   H. Chang, L.Tang, Y. Wang, J. Jiang, J. Li, Graphene fluorescence resonance energy transfer aptasensor for the thrombin detection, Analytical Chemistry, 82 (2010) 2341-2346. https://doi.org/10.1021/ac9025384

[2]   Y. Sun, B.Zhou, Y. Lin, W. Wang, K. Fernando, P.Pathak, M. Meziani, B. Harruff, X. Wang, H. Wang, P. Luo, H. Yang, M. Kose, B.Chen, L.Veca, S. Xie, Quantum-sized carbon dots for bright and colorful photoluminescence, Journal of the American Chemical Society, 21 (2006) 756-7757. https://doi.org/10.1021/ja062677d

[3]   N. Wang, M. Lin, H. Dai, H. Ma, Functionalized gold nanoparticles/reduced graphene oxide nanocomposites for ultrasensitive electrochemical sensing of mercury ions based on thyminemercury-thymine structure,Biosensors and Bioelectronics, 79 (2016) 320-326. https://doi.org/10.1016/j.bios.2015.12.056

[4]   N. Liu, Z. Ma, Au-ionic liquid functionalized reduced graphene oxide immunosensing platform for simultaneous electrochemical detection of multiple analytes,Biosensors and Bioelectronics, 51 (2014) 184-190. https://doi.org/10.1016/j.bios.2013.07.051

[5]   U.Jensen, E. Ferapontova, D.Sutherland,Quantifying protein adsorption and function at nanostructured materials: Enzymatic activity of glucose oxidase at GLAD structured electrodes, Langmuir, 28 (2012) 11106-11114. https://doi.org/10.1021/la3017672

[6]   P.Rafighi, M.Tavahodi, B.Haghighi, Fabrication of a thirdgeneration glucose biosensor using graphene-polyethyleneimine-gold nanoparticles hybrid,Sens. Actuators:B, 232 (2016) 454−461. https://doi.org/10.1016/j.snb.2016.03.147

[7]   H. Zhong, R. Yuan, Y.Chai, W. Li, X. Zhong, Y. Zhang, In situ chemo-synthesized multi-wall carbon nanotube-conductive polyaniline nanocomposites: Characterization and application for a glucose amperometric biosensor, Talanta, 85 (2011) 104-111. https://doi.org/10.1016/j.talanta.2011.03.040

[8]   J. Lu, I. Do, L. Drzal, R. Worden, I. Lee, Nanometal decorated exfoliated graphite nanoplatelet based glucose biosensors with high sensitivity and fast response, ACS Nano, 2 (2008) 1825-1832. https://doi.org/10.1021/nn800244k

[9]   M. Foglia, G. Alvarez, P. Catalano, A. Mebert, L.Diaz, T. Coradin, M. Desimone, Recent patents on the synthesis and application of silica nanoparticles for drug delivery, Recent Patents on Biotechnology, 5 (2011) 54-61. https://doi.org/10.2174/187220811795655887

[10] Y. Sun, B. Zhou, Y. Lin, W. Wang, K. Fernando, P. Pathak, M. Meziani, B. Harruff, X. Wang, H. Wang, P. Luo, H. Yang, M. Kose, B. Chen, L. Veca, S. Xie, Quantum-sized carbon dots for bright and colorful photoluminescence, Journal of the American Chemical Society, 21 (2006) 7756-7757. https://doi.org/10.1021/ja062677d

[11] H. Chang, L. Tang, Y. Wang, J. Jiang, J. Li, Graphene fluorescence resonance energy transfer aptasensor for the hrombin detection, Analytical Chemistry, 82 (2010) 2341-2346. https://doi.org/10.1021/ac9025384

[12] M. Ma, J. Gu,M. Yang, Z. Li, Z. Lu, Y. Zhang, P. Xing, S. Li, X. Chu, Y. Wang, Controllable self-assemblies of sodium benzoate in different solvent environments, RSC Advances, 5 (2015) 70178-70185. https://doi.org/10.1039/C5RA13026C

[13] S. Su, J.Wang, E. Vargas, J. Wei, R. Martínez-Zaguilán, S. Sennoune, M. Pantoya, S. Wang, J. Chaudhuri, J. Qiu, Porphyrin immobilized nanographene oxide for enhanced and targeted photothermal therapy of brain cancer, ACS Biomaterials Science and Engineering, 2 (2016) 1357-1366. https://doi.org/10.1021/acsbiomaterials.6b00290

[14] W. Grosse, J. Champavert, S. Gambhir, G. Wallace, S. Moulton, Aqueous dispersions of reduced graphene oxide and multi wall carbon nanotubes for enhanced glucose oxidase bioelectrode performance, Carbon, 61 (2013) 467-475. https://doi.org/10.1016/j.carbon.2013.05.029

[15] S. Iijima and T. Ichihashi, Single-shell carbon nanotubes of 1-nm diameter, Nature, 363 (1993) 603-605. https://doi.org/10.1038/363603a0

[16] D. S. Bethune, C. H. Kiang, M. S. Devries, G. Gorman, R. Savoy, J. Vazquez, R. Beyers, Cobalt-catalyzed growth of carbon nanotubes with single-atomic-layerwalls, Nature, 363 (1993) 605-607. https://doi.org/10.1038/363605a0

[17] Y. Liu, D. Yu, C. Zeng, Z. Miao, L. Dai, Biocompatible graphene oxide-based glucose biosensors, Langmuir, 26 (2010) 6158-6160. https://doi.org/10.1021/la100886x

[18] K. Atacan, B. Çakiroğlu, M. Özacar, Efficient protein digestion using immobilized trypsin onto tannin modified Fe_3O_4 magnetic nanoparticles, Colloids and Surfaces B: Biointerfaces, 156 (2017) 9-18. https://doi.org/10.1016/j.colsurfb.2017.04.055

[19]   S. Tadepalli, H. Hamper, S. Park, S. Cao, R. Naik, Adsorption behavior of silk fibroin on amphiphilic graphene oxide. ACS Biomaterials Science and Engineering, 2 (2016) 1084-1092. https://doi.org/10.1021/acsbiomaterials.6b00232

[20]   K. Atacan, M. Çakiroğlu, M. Ozacar, Covalent immobilization of trypsin onto modified magnetite nanoparticles and its application for casein digestion,International Journal of Biological Macromolecules, 97 (2017) 148-155. https://doi.org/10.1016/j.ijbiomac.2017.01.023

[21]   D.Wu,Y. Gao, W. Li, X. Zheng, Y. Chen, Q. Wang, Selective Adsorption of La3+ using a tough alginate-clay-poly (nisopropylacrylamide) hydrogel with hierarchical pores and reversible re-deswelling/swelling cycles, ACS Sustainable Chemistry and Engineering, 4 (2016) 6732-6743. https://doi.org/10.1021/acssuschemeng.6b01691

[22]   Y. Takemoto, H. Ajiro, M. Akashi, Hydrogen-bonded multilayer films based on poly(n-vinylamide) derivatives and tannic acid, Langmuir, 31 (2015) 6863-6869. https://doi.org/10.1021/acs.langmuir.5b00767

[23]   N. Kovtyukhova, P. Ollivier, B. Martin, T. Mallouk, S. Chizhik, E. Buzaneva, A. Gorchinskiy, Layer-by-layer assembly of ultrathin composite films from micron-sized graphite oxide sheets and polycations, Chemistry of Materials, 11 (1999) 771-778. https://doi.org/10.1021/cm981085u

[24]   M. Bradford, A rapid and sensitive method for the quantitation of microgram quantities of protein utilizing the principle of protein-dye binding, Analytical Biochemistry, 72 (1976) 248-254. https://doi.org/10.1016/0003-2697(76)90527-3

[25]   M. Franssen, P. Steunenberg, E. Scott, H. Zuilhof, J. Sanders, Immobilised enzymes in biorenewables production,Chemical Society Reviews, 42 (2013) 6491-6533. https://doi.org/10.1039/c3cs00004d

[26]   Z. S. Wu, W. Ren, L. Wen, L. Gao, J. Zhao, Z. Chen, Graphene anchored with Co3O4 nanoparticles as anode of lithium ion batteries with enhanced reversible capacity and cyclic performance, ACS Nano, 4 (2010) 3187-3194. https://doi.org/10.1021/nn100740x

[27]   J. Fowler, M. Allen, V. Tung, Y. Yang, R. Kaner, B. Weiller, Practical chemical sensors from chemically derived graphene, ACS Nano, 3 (2009) 301-306. https://doi.org/10.1021/nn800593m

[28]   Y. Dan, Y. Lu, N.Kybert, Z. Luo, A. Johnson, Intrinsic response of graphene vapor sensors, ACS Nano Letter, 9 (2009) 1472-1475. https://doi.org/10.1021/nl8033637

[29]   Y. Liu, D. Yu, C. Zeng, Z. Miao, L. Dai, Biocompatible graphene oxide-based glucose biosensors, Langmuir, 26 (2010) 6158-60. https://doi.org/10.1021/la100886x

[30]   W. Hu, C. Peng, W. Luo, M. Lv, X. Li, D. Li, Graphene-based antibacterial paper, ACS Nano, 4 (2010) 4317-4323. https://doi.org/10.1021/nn101097v

[31]   Q. He, S. Wu, S. Gao, X. Cao, Z. Yin, H. Li, Transparent, flexible, all-reduced graphene oxide thin film transistors, ACS Nano, 5 (2011) 5038-5044. https://doi.org/10.1021/nn201118c

[32]   X. Sun, Z. Liu, K. Welsher, J. Robinson, A. Goodwin, S. Zaric, Nano-graphene oxide for cellular imaging and drug delivery, Nano Research, 1 (2008) 203-212. https://doi.org/10.1007/s12274-008-8021-8

[33]   G. Zhou, D. Wang, F. Li, L. Zhang, N. Li, Z. Wu. Graphene-wrapped Fe3O4 anode material with improved reversible capacity and cyclic stability for lithium ion batteries, Chemistry of Materials, 22 (2010) 5306-5313. https://doi.org/10.1021/cm101532x

[34]   C. Guo, H. Yang, Z. Sheng, Z. Lu, Q. Song, C. Li, Layered graphene/quantum dots for photovoltaic devices, Angewandte Chemie International Edition, 49 (2010) 3014-3027. https://doi.org/10.1002/anie.200906291

[35]   C. Peng, W. Hu, Y. Zhou, C. Fan, Q. Huang, Intracellular imaging with a graphenebased fluorescent probe, Small, 6 (2010) 1686-1692. https://doi.org/10.1002/smll.201000560

[36]   J. Williams, L. DiCarlo, C. Marcus, Quantum-Hall effect in a gate-controlled pn junction of graphene, Science, 317 (2007) 638-641. https://doi.org/10.1126/science.1144657

[37]   C. XianáGuo, C. MingáLi, A self-assembled hierarchical nanostructure comprising carbon spheres and graphene nanosheets for enhanced supercapacitor performance, Energy, Energy & Environmental Science, 4 (2011) 4504-4517. https://doi.org/10.1039/c1ee01676h

[38]   F. Bonaccorso, Z. Sun, T. Hasan, A. Ferrari, Graphene photonics and optoelectronics, Nature Photonics, 4 (2010) 611-622. https://doi.org/10.1038/nphoton.2010.186

[39]   V.Eswaraiah, V.Sankaranarayanan, S.Ramaprabhu, Graphene-based engine oil nanofluids for tribological applications, ACS Applied Materials & Interfaces, 3 (2011) 4221-4227. https://doi.org/10.1021/am200851z

[40]   Z. Sui, Q. Meng, X. Zhang, R. Ma, B. Cao, Green synthesis of carbon nanotube-graphene hybrid aerogels and their use as versatile agents for water purification, Journal Material Chemistry, 22 (2012) 8767-8771. https://doi.org/10.1039/c2jm00055e

[41]   Y. Zhao, C. Jiang, C. Hu, Z. Dong, J. Xue, Y. Meng. Large-scale spinning assembly of neat, morphology-defined, graphene-based hollow fibers, ACS Nano, 7 (2013) 2406-2412. https://doi.org/10.1021/nn305674a

[42]   H. Sun, Z. Xu, C. Gao, Multifunctional, ultraflyweight, synergistically assembled carbon aerogels, Advanced Materials, 25 (2013) 2554-2560. https://doi.org/10.1002/adma.201204576

[43]   M. Shin, K. Kim, W. Shim, J. Yang, H. Lee, Tannic acid as a degradable mucoadhesive compound, ACS Biomaterials Science and Engineering, 2 (2016) 687-696. https://doi.org/10.1021/acsbiomaterials.6b00051

[44]   K. Xiong, P. Qi, Y. Yang, X. Li, H. Qiu, X. Li, R. Shen, Q. Tu, Z. Yang, N. Huang, Facile immobilization of vascular endothelial growth factor on a tannic acid-functionalized plasma-polymerized allylamine coating rich in quinone groups, RSC Advances, 6 (2016) 17188-17195. https://doi.org/10.1039/C5RA25917G

[45]   S. Çakar, N. Güy, M. Özacar, F. Findik, Investigation of vegetable tannins and their iron complex dyes for dye sensitized solar cell applications, Electrochimica Acta, 209 (2016) 407-422. https://doi.org/10.1016/j.electacta.2016.05.024

[46]   T. Terse-Thakoor, K. Komori, P. Ramnani, I. Lee, A. Mulchandani, Electrochemically functionalized seamless threedimensional graphene-carbon nanotube hybrid for direct electron transfer of glucose oxidase and bioelectrocatalysis, Langmuir, 31 (2015) 13054-13061. https://doi.org/10.1021/acs.langmuir.5b03273

[47]   V. Mani, B. Devadas, S. Chen, Direct electrochemistry of glucose oxidase at electrochemically reduced graphene oxide-multiwalled carbon nanotubes hybrid material modified electrode for glucose biosensor, Biosensors and Bioelectronics, 41 (2013) 309-315. https://doi.org/10.1016/j.bios.2012.08.045

[48]   M. Jose, S. Marx, H. Murata, R. Koepsel, A.Russell, Direct electron transfer in a mediator-free glucose oxidase-based carbon nanotube-coated biosensor, Carbon, 50 (2012) 4010-4020. https://doi.org/10.1016/j.carbon.2012.04.044

[49]   S. Sali, H. Mackey, A. A. Abdala, Effect of graphene oxide synthesis method on properties and performance of polysulfone-graphene oxide mixed matrix

membranes, Nanomaterials, 9 (2019) 769-775.
https://doi.org/10.3390/nano9050769

[50] X. Ji, P. Herle, Y. Rho, L. Nazar, Carbon/MoO2 composite based on porous semi-graphitized nanorod assemblies from in situ reaction of Tri-Block polymers, Chemistry of Materials,19 (2007) 374-383. https://doi.org/10.1021/cm060961y

[51] A. N. Ahmad, A. Kausar, B. Muhammad, An investigation on 4-aminobenzoic acid modified polyvinyl chloride/graphene oxide and PVC/graphene oxide based nanocomposite membranes, Journal of Plastic Film & Sheeting, 32 (2016) 419-448. https://doi.org/10.1177/8756087915616434

[52] N. Sorokina, M. Khaskov, V. Avdeev, I. Nikol Skaya, Reaction of graphite with sulfuric acid in the presence of KMnO4, Russian Journal of General Chemistry, 75 (2005) 162-168. https://doi.org/10.1007/s11176-005-0191-4

[53] S. Y. Toh, K. S. Loh, S. K. Kamarudin, W. R. Daud, Graphene production via electrochemical reduction of graphene oxide: Synthesis and characterisation, Chemical Engineering Journal, 251 (2014) 422-434.
https://doi.org/10.1016/j.cej.2014.04.004

[54] M. J. McAllister, J. L. Li, D. H. Adamson, H. C. Schniepp, A. A. Abdala, J. Liu, M. Herrera-Alonso, D. L. Milius, R. Car, R. K. Prud'homme, Single sheet functionalized graphene by oxidation and thermal expansion of graphite, Chemistry of Materials, 19(2007) 4396-4404. https://doi.org/10.1021/cm0630800

[55] D. Marcano, D. Kosynkin, J. Berlin, A. Sinitskii, Z. Sun, A. Slesarev, L. Alemany, W. Lu, J. Tour, Improved synthesis of graphene oxide, ACS Nano, 4 (2010) 4806-4814. https://doi.org/10.1021/nn1006368

[56] C.R.Yang, S. F. Tseng, Y. T. Chen, Laser-induced reduction of graphene oxide powders by high pulsed ultraviolet laser irradiations, Applied Surface Science, 444 (2018) 578-583. https://doi.org/10.1016/j.apsusc.2018.03.090

[57] Y. Xu, K. Sheng, C. Li, G. Shi, Highly conductive chemically converted graphene prepared from mildly oxidized graphene oxide, Journal of Material Chemistry, 21 (2011) 7376-7380. https://doi.org/10.1039/c1jm10768b

[58] Y. Q. Niu, Fang, X. Zhang, P. Zhang, Y. Li, Reduction and structural evolution of graphene oxide sheets under hydrothermal treatment, Physics Letter A, 380 (2016) 3128-3132. https://doi.org/10.1016/j.physleta.2016.07.027

[59] B. C. Brodie, XIII. On the atomic weight of graphite. Philosophical Transactions of the Royal Society of London. 149, (1859) 249-259.
https://doi.org/10.1098/rstl.1859.0013

[60]   H. Ma, M. Ma, J. Zeng, X. Guo, Y. Ma, Hydrothermal synthesis of graphene nanosheets and its application in electrically conductive adhesives, Materials Letters,178(2016) 181-184. https://doi.org/10.1016/j.matlet.2016.05.008

[61]   S. P. Lonkar, V. Pillai, A. Abdala, V. Mittal, In situ formed graphene/ZnO nanostructured composites for low temperature hydrogen sulfide removal from natural gas, RSC Advances, 6 (2016) 81142-81150. https://doi.org/10.1039/C6RA08763A

[62]   S. P. Lonkar, V. V. Pillai, S. Stephen, A. Abdala, V. Mittal, Facile in situ fabrication of nanostructured graphene-CuO hybrid with hydrogen sulfide removal capacity, Nano-Micro Letters, 8 (2016) 312-319. https://doi.org/10.1007/s40820-016-0090-8

[63]   R. K. Prud'Homme, I. A. Aksay, A. Abdala, Thermally exfoliated graphite oxide, (2011) Patent No. 8066964.

[64]   H. He, J. Klinowski, M. Forster, A. Lerf, A new structural model for graphite oxide, Chemical Physics Letters, 287 (1998) 53-56. https://doi.org/10.1016/S0009-2614(98)00144-4

[65]   D. R. Dreyer, S. Park, C. W. Bielawski, R. S. Ruoff, The chemistry of graphene oxide, Chemical Society Reviews, 39 (2010) 228-240. https://doi.org/10.1039/B917103G

[66]   J. Paredes, S. Villar-Rodil, A. Martinez-Alonso, J. Tascon, Graphene oxide dispersions in organic solvents, Langmuir, 24 (2008) 10560-10564. https://doi.org/10.1021/la801744a

[67]   G. Shao, Y. Lu, F. Wu, C. Yang, F. Zeng, Q. Wu, Graphene oxide: the mechanisms of oxidation and exfoliation, Journal Material Science, 47 (2012) 4400-4409. https://doi.org/10.1007/s10853-012-6294-5

[68]   P. Chouke, A. Potbhare, G. Bhusari, S. Somkuwar, Dadamia PMD Shaik, R. Mishra, R.G. Chaudhary, Green fabrication of zinc oxide nanospheres by Aspidopterys cordata for effective antioxidant and antibacterial activity, Advanced Materials Letters, 10 (2018) 355-360. https://doi.org/10.5185/amlett.2019.2235

[69]   M. Khan, A.H. Al-Marri, M. Khan, N. Mohri, S.F. Adil, A. Al-Warthan, Pulicaria glutinosa plant extract: a green and eco-friendly reducing agent for the preparation of highly reduced graphene oxide, RSC Advances, 4 (2014) 24119-24125. https://doi.org/10.1039/C4RA01296H

[70] F. Tavakoli, M. Salavati-Niasari, F. Mohandes, Green synthesis and characterization of graphene nanosheets, Materials Research Bulletin, 63 (2015) 51-57. https://doi.org/10.1016/j.materresbull.2014.11.045

[71] R. K. Upadhyay, N. Soin, G. Bhattacharya, S. Saha, A. Barman, S. S. Roy, Grape extract assisted green synthesis of reduced graphene oxide for water treatment application, Materials Letter,160(2015) 355-358. https://doi.org/10.1016/j.matlet.2015.07.144

[72] M. Nasrollahzadeh, M. Maham, A. Rostami-Vartooni, M. Bagherzadehc, S. Mohammad Sajadi, Barberry fruit extract assisted in situ green synthesis of Cu nanoparticles supported on a reduced graphene oxide-Fe3O4 nanocomposite as a magnetically separable and reusable catalyst for the O-arylation of phenols with aryl halides under ligand-free conditions, RSC Advances, 5 (2015) 64769. https://doi.org/10.1039/C5RA10037B

[73] J. Li, G. Xiao, C. Chen, R. Li, D. Yan, Superior dispersions of reduced graphene oxide synthesized by using gallic acid as a reductant and stabilizer, Journal of Material Chemistry A, 1 (2013) 1481-1487. https://doi.org/10.1039/C2TA00638C

[74] M. Mitra, K. Chatterjee, K. Kargupta, S. Ganguly, D. Banerjee, Reduction of graphene oxide through a green and metal-free approach using formic acid, Diamond and Related Materials, 37 (2013) 74-79. https://doi.org/10.1016/j.diamond.2013.05.003

[75] Z. Bo, X. Shuai, S. Mao, H. Yang, J. Qian, J. Chen, J. Yan, K. Cen, Green preparation of reduced graphene oxide for sensing and energy storage applications, Scientic Reports, 4 (2014) 2525-2535. https://doi.org/10.1038/srep04684

[76] N. Kim, P. Khanra, T. Kuila, D. Jung, J. Lee, Efficient reduction of graphene oxide using Tin-powder and its electrochemical performances for use as an energy storage electrode material, Journal of Material Chemistry A, 1 (2013) 11320-11328. https://doi.org/10.1039/c3ta11987d

[77] Z. Fan, K. Wang, T. Wei, J. Yan, L. Song, B. Shao, An environmentally friendly and efficient route for the reduction of graphene oxide by aluminum powder, Carbon, 48 (2010) 1686-1689. https://doi.org/10.1016/j.carbon.2009.12.063

[78] T. A. Pham, J. S. Kim, J. S. Kim, Y. T. Jeong, One-step reduction of graphene oxide with L-glutathione, Colloids and Surfaces A: Physicochemical and Engineering Aspects, 384 (2011) 543-548. https://doi.org/10.1016/j.colsurfa.2011.05.019

[79]   J. Ma, X. Wang, Y. Liu, T. Wu, Y. Liu, Y. Guo, R. Li, X. Sun, F.Wu, C. Lia J.
        Gao, Reduction of graphene oxide with l-lysine to prepare reduced graphene oxide
        stabilized with polysaccharide polyelectrolyte, Journal of Material Chemistry A, 1
        (2013) 2192-2201. https://doi.org/10.1039/C2TA00340F

[80]   X. Kanga, J. Wanga, H. Wua, Ilhan A. Aksayc, J. Liua, Y. Lina, Glucose Oxidase-
        graphene-chitosan modified electrode for direct electrochemistry and glucose
        sensing, Biosensors and Bioelectronics, 25 (2009) 901-905.
        https://doi.org/10.1016/j.bios.2009.09.004

[81]   C. Liu, J. Zhang, E.Yifeng, J.Yue , L. Chen, D.Li,One-pot synthesis of graphene-
        chitosan nanocomposites modified carbon paste electrode for selective
        determination of dopamine, Electronic Journal of Biotechnology,17 (2014) 183-
        188. https://doi.org/10.1016/j.ejbt.2014.04.013

[82]   H. Peng, L. Meng, L. Niu, Q. Lu, Simultaneous reduction and surface
        functionalization of graphene oxide by natural cellulose with the assistance of the
        ionic liquid, The Journal of Physical Chemistry C, 116 (2012) 16294-16299.
        https://doi.org/10.1021/jp3043889

[83]   B. Kartick, S. Srivastava, Green synthesis of graphene,Journal of Nanoscience
        Nanotechnology, 13 (2013) 4320-4324. https://doi.org/10.1166/jnn.2013.7461

[84]   C. Zhu, S. Guo, Y. Fang, S. Dong, Reducing sugar: new functional molecules for
        the green synthesis of graphene nanosheets, ACS Nano,4 (2010) 2429-2437.
        https://doi.org/10.1021/nn1002387

[85]   O. Akhavan, E. Ghaderi, Escherichia coli bacteria reduce graphene oxide to
        bactericidal graphene in a self-limiting manner, Carbon, 50 (2012) 1853-1860.
        https://doi.org/10.1016/j.carbon.2011.12.035

[86]   S. Gurunathan, J. Han, E. Kim, D. N. Kwon, J. K. Park, J. H. Kim, Enhanced
        green fluorescent protein-mediated synthesis of biocompatible graphene, Journal
        of Nanobiotechnology, 12 (2014) 41. https://doi.org/10.1186/s12951-014-0041-9

[87]   S. Gurunathan, J. W. Han, J. H. Park, V. Eppakayala, J. H. Kim, Ginkgo biloba: a
        natural reducing agent for the synthesis of cytocompatible graphene, International
        Journal of Nanomedicine, 9 (2014) 363-377. https://doi.org/10.2147/IJN.S53538

[88]   O. Akhavan, M. Kalaee, Z. Alavi, S. Ghiasi, A. Esfandiar, Increasing the
        antioxidant activity of green tea polyphenols in the presence of iron for the
        reduction of graphene oxide, Carbon, 50 (2012) 3015-3025.
        https://doi.org/10.1016/j.carbon.2012.02.087

[89]   S. Bose, T. Kuila, A. K. Mishra, N. H. Kim, J. H. Lee, Dual role of glycine as a
       chemical functionalizer and a reducing agent in the preparation of graphene: an
       environmentally friendly method, Journal of Material Chemistry, 22 (2012) 9696-
       9703. https://doi.org/10.1039/c2jm00011c

[90]   H. J. Chu, C. Y. Lee, N. H. Tai, Green reduction of graphene oxide by Hibiscus
       sabdariffa L. to fabricate flexible graphene electrode, Carbon,80 (2014) 725-733.
       https://doi.org/10.1016/j.carbon.2014.09.019

[91]   J. Zhang, H. Yang, G. Shen, P. Cheng, J. Zhang, S. Guo, Reduction of graphene
       oxide via L-ascorbic acid, Chemical Communication, 46 (2010) 1112-1114.
       https://doi.org/10.1039/B917705A

[92]   J. Ma, X. Wang, Y. Liu, T. Wu, Y. Liu, Y. Guo, R. Li, X. Sun, F.Wu, C. Lia J.
       Gao, Reduction of graphene oxide with l-lysine to prepare reduced graphene oxide
       stabilized with polysaccharide polyelectrolyte, Journal of Material Chemistry A, 1
       (2013) 2192-2201. https://doi.org/10.1039/C2TA00340F

[93]   A. Esfandiar, O. Akhavan, A. Irajizad, Melatonin as a powerful bio-antioxidant for
       reduction of graphene oxide, Journal of Material Chemistry, 21 (2011) 10907-
       10914. https://doi.org/10.1039/c1jm10151j

[94]   Q. Zhuo, J. Gao, M. Peng, L. Bai, J. Deng, Y. Xia, Y. Ma, J. Zhong, X. Sun.,
       Large-scale synthesis of graphene by the reduction of graphene oxide at room
       temperature using metal nanoparticles as catalyst, Carbon, 52 (2013) 559-564.
       https://doi.org/10.1016/j.carbon.2012.10.014

[95]   X. Li, X. Xu, F. Xia, L. Bu, H. Qiu, M. Chen, L. Zhang, J. Gao.,
       Electrochemically active MnO2/rGO nanocomposites using Mn powder as the
       reducing agent of GO and MnO2 precursor, Electrochimica Acta, 130 (2014) 305-
       313. https://doi.org/10.1016/j.electacta.2014.03.040

[96]   S. Yang, W. Yue, D. Huang, C. Chen, H. Lin, X. Yang, A facile green strategy for
       rapid reduction of graphene oxide by metallic zinc, RSC Advances, 2 (2012)
       8827-8832. https://doi.org/10.1039/c2ra20746j

[97]   C. Li, X. Wang, Y. Liu, W. Wang, J. Wynn, J. Gao, Using glucosamine as a
       reductant prepare reduced graphene oxide and its nanocomposites with metal
       nanoparticles, Journal of Nanoparticle Research, 14 (2012) 1-11.
       https://doi.org/10.1007/s11051-012-0875-8

[98]   Y. Wang, Z. Shi, J. Yin, Facile synthesis of soluble graphene via a green reduction
       of graphene oxide in tea solution and its biocomposites, ACS Applied Materials &
       Interfaces, 3 (2011) 1127-1133. https://doi.org/10.1021/am1012613

[99]   S. Gurunathan, J. Han, J. Kim, Humanin: a novel functional molecule for the
       green synthesis of graphene, Colloids and Surfaces B: Biointerfaces,11 (2013)
       376-383. https://doi.org/10.1016/j.colsurfb.2013.06.018

[100] P. Khanra, T. Kuila, N. Kim, S. Bae, D.-s. Yu, J. Lee, Simultaneous
       biofunctionalization and reduction of graphene oxide by baker's yeast, Chemical
       Engineering Journal,183 (2012) 526-533. https://doi.org/10.1016/j.cej.2011.12.075

[101] D. Dreyer, S. Murali, Y. Zhu, R. Ruoff, C. Bielawski, Reduction of graphite oxide
       using alcohols, Jornal Material Chemistry, 21(2011) 3443-3457.
       https://doi.org/10.1039/C0JM02704A

[102] S. Liu, J. Tian, L. Wang, X. Sun, A method for the production of reduced
       graphene oxide using benzylamine as a reducing and stabilizing agent and its
       subsequent decoration with Ag nanoparticles for enzymeless hydrogen peroxide
       detection, Carbon, 49(2011) 3158-3164.
       https://doi.org/10.1016/j.carbon.2011.03.036

[103] D. Chen, L. Li, L. Guo, An environment-friendly preparation of reduced graphene
       oxide nanosheets via amino acid, Nanotechnology, 22 (2011) 325601.
       https://doi.org/10.1088/0957-4484/22/32/325601

[104] D.Tran, S. Kabiri, D. Losic, A green approach for the reduction of graphene oxide
       nanosheets using non-aromatic amino acids, Carbon. 76 (2014) 193-202.
       https://doi.org/10.1016/j.carbon.2014.04.067

[105] M. Fernandez-Merino, L. Guardia, J. Paredes, S. Villar-Rodil, P. Solis-Fernandez,
       A. Martinez-Alonso, J. Tascón, Vitamin C is an ideal substitute for hydrazine in
       the reduction of graphene oxide suspensions, The Journal of Physical Chemistry
       C, 114 (2010) 6426-6432. https://doi.org/10.1021/jp100603h

[106] S. Thakur, N. Karak, Green reduction of graphene oxide by aqueous phytoextracts,
       Carbon. 50 (2012) 5331-5339. https://doi.org/10.1016/j.carbon.2012.07.023

[107] R. Zaid, F.Chong, E. Y. L. Teo, E. P. Ng, K. F. Chong, Reduction of graphene
       oxide nanosheets by natural beta carotene and its potential use as supercapacitor
       electrode, Arabian Journal of Chemistry, 8 (2015) 560-569.
       https://doi.org/10.1016/j.arabjc.2014.11.036

[108] T. Kuila, S. Bose, P. Khanra, A.K. Mishra, N.H. Kim, J. H. Lee, A green approach
       for the reduction of graphene oxide by wild carrot root, Carbon, 50 (2012) 914-
       921. https://doi.org/10.1016/j.carbon.2011.09.053

[109] D. Suresh, M. Kumar, H. Nagabhushana, S. Sharma, Cinnamon supported facile green reduction of graphene oxide, its dye elimination and antioxidant activities, Materials Letters, 151 (2015) 93-95. https://doi.org/10.1016/j.matlet.2015.03.035

[110] D. Suresh, H.Nagabhushana, S. Sharma, Clove extract mediated facile green reduction of graphene oxide, its dye elimination and antioxidant properties, Materials Letters, 142 (2015) 4-6. https://doi.org/10.1016/j.matlet.2014.11.073

[111] D. Li, M. Müller, S. Gilje, R. Kaner, G. Wallace, Processable aqueous dispersions of graphene nanosheets, Nature Nanotechnology,3 (2008) 101-105. https://doi.org/10.1038/nnano.2007.451

[112] A. Khaled, M. Ali, K. Hantanasirisakul, A. Abdala, P. Urbankowski, M. Zhao, B. Anasori, Y.Gogotsi, B. Aissa, Effect of synthesis on performance of iron oxide anode material for lithium-ion batteries, Langmuir, 34 (2018) 11325-11334. https://doi.org/10.1021/acs.langmuir.8b01953

[113] L. Zhang, G. Shi, Preparation of highly conductive graphene hydrogels for fabricating supercapacitors with high rate capability, The Journal of Physical Chemistry C,115, (2011) 17206-17212. https://doi.org/10.1021/jp204036a

[114] C. Wang, Y. Zhou, M. Ge, X. Xu, Z. Zhang, J. Z. Jiang, Large-scale synthesis of SnO2nanosheets with high lithium storage capacity, Journal of the American Chemical Society, 132 (2010)46-47. https://doi.org/10.1021/ja909321d

[115] L. Li, X. Yin, S. Liu, Y. Wang, L. Chen, T. Wang, Electrospun porous SnO2 nanotubes as high capacity anode materials for lithium ion batteries, Electrochemistry Communications, 12 (2010)1383-1386. https://doi.org/10.1016/j.elecom.2010.07.026

[116] K. Fukuda, K. Kikuya, K. Isono, M. Yoshio, Foliated natural graphite as the anode material for rechargeable lithium-ion cells, Journal of Power Sources, 69 (1997) 165-168. https://doi.org/10.1016/S0378-7753(97)02568-8

[117] M. Goh, M.Pumera, Multilayer graphene nanoribbons exhibit larger capacitance than their few-layer and single-layer graphene counterparts, Electrochemistry Communications, 12 (2010) 1375-1377. https://doi.org/10.1016/j.elecom.2010.07.024

[118] L. Buglione, E.Chng, A. Ambrosi, Z. Sofer, M. Pumera,Graphene materials preparation methods have dramatic influence upon their capacitance, Electrochemistry Communications,14 (2012) 5-8. https://doi.org/10.1016/j.elecom.2011.09.013

[119] S. Cho, J. Ung, C. Kim, I.Kim,Rational design of 1-D Co3O4nanofibers@low content graphene composite anode for high performance Li-Ion batteries, Scientific Reports, 7 (2017) 45105 https://doi.org/10.1038/srep45105

[120] M. Jing, M. Zhou, G. Li, Z. Chen, W. Xu, X. Chen, and Z. Hou, Graphene-embedded Co3O4 rose-spheres for enhanced performance in Lithium Ion batteries, ACS Applied Materials & Interfaces, 9 (2017) 9662-9668. https://doi.org/10.1021/acsami.6b16396

[121] D.Su, S. Dou, G.Wang, Mesocrystal Co3O4 nanoplatelets as high capacity anode materials for Li-ion batteries, Nano Research, 7 (2014) 794-803. https://doi.org/10.1007/s12274-014-0440-0

[122] Y. Gua, M. Xinga, J. Zhang, Synthesis and photocatalytic activity of graphene based doped TiO2 nanocomposites, Applied Surface Science, 319 (2014) 8-15. https://doi.org/10.1016/j.apsusc.2014.04.182

[123] M. Fagnoni, D. Dondi, D. Ravelli, A. Albini, Photocatalysis for the formation of the C−C Bond,Chemical Reviews, 107 (2007) 2725-2756. https://doi.org/10.1021/cr068352x

[124] G. Colon, J. M. Sanchez-Espana, M.C. Hidalgo, J.A. Navıo, Effect of TiO2 acidic pre-treatment on the photocatalytic properties for phenol degradation, Journal of Photochemistry and Photobiology A: Chemistry,179 (2006) 20-27. https://doi.org/10.1016/j.jphotochem.2005.07.007

[125] M. Z. Iqbal, P. Pal, M. Shoaib, A. A. Abdala, Efficient removal of different basic dyes using graphene, Desalination and Water Treatment, 68 (2017) 226-235. https://doi.org/10.5004/dwt.2017.20213

[126] V. Sonkusare, R.G. Chaudhary, G. Bhusari, A. Rai, H. Juneja, Microwave-mediated synthesis, photocatalytic degradation and antibacterial activity of α-Bi2O3 microflowers/novel γ-Bi2O3 microspindles, Nano-Structure & Nano-Objects, 13 (2018) 121-131. https://doi.org/10.1016/j.nanoso.2018.01.002

[127] N. Serpone, D. Lawless, R. Khairutdinov,Subnanosecond relaxation dynamics in TiO2 Colloidal Sols (Particle Sizes R, = 1.0- 13.4 nm) relevance to heterogeneous photocatalysis, Journal Physical Chemistry, 99 (1995) 16655-16661. https://doi.org/10.1021/j100045a027

[128] Q. Rahman , M. Ahmad, S.Misra, M. Lohani, Effective photocatalytic degradation of rhodamine B dye by ZnO nanoparticles, Materials Letters 91 (2013) 170-174 https://doi.org/10.1016/j.matlet.2012.09.044

[129] M. Iqbal, A. Abdala, Thermally reduced graphene: synthesis, characterization and
       dye removal applications, RSC Advances, 3 (2013) 24455-24464.
       https://doi.org/10.1039/c3ra43914c

[130] N. Daneshvar,A. Oladegaragoze, N. Djafarzadeh, Decolorization of basic dye
       solutions by electrocoagulation: An investigation of the effect of operational
       parameters. Journal of Hazardous Materials,129 (2006) 116-122.
       https://doi.org/10.1016/j.jhazmat.2005.08.033

[131] T. Sauer, G. Cesconeto Neto, H. José, R. Moreira. Kinetics of Photocatalytic
       Degradation of Reactive Dyes in a TiO2 Slurry Reactor,Journal of Photochemistry
       and Photobiology A: Chemistry, 149 (2002) 147-154.
       https://doi.org/10.1016/S1010-6030(02)00015-1

[132] I. Poulios, M. Kositzi, A. Kouras, Photocatalytic decomposition of triclopyr over
       aqueous semiconductor suspensions, Journal of Photochemistry and Photobiology
       A: Chemistry,115 (1998) 175-183. https://doi.org/10.1016/S1010-
       6030(98)00259-7

[133] D. Dionysiou , A. Khodadoust, A. Kern, M. Suidan, Isabelle Baudin, Jean-Michel
       Laîné, Continuous-mode photocatalytic degradation of chlorinated phenols and
       pesticides in water using a bench-scale TiO2 rotating disk reactor, Applied
       Catalysis B: Environmental, 24 (2000) 139-155. https://doi.org/10.1016/S0926-
       3373(99)00103-4

[134] Z. Ghouri, K. Elsaid, A. Abdala, S.Meer, N. Barakat, Surfactant/organic solvent
       free single-step engineering of hybrid graphene-Pt/TiO2 nanostructure: Efficient
       photocatalytic system for the treatment of wastewater coming from textile
       industries,Scientific Reports, 8 (2018) 14656. https://doi.org/10.1038/s41598-018-
       33108-4

[135] J. Zhang, Z. Xiong, X. Zhao, Graphene-metal-oxide composites for the
       degradation of dyes under visible light irradiation, Journal of Materials Chemistry,
       21 (2011) 3634-3640 https://doi.org/10.1039/c0jm03827j

[136] R. Bunsen, Ueber eine neue Construction der galvanischen Säule, Justus Liebigs
       Annalen der Chemie, 38 (1841) 311-313.
       https://doi.org/10.1002/jlac.18410380307

[137] H. W. Kroto, J. R. Heath, S. C. O'Brien, R. F. Curl, R. E. Smalley, C60:
       Buckminsterfullerene, Nature, 318 (1985) 162-163.
       https://doi.org/10.1038/318162a0

[138] K. Drew, G. Girishkumar, K. Vinodgopal, Prashant V. Kamat, Boosting fuel cell
       performance with a semiconductor photocatalyst: TiO2/Pt−Ru hybrid catalyst for

methanol oxidation, The Journal of Physical Chemistry B,109 (2005) 11851-11857. https://doi.org/10.1021/jp051073d

[139] V.N. Sonkusare, R.G. Chaudhary, G.S. Bhusari, A. Mondal, A.K. Potbhare, R. Kumar Mishra, A.A. Abdala, H.D. Juneja,Mesoporous Octahedron-Shaped Tricobalt Tetroxide Nanoparticles for Photocatalytic Degradation of Toxic Dyes, ACS Omega, 5 (2020) 7823-7835. https://doi.org/10.1021/acsomega.9b03998

[140] A. Abu-Nada, G. McKay, A. Abdala, Recent Advances in Applications of Hybrid Graphene Materials for Metals Removal from Wastewater, Nanomatrials, 10 (2020) 595. https://doi.org/10.3390/nano10030595

[141] S. P. Lonkar, V. Pillai, A. Abdala, Solvent-free synthesis of ZnO-graphene nanocomposite with superior photocatalytic activity, Applied Surface Science, 465, (2019) 1107-1113. https://doi.org/10.1016/j.apsusc.2018.09.264

[142] R.G. Chaudhary, G.S. Bhusari, A.D. Tiple, A.R. Rai, S.R. Somkuvar, A.K. Potbhare, T.L. Lambat, P.P. Ingle, A. A. Abdala, Metal/Metal Oxide Nanoparticles: Toxicity, Applications, and Future Prospects, Current Pharmaceutical Design, 25 (2019) 4013-4029. https://doi.org/10.2174/1381612825666191111091326

[143] Y. Zhang, N. Zhang, Z. R. Tang, Y-J Xu. Graphene transforms wide band gap ZnS to a visible light photocatalyst. The new role of graphene as a macromolecular photosensitizer. ACS Nano, 6 (2012) 9777-9789. https://doi.org/10.1021/nn304154s

[144] R. Bagade, V. Sonkusare, A. Potbhare, R. Chaudhary, R. Husain, H. Juneja, Fabrication of microflower-shaped mesoporous Fe (II) chelate polymer for photocatalytic performance under visible light, Material Today: Proceedings, 15 (2019) 566-577. https://doi.org/10.1016/j.matpr.2019.04.122

[145] H. Wang, L. Zhang, Z. Chen, J. Hu, S. Li, Z. Wang, J. Liu, X. Wang, Semiconductor heterojunction photocatalysts: design, construction and photocatalytic performances, Chemical Society Reviews,43 (2014) 5234-5244. https://doi.org/10.1039/C4CS00126E

[146] J. Tanna, R.G. Chaudhary, N. Gandhare, A. Rai, S. Yerpude, H. Juneja, Copper nanoparticles catalysed an efficient one-pot multicomponents synthesis of chromenes derivatives and its antibacterial activity, Journal of Experimental Nanoscience,11 (2016) 884-900. https://doi.org/10.1080/17458080.2016.1177216

[147] N. Gandhare, R.G. Chaudhary, M.P. Gharpure, V.P. Meshram, H. Juneja, An efficient and one-pot synthesis of 2, 4, 5-trisubstituted imidazole compounds

catalyzed by copper nanoparticles, Journal of the Chinese Advanced Material Society, 3 (2015) 270-279. https://doi.org/10.1080/22243682.2015.1068134

[148] J. Tanna, R.G. Chaudhary, N. Gandhare, A. Mondal, H. Juneja, Silica-coated nickel oxide a core-shell nanostructure: synthesis, characterization and its catalytic property in one-pot synthesis of malononitrile derivative, Journal of the Chinese Advanced Material Society, 5 (2017) 103-117. https://doi.org/10.1080/22243682.2017.1296371

[149] N. Gandhare, V. Meshram, R.G. Chaudhary, H.D. Juneja,Cu-nanoparticles as catalyst for synthesis of 1H-imidazoles under microwave irradiation, Bionanomaterial Frontier, 5 (2012) 12-14.

[150] J. Tanna, R.G. Chaudhary, V. Sonkusare, H. Juneja, CuO nanoparticles: synthesis, characterization and reusable catalyst for polyhydroquinoline derivatives under ultrasonication, Journal of the Chinese Advanced Materials Society, 4 (2016) 110-122. https://doi.org/10.1080/22243682.2016.1164618

[151] C. Hu, T. Lu, F. Chen, R. Zhang, A brief review of graphene-metal oxide composites synthesis and applications in photocatalysis, Journal of the Chinese Advanced Materials Society, 1(2013). 21-39. https://doi.org/10.1080/22243682.2013.771917

[152] Q. Xiang, J. Yu, M. Jaroniec, Graphene-based semiconductor photocatalysts, Chemical Society Reviews, 41 (2012) 782-796. https://doi.org/10.1039/C1CS15172J

[153] M. Z. Iqbal, P. Pal, M. Shoaib, A. A. Abdala, Efficient removal of different basic dyes using graphene, Desalination and Water Treatment, 68 (2017) 226-235. https://doi.org/10.5004/dwt.2017.20213

[154] K. Sauer, A. Camper, G. Ehrlich, J. Costerton, D. Davies, Pseudomonas aeruginosa displays multiple phenotypes during development as a biofilm, Journal of Bacteriology, 184 (2002) 1140-1154. https://doi.org/10.1128/jb.184.4.1140-1154.2002

[155] R. Bagade, R.G. Chaudhary, A. Potbhare, A. Mondal, M. Desimone, K. Dadure, R. Mishra, H. Juneja,Microspheres/custard apples copper (II) chelate polymer: characterization, docking, antioxidant and antibacterial assay, ChemistrySelect, 4 (2019) 6233-6244. https://doi.org/10.1002/slct.201901115

[156] C. Vecitis, K. Zodrow, S. Kang, M. Elimelech, Electronic-structure-dependent bacterial cytotoxicity of single-walled carbon nanotubes,ACS Nano, 4 (2010) 5471-7479. https://doi.org/10.1021/nn101558x

[157] A. Potbhare, P. Chauke, S. Zahra, V. Sonkusare, R. Bagade, M. Umekar, R. G. Chaudhary, Microwave-mediated fabrication of mesoporous Bi-doped CuAl2O4 nanocomposites for antioxidant and antibacterial performances, Materials Today: Procdeengs, 15 (2019) 454-463. https://doi.org/10.1016/j.matpr.2019.04.107

[158] R.G. Chaudhary, J. Tanna, N. Gandhare, A. R. Rai, H. Juneja, Synthesis of nickel nanoparticles: Microscopic investigation, an efficient catalyst and effective antibacterial activity,Advanced Materials Letter, 6 (2015) 990-998. https://doi.org/10.5185/amlett.2015.5901

[159] A. Potbhare, R.G. Chaudhary, P. Chouke, S. Yerpude, A. Mondal, V. Sonkusare, A. Rai, H. Juneja,Phytosynthesis of nearly monodisperse CuOnanospheres usingPhyllanthusreticulatus/Conyzabonariensis and its antioxidant/antibacterial assays, Material Science & Engineering C, 99 (2019) 783-793. https://doi.org/10.1016/j.msec.2019.02.010

[160] S. Liu, T. H Zeng, M. Hofmann, E. Burcombe, J. Wei, R, Jiang, J. Kong, Y. Chen, Antibacterial Activity of Graphite, Graphite Oxide, Graphene Oxide, and Reduced Graphene Oxide: Membrane and Oxidative Stress, ACS Nnao, 5 (2011) 6971-6980 ' https://doi.org/10.1021/nn202451x

[161] H. Jung, P. Verwilst, A. Sharma, J. Shin, J. Sessler, J. Kim, Organic molecule-based photothermal agents: an expanding photothermal therapy universe. Chemical Society Reviews, 47 (2018) 2280. https://doi.org/10.1039/C7CS00522A

[162] H. Lin, S.Liao, S.Hung,The DC thermal plasma, synthesis of ZnOnanoparticles for visible-light photocatalyst. Journal of Photochemistry and Photobiology A: Chemistry, 174 (2005) 82-87. https://doi.org/10.1016/j.jphotochem.2005.02.015

[163] C. Thomas, N. Saleh, R. Tilton, Lowry, B. Veronesi,Titanium dioxide (P25) produces reactive oxygen species in immortalized brain microglia (BV2): implications for nanoparticles neurotoxicity, Environment Science Technology, 40 (2006) 4346-4352. https://doi.org/10.1021/es060589n

[164] S. Liu, M. Hu, T. H. Zeng, Ran Wu, R.Jiang, J. Wei, L.Wang, J. Kong, Y.Chen, Lateral Dimension-Dependent Antibacterial Activity of Graphene Oxide Sheets, Langmuir, 28 (2012) 12364-12372 https://doi.org/10.1021/la3023908

[165] D. Schubert, R. Dargusch, J. Raitano, S Chan, Cerium and yttrium oxide nanoparticles are neuroprotective, Biochemical and Biophysical Research Communications, 342 (2006) 86-91. https://doi.org/10.1016/j.bbrc.2006.01.129

[166] A. Thill, O. Zeyons, O. Spalla, F. Chauvat, J. Rose, M. Auffan, A. Flank,Cytotoxicity of CeO2nanoparticles for Escherichia coliphysico-chemical

insight of the cytotoxicity mechanism. Environmental Science & Technology, 40 (2006) 6151-6156. https://doi.org/10.1021/es060999b

[167] J. Tanna, R.G. Chaudhary, H.D. Juneja, N. Gandhare, A. Rai, Histidine-capped ZnO nanoparticles: an efficient synthesis, spectral characterization and effective antibacterial activity, BioNanoScience, 5 (2015) 123-134. https://doi.org/10.1007/s12668-015-0170-0

[168] K. Krishnamoorthy, N. Umasuthan, R. Mohan, J. Lee, S-J. Kim, Science of Advanced Materials, 4 (2012) 1-7. https://doi.org/10.1166/sam.2012.1402

[169] X. Zeng, G. Wang, Y. Liua, X. Zhang, Graphene-based antimicrobial nanomaterials: rational design and applications for water disinfection and microbial control, Environmental Science: Nano, 4 (2017) 2248-2266. https://doi.org/10.1039/C7EN00583K

[170] T. Long, J. Tajuba, P. Sama, N. Saleh, C. Swartz, J. Parker, S, Hester, G. Lowry, B. Veronesi, Nanosize titanium dioxide stimulates reactive oxygen species in brain microglia and damages neurons in vitro, Environmental Health Perspectives, 115 (2007) 1631-1637. https://doi.org/10.1289/ehp.10216

[171] S. Wang, R. Gao, F. Zhou, M. Selke, Nanomaterials and singlet oxygen photosensitizers: potential applications in photodynamic therapy, Journal of Material Chemistry, 14 (2014) 487-493. https://doi.org/10.1039/b311429e

[172] R. G. Chaudhary, V. Sonkusare, G. Bhusari, A. Mondal, D. Shaik, H. Juneja, Microwave-mediated synthesis of spinel-CuAl2O4 nanocomposites for enhanced electrochemical and catalytic performance, Research on Chemical Intermediates, 44 (2017) 2039-2060. https://doi.org/10.1007/s11164-017-3213-z

[173] F. Irwin, Biological effects of superoxide radical, Archives of Biochemistry and Biophysics, 247 (1986) 1-11. https://doi.org/10.1016/0003-9861(86)90526-6

Materials Research Forum LLC
https://doi.org/10.21741/9781644900970-5

Chapter 5

Synthesis and Antimicrobial Study of Co-Ni-Cd Nanoferrites

M. Raghasudha

Department of Chemistry, National Institute of Technology Warangal 506 004 Telangana, India

raghas13chem@nitw.ac.in

Abstract

$Co_xNi_{0.5}Cd_{0.5-x}Fe_2O_4$ (x = 0.0, 0.1, 0.2, 0.3, 0.4, 0.5) nanoferrites were synthesized by citrate-gel auto combustion method. Antimicrobial study was tested on Gram-negative bacteria and Gram-positive bacteria by adopting disc-diffusion agar method. The prepared nanoferrites show excellent antimicrobial function with a zone of inhibition of 10-12 mm against E. coli, 5-7 mm against K. pneumonia, 9-12 mm against B. subtilis and 5-7 mm against S. aureus. Inhibition of the growth was ascertained with reference to Ampicillin, a standard antibacterial drug, that acts as positive control. The antibacterial properties of the prepared ferrites bid great challenges in biomedical and pharmaceutical applications.

Keywords

Nanoferrites, Structural Properties, Citrate-Gel Auto Combustion Method, Anti-Microbial Study

Contents

1.   Introduction

Ferrites are very significant inorganic materials that possess extraordinary electrical, magnetic and biological properties. Due to these properties, they find excellent applications in distinct fields such as electronic industry, magnetic storage, catalytic purposes, biomedical field etc. Magnetic ferrite nanoparticles extend their possible applications in various biomedical fields such as targeted drug delivery, hyperthermia and MRI (Magnetic Resonance Imaging) which are all guided by magnetic field [1, 2]. Magnetic nanoparticles are used as anti-microbial agents because of the fact that they can be easily influenced by an external magnetic field.

The properties of the ferrites are influenced by the synthesis process and particle size [3]. Ferrites prepared in the nano regime have fascinating properties consequently wonderful applications. Some of the common synthesis methods used for the synthesis of ferrite nano particles are mechanical milling [4], co-precipitation [5], hydrothermal reaction [6], micro emulsion method [7] and sol–gel technique [8]. The properties are also influenced by the type of substituting metal ion i.e. dopant. In present work, citrate sol-gel method is used to prepare ferrites.

Day by day infectious diseases are emerging out with great speed and the pathogenic bacteria are becoming drug resistant. Under these conditions, there is a great need for novel antimicrobial agents. Nanoparticles possessing anti-microbial activity can find gigantic application in various biomedical fields. Earlier reports also show that nanoparticles with good antimicrobial activity can be used as effective antimicrobial agent [9]. Mixed spinel ferrites with different cations, are important materials for both the basic studies and application point of view [10].

The nano ferrites with nickel as the dopant were found to have extensive applications in the electronic industry due to their noteworthy properties such as high electrical resistivity, low dielectric and eddy current losses [11]. It is described in the literature that Ni-Cd ferrites showcase good coercivity, remanence and saturation magnetization due to which they display excellent microwave applications [12]. Non-magnetic Cd is substituted with magnetic Cobalt in Ni-cd nano ferrites with an expectation that the magnetic properties were improved. It is also reported that the metal oxide nano particles such as nano ferrites have been used as effective antimicrobial units. It is found that many

Magnetic Oxides and Composites II   Materials Research Forum LLC
Materials Research Foundations **83** (2020) 117-133   https://doi.org/10.21741/9781644900970-5

of them show good efficiency against resistant micro-organisms. Kim et al have reported that cobalt nano particles possessing good coercivity, reasonable saturation magnetization and remarkable chemical stability can be efficiently used in MRI and biomedical drug delivery, hyperthermia cancer treatment [13]. Recently it is also reported by Sanpo et al. that Cobalt nano ferrite particles display outstanding antibacterial activity against E. Coli and S. Aureus [14]. All these facts made the author to use the cobalt ion as dopant in Ni-Cd ferrites and to prepare in nano regime for the effective anti-microbial property of the material. For the synthesis of proposed ferrites under investigation, sol-gel method with citric acid as fuel was selected because this method has a good control over the stoichiometry and result in the formation of very low sized nano particles [15]. The present paper reports the synthesis, characterization and antimicrobial studies of cobalt doped Ni-Cd nano ferrites. The microbes considered for the study are Gram-negative strains of Escherichia coli, Klebsiella pneumonia and Gram-positive strains of Staphylococcus aureus and Bacillus subtilis.

2.   Materials and methods

Citrate-gel auto combustion method was used to synthesize cobalt doped Ni-Cd nano ferrites [16]. Appropriate metal nitrates such as cobalt nitrate, nickel nitrate, cadmium nitrate, ferric nitrate and citric acid (all AR grade of high purity) were used as initial materials during the synthesis. Required quantities of chosen metal nitrates were dissolved in proper quantity of distilled water and were mixed well and then citric acid solution was added. The molar ratio of the metal nitrate to the citric acid was maintained as 1:3. Magnetic stirrer was used to get a homogenous solution by continuous stirring of the obtained solution. To maintain the pH at 7, Ammonia solution (25 % w/v) was added under constant stirring. The resulting solution was then heated at 80 °C on magnetic hot plate till the water evaporates from the solution resulting in dry solid. On further heating, the solution turns into gel and the heating is continued up to 180 °C where the gel self-ignites and undergoes auto combustion from bottom to top like a volcanic eruption. This results loose puffy mass of the proposed ferrites. The puffy mass was crushed using an Agate mortar-pastel to result in very fine powder of cobalt substituted Ni-Cd nano ferrites. The obtained powder was subjected to sintering in muffle furnace at 500 °C for duration of four hours. Fig.1 represents flow chart to prepare cobalt substituted Ni-Cd ferrites.

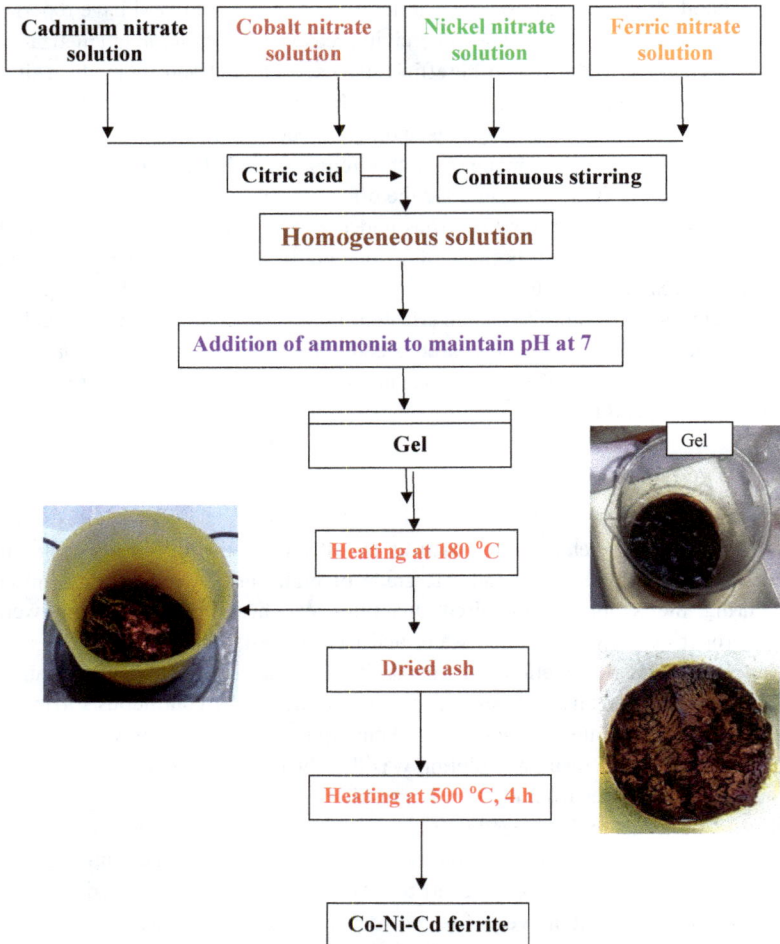

Fig. 1 Flow chart for the synthesis of Co-Ni-Cd nano ferrite powder.

The resulting ferrite powder was subjected to structural characterization by XRD using Phillips X ray diffractometer (model 3710) with Cu-Kα radiation of wave length λ = 1.5405 Å. The XRD analysis was carried out to identify crystal structure and the phase. XRD patterns were recorded at room temperature by continuous scanning in the range of

2θ from 5° to 85°. Infrared spectra of prepared ferrites were recorded by IR Prestige-21 Shimadzu Fourier Transform Infrared spectrophotometer in wave number range of 800-250 cm^{-1} with a resolution of 1 cm^{-1}. Antimicrobial studies on Gram-negative strains of Escherichia coli, Klebsiella pneumonia and Gram-positive strains of Staphylococcus aureus and Bacillus subtilis of prepared nano ferrites were performed using agar diffusion method. For this, synsized ferrites with various compositions were dissolved in DMSO at 30 μg/μL concentration and tested against Gram-negative strains of (1) Escherichia coli, (2) Klebsiella. Pneumonia and Gram-positive strains of (3) Staphylococcus aureus and (4) Bacillus subtilis using agar well diffusion method according to the literature protocol [17]. Ampicillin, a standard antibacterial drug was used as the reference antibiotic. With the help of zones showing complete inhibition (mm), the activity was determined. Growth inhibition was measured with reference to the positive control. All the samples were used in triplicates.

3.   Results and discussion

3.1   XRD analysis

Fig. 2 represents X-ray diffraction patterns of $Co_xNi_{0.5}Cd_{0.5-x}Fe_2O_4$ (x = 0.0, 0.1, 0.2, 0.3, 0.4, 0.5) nanoferrites, heated at 500 °C for 4 h. The XRD peaks of all the compositions were compared with reference data from ICSD card no. 96-591-0065 of $NiFe_2O_4$ and ICSD card no. 00-022-1063 of $CdFe_2O_4$. All the Bragg's reflection peaks have been indexed as (220), (311), (222), (400), (511) and, (440) as shown in Fig. 2. X-ray analysis confirmed the formation of cubic spinel structure with single phase without any impurity peak. Cobalt doped Ni-Cd nano ferrites displayed a homogeneous cubic spinel with single phase, which is belong to the Fd_3m space group as confirmed by the ICSD reference data. All the obtained peaks are allowed peaks with the maximum reflection exhibited by (311) plane, which indicates the spinel phase.

Debye-Scherrer's formula was used to determine the crystallite size (D_{xrd}) of the prepared ferrite compositions from maximum intensity peak (311) [18].

$$D_{xrd} = \frac{0.91\,\lambda}{\beta\cos\theta} \tag{1}$$

Where, λ = X-rays wavelength

β = FWHM (Full width half maxima) in radians,

θ = Bragg's angle at the peak position.

The crystallite size was found to be in the range of 11.4 nm to 26.7 nm for various compositions with a notable increase in size corresponding to the rise in Cobalt content (x) as shown in the Table 1.

Table 1 Sample code, compositions and crystallite size prepared ferrites.

Sample code	Co content (x)	composition	Crystallite size (nm)
I-1	0.0	$Ni_{0.5}Cd_{0.5}Fe_2O_4$	11.4
I-2	0.1	$Co_{0.1}Ni_{0.5}Cd_{0.4}Fe_2O_4$	25.1
I-3	0.2	$Co_{0.2}Ni_{0.5}Cd_{0.3}Fe_2O_4$	24.6
I-4	0.3	$Co_{0.3}Ni_{0.5}Cd_{0.2}Fe_2O_4$	22.9
I-5	0.4	$Co_{0.4}Ni_{0.5}Cd_{0.1}Fe_2O_4$	23.7
I-6	0.5	$Co_{0.5}Ni_{0.5}Fe_2O_4$	26.6

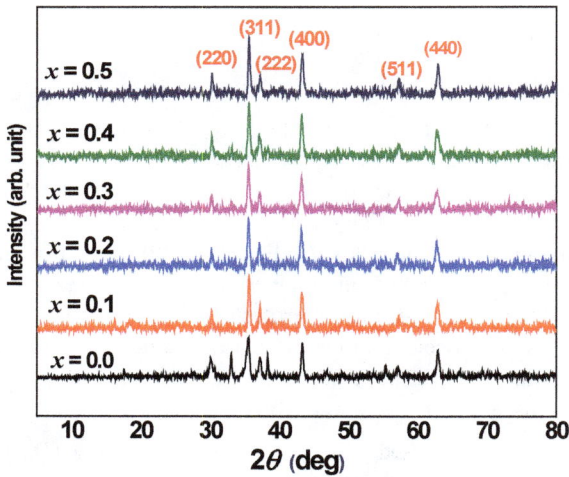

Fig. 2 XRD patterns of $Co_xNi_{0.5}Cd_{0.5-x}Fe_2O_4$ (x = 0.0, 0.1, 0.2, 0.3, 0.4, 0.5) nanoferrites, heated at 500 °C for 4 h.

Lattice parameter 'a' of all heated samples was calculated using the formula

$$a = d\sqrt{h^2 + k^2 + l^2} \qquad (2)$$

Materials Research Forum LLC
https://doi.org/10.21741/9781644900970-5

Where, a = Lattice parameter,

d = distance between the planes,

h, k, l = Miller indices of the plane

X-ray density (d_x) of the prepared samples was calculated using the formula [19]:

$$d_x = \frac{8M}{Na^3}\tag{3}$$

Where, d_x = X-ray density in gm/cc

M = Molecular weight of the sample

N = Avogadro's number

a = Lattice parameter

The specific surface area of the particles (S) can be defined as the summation of exposed particle surface area per unit mass. It was calculated using the relation [20]

$$S = \frac{6}{D_{xrd}\, d_x}\tag{4}$$

Where, D_{xrd} = Crystallite size obtained from Scherrer's formula (nm),

d_x = X-rays density (g/m^3)

The XRD data can be used to confirm the formation of single phase by determining the tetrahedral and octahedral site radii viz., r_A and r_B respectively using the following relations [21]:

$$r_A = \left(\mu - \frac{1}{4}\right)a\sqrt{3} - R_O\tag{5}$$

$$r_B = \left(\frac{5}{8} - \mu\right)a - R_O\tag{6}$$

Where, μ = Oxygen parameter its ideal value is 3/8

a = Lattice parameter

Materials Research Forum LLC
https://doi.org/10.21741/9781644900970-5

R_O = radius of oxygen ion (1.32 Å)

The hopping length (L) between magnetic ions in the tetrahedral A site (L_A) and in the octahedral B-site (L_B) was determined using the relations given in equations 7 and 8 respectively.

$$L_A = \frac{a\sqrt{3}}{4} \tag{7}$$

$$L_B = \frac{a\sqrt{2}}{4} \tag{8}$$

Where, 'a' is the lattice parameter.

For cubic samples, bond length values of tetrahedral and octahedral site (R_A and R_B) were determined using the equations (9) and (10) respectively.

$$R_A = r_A + R_o \tag{9}$$

$$R_B = r_B + R_o \tag{10}$$

Where r_A and r_B corresponds to the tetrahedral and octahedral site radii respectively and R_O is the oxygen ion radius. The obtained values of lattice parameter, X-ray density, specific surface area, site radii, hopping length and bond length for all compositions are in Table 2.

Table 2 Lattice parameter, X-ray density, specific surface area, site radii, hopping length and bond length of $Co_xNi_{0.5}Cd_{0.5-x}Fe_2O_4$ (x = 0.0, 0.1, 0.2, 0.3, 0.4, 0.5) ferrites, heated at 500 °C for 4 h.

Cobalt content (x)	Lattice parameter a (Å)	X-ray density d_x (g/cc)	Specific surface area (S) (m^2)	Site radii (Å)		Hopping length (Å)		Bond length (Å)	
				r_A (Å)	r_B (Å)	L_A	L_B	R_A	R_B
0.0	8.359	5.9408	88.75	0.489	0.7697	3.6194	2.9549	1.809	2.0897
0.1	8.353	5.8318	40.97	0.488	0.7682	3.6168	2.9527	1.808	2.0882
0.2	8.373	5.6691	42.92	0.493	0.7732	3.6255	2.9598	1.813	2.0932
0.3	8.374	5.5461	47.16	0.493	0.7735	3.6259	2.9602	1.813	2.0935
0.4	8.360	5.4524	46.47	0.490	0.7700	3.6198	2.9552	1.810	2.0900
0.5	8.359	5.3328	42.22	0.489	0.7697	3.6194	2.9549	1.809	2.0897

Magnetic Oxides and Composites II   Materials Research Forum LLC
Materials Research Foundations **83** (2020) 117-133   https://doi.org/10.21741/9781644900970-5

From the data recorded in the Table 2, irregularity in lattice parameter, X-ray density, bond length, site radii and hopping length with cobalt content was observed, which can be ascribed to the irregularity in the value of lattice parameter with cobalt content.

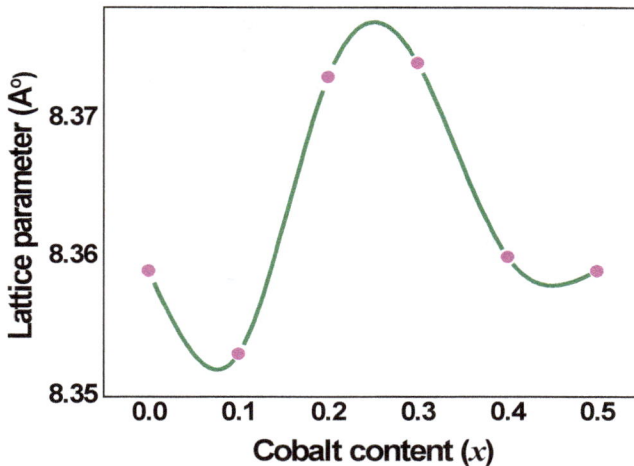

Fig. 3 The variation of lattice parameter with cobalt content (x).

Fig. 3 depicts the variation of lattice parameter of Co-Ni-Cd ferrite system with cobalt content (x). It is clear from Fig.3 that an irregular trend in the lattice parameter is noticed with an increase in cobalt content. The lattice parameter was observed to decrease initially up to $x = 0.1$ and later increased up to $x = 0.3$. Thereafter, a fall in the lattice parameter was noticed to further rise with cobalt content. This decline in lattice parameter with rise in cobalt composition is likely because of small ionic radius of Co^{2+} ion (0.73 Å) with reference to that of Cd^{2+} ion (0.97A°). It is reported that Cd^{2+} ions tend to occupy on A-site [22, 23] that push Fe^{3+}ions on A site to go to B-site. As the ionic radius of Cd^{2+} ions (0.97 Å) is higher than Fe^{3+} (0.67 Å), the four oxygen atoms comprising tetrahedral site have to step outward its position in the lattice to offer more space for Cd^{2+} ion. It is reported that the Ni^{2+} ions have strong site preference to octahedral site [24, 25]. Site occupancy in spinel is influenced majorly by ionic radius and CFSE (Crystal Field Stabilization Energy). Radius of A-site (tetrahedral) is 0.5-0.67 Å and that of B-site (octahedral) is 0.70-0.75 Å. Hence, irrespective of the crystal field effects, an ion will be accommodated in octahedral hole if it cannot be accommodated into tetrahedral hole. Fe^{2+} ions passes into B –site easier than Fe^{3+}, whereas Co^{2+} ions and Ni^{2+} ions enters B-site even easier than Fe^{2+} ions. When Co^{2+} ions and Ni^{2+} ions enters the ferrite lattice they not only substitute Fe^{2+} ions on B-site but also force a part of Fe^{3+} ions into A site and results in an increment in lattice parameter up to $x = 0.4$. But, with further increase in Co^{2+} ions lattice constant will decrease instead of increasing because of small Co^{2+} ions (0.73Å) compared to large Fe^{2+} ions (0.81Å). As far as Ni^{2+}ions are concern, it seems that it can only substitute for Fe^{2+} ions on B-site. In fact, it can substitute for both Fe^{2+} and Fe^{3+} ions equally, as both the ions can be interchanged via charge transfer. Thus, the variation in the lattice parameter is the outcome of combined effect of all the Fe^{3+}ions, Fe^{2+} ions, Ni^{2+} ions and Co^{2+} ions.

3.2   FT-IR study

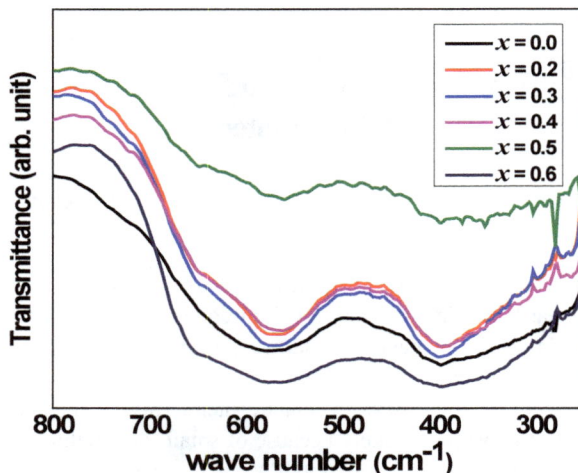

Fig. 4 FT-IR spectra of $Co_xNi_{0.5}Cd_{0.5-x}Fe_2O_4$ (x = 0.0, 0.1, 0.2, 0.3, 0.4, 0.5) nanoferrites, heated at 500 °C for 4 h.

FT-IR spectral studies is an important tool to check the formation of ferrite in the synthesized ferrite samples. The FT-IR spectra of $Co_xNi_{0.5}Cd_{0.5-x}Fe_2O_4$ (x = 0.0, 0.1, 0.2,

0.3, 0.4, 0.5) nanoferrites, heated at 500 °C for 4 h. recorded in wave number range of 800 cm^{-1} to 250 cm^{-1}, are depicted in Fig. 4. It is clear from Fig.4 that two absorption bands and are observed at around 600 cm^{-1} (v_1) and 400 cm^{-1} (v_2) respectively. It is a known fact that IR spectra of spinel ferrites are expected to show two major absorption bands in wave number range of 400-600 cm^{-1} [26] . Hence, the ferrite samples under study show formation of ferrite. From the figure, the higher frequency band (v_1) is observed in the range 559-579 cm^{-1} and is due to the stretching vibrations of the tetrahedral (A) metal- oxygen bond. The lower frequency band (v_2) is in the range 393-399 cm^{-1}, and is caused by the metal- oxygen vibrations in the octahedral (B) sites. The band positions of (v_1) and (v_2) of the prepared Co-Ni-Cd ferrites is shown in Table 3. The change in band position is ascribed to the difference in Fe^{3+} - O^{2-} for tetrahedral and octahedral complexes [27]. These band positions are influenced by mainly the bond lengths and nature of the cations present in the sample.

Table 3 The band position assignments of (v_1) and (v_2) for prepared compositions.

Sample code	Cobalt content (x)	Composition	v_1 (cm^{-1})	v_2 (cm^{-1})
I-1	0.0	$Ni_{0.5}Cd_{0.5}Fe_2O_4$	399.26	578.64
I-2	0.1	$Co_{0.1}Ni_{0.5}Cd_{0.4}Fe_2O_4$	397.60	573.60
I-3	0.2	$Co_{0.2}Ni_{0.5}Cd_{0.3}Fe_2O_4$	397.33	570.92
I-4	0.3	$Co_{0.3}Ni_{0.5}Cd_{0.2}Fe_2O_4$	396.53	563.73
I-5	0.4	$Co_{0.4}Ni_{0.5}Cd_{0.1}Fe_2O_4$	395.40	561.28
I-6	0.5	$Co_{0.5}Ni_{0.5}Fe_2O_4$	393.47	559.35

3.2 Antibacterial studies

The synthesized Co-Ni-Cd ferrite NPs (I-1 to I-6 samples) were assessed for their antimicrobial activity along with the standard drug Ampicillin. The bacteria used in this study are viz., Gram-negative strains of Escherichia coli, Klebsiella pneumonia and Gram-positive strains of Staphylococcus aureus and Bacillus subtilis. The disc diffusion agar method was used for the evaluation of antimicrobial activities of the nanoparticles of the prepared Co-Ni-Cd ferrites by determining the presence of inhibition zones. The disc diffusion assay showed that the prepared ferrite NP's are inhibiting the growth of the bacteria efficiently. It is due to the fact that the prepared nanoferrites have enormously large surface area that provides good interaction with micro-organisms.

0.1 ml from 10^8 CFU/ml of the four disease causing bacteria viz., Gram-negative strains of Escherichia coli, Klebsiella pneumonia and Gram-positive strains of Staphylococcus aureus and Bacillus subtilis suspensions were outspread on individual agar plates that are nourished with LB (Luria Bertani) media. Over the agar plates, filter paper discs (of 5 mm in diameter) were positioned and then nano particles of various compositions of the prepared Co-Ni-Cd ferrites (10 μl.) were impregnated onto the discs in different amounts (concentrations). To measure the antimicrobial activity, 5 μl (10 μg/10 μl. concentration) of Ampicillin has served as the standard. Then, the agar plates were incubated for 24 h. at 37 °C. Around each disc, the inhibition zones were measured in (mm).

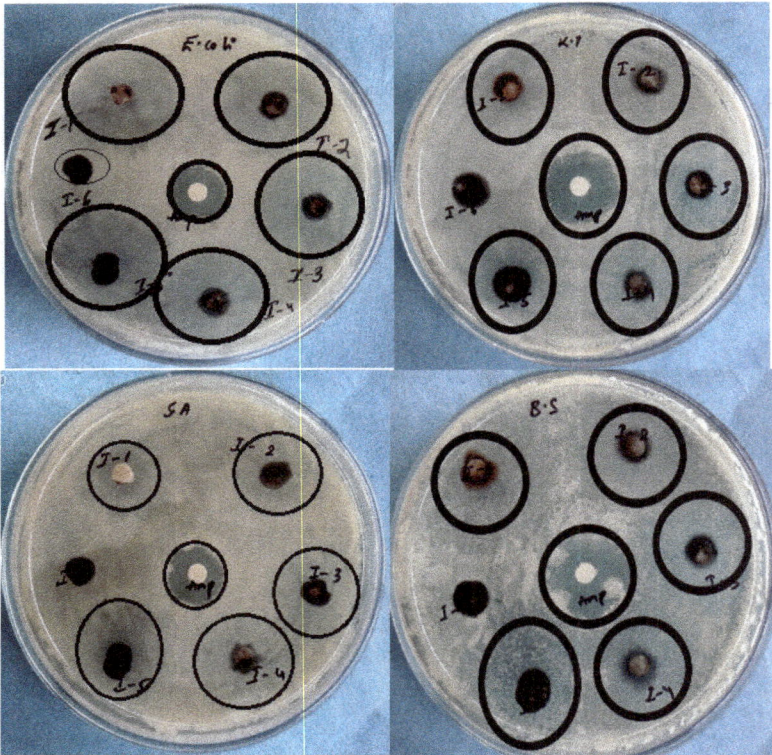

Fig. 5 An antibacterial activity and zone of Inhibition of $Co_xNi_{0.5}Cd_{0.5-x}Fe_2O_4$ (I-1 to I-6 samples) ferrites on Escherichia coli, Klebsiella pneumonia (Gram-negative strains) and Staphylococcus aureus and Bacillus subtilis (Gram-positive strains).

Table 4 Zone of inhibition (in mm) of $Co_xNi_{0.5}Cd_{0.5-x}Fe_2O_4$ system against Escherichia coli, Klebsiella pneumonia (Gram negative strains) and Staphylococcus aureus and Bacillus subtilis (Gram positive strains).

Name of Micro Organism	I-1 (30µg/30 µl)	I-2 (30µg/30 µl)	I-3 (30µg/30 µl)	I-4 30µg/30 µl)	I-5 (30µg/30 µl)	I-6 (30µg/30 µl)
Escherichia coli	10	11	11	11	12	0
Klebsiella pneumonia	5	4	5	5	7	0
Bacillus subtilis	9	10	10	11	12	0
Staphylococcus aureus	5	6	4	6	7	0

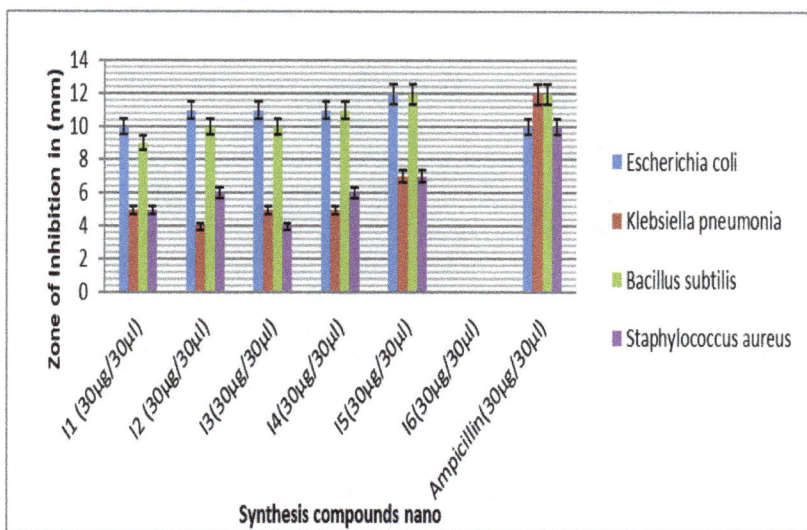

Fig. 6 Graphical representation of Zone of inhibition exhibited by $Co_xNi_{0.5}Cd_{0.5-x}Fe_2O_4$ system and standard drug Ampicillin against Escherichia coli, Klebsiella pneumonia (Gram-negative strains) and Staphylococcus aureus and Bacillus subtilis (Gram-positive strains).

The result of antibacterial growth around the ferrite NP's is an estimate of the potential of NP's to inhibit the growth [28]. The results of antibacterial study were shown in the form of zone of inhibition in Fig.5. The values in mm are recorded in Table 4 and the results were graphically represented in Fig.6.

Ampicillin was used as positive control. It is noticed that compounds I-1 to I-5 revealed outstanding antimicrobial activity with a zone of inhibition of 10-12 mm against E. coli, 4-7 mm against K. pneumonia (Gram-negative bacteria) and 9-12 mm against B. Subtilis, 5-7mm against S. aureus (Gram-positive bacteria). Compound $Co_{0.4}Ni_{0.5}Cd_{0.1}Fe_2O_4$ (I-5) exhibited remarkable antimicrobial activity in case of Gram-negative bacteria and Gram-positive bacteria, with a zone of inhibition 12 mm and 7 mm (Table 4) respectively. Zone of inhibition of standard drug Ampicillin against bacteria stains under test was shown in table 5.

Table 5 Zone of inhibition of standard drug Ampicillin against Escherichia coli, Klebsiella pneumonia (Gram-negative strains) and Staphylococcus aureus and Bacillus subtilis (Gram-positive strains).

Name of Micro Organism	Ampicillin ($30\mu g/30\mu l$)
Escherichia coli	10
Klebsiella pneumonia	12
Bacillus subtilis	12
Staphylococcus aureus	10

From Table 4 and Table 5, it is clear that the synthesized nano ferrites (I-1 to I-5) show excellent antimicrobial activity against E. Coli, K. pneumonia, B. subtilis and S. aureus which is almost equivalent to the standard drug Ampicillin. The activity exhibited by composition I-5 (zone of inhibition 12 mm) is more than that (10 mm) of the standard drug Ampicillin. Ferrite NPs of I-1 to I-5 exhibited comparable antimicrobial activity against K. pneumonia and S. Aureus with that of standard drug Ampicillin. It is clear that with increase in cobalt content there is a gradual increase in zone of inhibition (antimicrobial activity) up to $x = 0.5$ and there is a fall in the activity. The efficient antimicrobial activity of these ferrites is ascribed to an ample surface to volume ratio of the ferrite nanoparticles which improves the contact area with the microbes. Among all the prepared ferrites, composition $Co_{0.4}Ni_{0.5}Cd_{0.1}Fe_2O_4$ (I-5) shows highest antimicrobial activity against the microbes.

From the above studies it is clear that Co-Ni-Cd nanoferrites show considerable antibacterial effect against Escherichia coli (Gram negative) and B. Subtilis (Gram

Magnetic Oxides and Composites II
Materials Research Foundations **83** (2020) 117-133

Materials Research Forum LLC
https://doi.org/10.21741/9781644900970-5

positive). Thus the cobalt substituted Ni-Cd nano ferrite particles with notable growth of inhibition against microbes can be considered for probable applications in the field of medicine as antimicrobials.

Conclusions

Cobalt substituted $Co_xNi_{0.5}Cd_{0.5-x}Fe_2O_4$ ($0.0 \leq x \leq 0.5$) ferrites were synthesized via citrate-gel auto-combustion method. X-ray diffraction analysis confirmed the formation of cubic spinel structure in single phase. The crystallite size of prepared compositions is found from 11.4 nm to 26.7 nm. The antimicrobial activity was assessed against Gram negative (Escherichia coli, Klebsiella pneumonia) and Gram positive (Bacillus subtilis, Staphylococcus aureus) bacterial strains and the results show an augmentation in the activity by doping cobalt into Ni-Cd ferrite. The efficient antimicrobial activity is ascribed to an ample surface to volume ratio of the prepared ferrite nanoparticles, which improves the area of contact with the microbes. Accordingly, the prepared cobalt substituted Ni-Cd ferrite nanoparticles with noteworthy structural and antimicrobial properties bid great challenges in biomedical and pharmaceutical applications.

Acknowledgement

The author thanks Dr. K. Karunakar Rao, Department of Biochemistry, Osmania University, Hydrabad, Telangana, India for providing antimicrobial measurements facility.

References

[1] Goldman A. Modern Ferrite Technology, Van Nostrand, New York 1990.

[2] Amiri S, Shokrollahi H, The role of cobalt ferrite magnetic nanoparticles in medical science, Materials Science Engineering C, 33 (2013) 1– 8. https://doi.org/10.1016/j.msec.2012.09.003

[3] M.A. Fischbach, C. T. Walsh, Antibiotics for emerging pathogens, Science, 325 (5944): (2009) 1089-1093. https://doi.org/10.1126/science.1176667

[4] S.K. Pradhan, S. Bid, M. Gateshki, V. Petkov, Microstructure characterization and cation distribution of nanocrystalline magnesium ferrite prepared by ball milling, Materials Chemistry Physics, 93 (2005) 224–230. https://doi.org/10.1016/j.matchemphys.2005.03.017

[5] R. Arulmurugan, G. Vidyanathan, S. Sendhilnathan, B. Jeyadevan, Thermomagnetic properties of $Co_{1-x}Zn_xFe_2O_4$ ($x = 0.1$–0.5) nanoparticles, Journal of Magnetism and Magnetic Materials, 303 (2006)131–137. https://doi.org/10.1016/j.jmmm.2005.10.237

[6] C.K. Kim, J.H. Lee, S. Katoh, R. Murakami, M. Yoshimura, Material Research Bulletin, 36 (2001) 2241–2250. https://doi.org/10.1016/S0025-5408(01)00703-6

[7] X. Gao, U. Du, X. Liu, P. Xu, X. Han, Synthesis and characterization of Co-Sn substituted barium ferrite particles by a reverse micro emulsion technique, Materials Research Bulletin, 46 (2011) 643–648. https://doi.org/10.1016/S0025-5408(01)00703-6

[8] M. Atif, M. Nadeem, R. Grossinger, R. S. Turtelli, Studies on the magnetic, magnetostrictive and electrical properties of sol–gel synthesized Zn doped nickel ferrite, Journal of Alloys and Compounds, 509 (2011)5720–5724. https://doi.org/10.1016/j.jallcom.2011.02.163

[9] Linlin Wang, Chen Hu, and Longquan Shao, The antimicrobial activity of nanoparticles: Preent situation and prospects for the future, International Journal of Nanomedicine, 12 (2017) 1227–1249. https://doi.org/10.2147/IJN.S121956

[10] M. Satalkar , S. N. Kane, On the study of Structural properties and cation distribution of $Zn_{0.75-x}Ni_xMg_{0.15}Cu_{0.1}Fe_2O_4$ nano ferrite: Effect of Ni addition, Journal of Physics: Conference Series, 755 (2016) 012050. https://doi.org/10.1088/1742-6596/755/1/012050

[11] K. S. Lohar, S. M. Patange, M. L. Mane, and Sagar E. Shirsath, Cation distribution investigation and characterizations of $Ni_{1-x}Cd_xFe_2O_4$ nanoparticles synthesized by citrate gel process, Journal of Molecular Structure,1032 (2013) 105–110. https://doi.org/10.1016/j.molstruc.2012.07.055

[12] M. B. Shelar, P. A. Jadhav, S. S. Chougule, M. M. Mallapur, and B. K. Chougule, Structural and electrical properties of nickel cadmium ferrites prepared through self-propagating auto combustion method, Journal of Alloys and Compounds, 476 (2009) 760–764. https://doi.org/10.1016/j.jallcom.2008.09.107

[13] D. H. Kim, D. E. Nikles, D. T. Johnson, C. S. Brazel, Heat generation of aqueously dispersed $CoFe_2O_4$ nanoparticles as heating agents for magnetically activated drug delivery and hyperthermia, Journal of Magnetism and Magnetic Materials, 320 (2008) 2390–2396. https://doi.org/10.1016/j.jmmm.2008.05.023

[14] N. Sanpo, C. C. Berndt, C. Wen, J. Wang, Transition metal substituted cobalt ferrite nanoparticles for biomedical applications, Acta Biomaterialia, 9 (3) (2013) 5830–5837. https://doi.org/10.1016/j.actbio.2012.10.037

[15] T. Smitha, P. J. Binu, X. Sheena, E. M. Mohammed, Effect of samarium substitution on structural and magnetic properties of magnesium ferrite nanoparticles, Journal of Magnetism and Magnetic Materials, 348 (2013)140 – 145. https://doi.org/10.1016/j.jmmm.2013.07.065

[16] M. Raghasudha, D. Ravinder, P. Veerasomaiah, Characterization of nanostructured magnesium-chromium ferrites synthesized by citrate gel auto combustion method, Advanced Material letters, 4 (12) (2013) 910– 916. https://doi.org/10.5185/amlett.2013.5479

[17] C. O. Ehi-Eromosele, J. A. O. Olugbuyiro, O. S. Taiwo, O. A. Bamgboye, C. E. Ango, Synthesis and evaluation of the antimicrobial potentials of cobalt doped- and magnesium ferrite spinel nanoparticles, Bulletin of Chemical Society of Ethiopia, 32(3) (2018) 451–458. https://doi.org/10.4314/bcse.v32i3.4

[18] B. D. Cullity, Elements of X-Ray Diffraction, Addison Wesley publishing Company, Inc., Massachusetts, 1(1959)132–136.

[19] R. C. Kumbale, P. A. Sheikh, S. S. Kamble and Y. D. Kolekar, Effect of cobalt substitution on structural, magnetic and electric properties of nickel ferrite, Journal of Alloys and Compounds, 478(1-2) (2009) 599–603. https://doi.org/10.1016/j.jallcom.2008.11.101

[20] X. Qi, J. Zhou, Z. Yue, Z. Gui, L. Li, Permeability and microstructure of manganese modified lithium ferrite prepared by sol-gel auto-combustion method, Materials Science Engineering B, 99 (2003) 278–281. https://doi.org/10.1016/S0921-5107(02)00524-X

[21] P. P. Hankare, V.T. Vader, U.B. Sankpal, L.V. Gavali, I.S. Mulla and R. Sashikala, Effect of sintering temperature and thermoelectric power studies of the system Mg $Fe_{2-x}Cr_xO_4$, Solid State Sciences, 11(2009) 2075–2079. https://doi.org/10.1016/j.solidstatesciences.2009.09.005

[22] K. S. Lohar, S. M. Patange, M. L. Mane, and Sagar E. Shirsath, Cation distribution investigation and characterizations of $Ni_{1-x}Cd_xFe_2O_4$ nano particles synthesized by citrate gel process, Journal of Molecular Structure, 1032 (2013) 105–110. https://doi.org/10.1016/j.molstruc.2012.07.055

[23] A. B. Gadkari, T. S. Shinde and P. N. Vasambekar, Magnetic properties of rate earth ion (Sm^{3+}) added nanocrystalline Mg-Cd ferrites, prepared by oxalate co-precipitation method, Journal of Magnetism and Magnetic Materials, 322 (2010) 3823–3827. https://doi.org/10.1016/j.jmmm.2010.06.021

[24] M.U. Rana, M.U. Islam, T. Abbas, X-ray diffraction and site preference analysis of Ni-substituted $MgFe_2O_4$ ferrites, Pakistan Journal of Applied Science, 2 (2002) 1110–1114. https://doi.org/10.3923/jas.2002.1110.1114

[25] E. Rezlescu, L. Sachelarie, P. D. Popa and N. Rezlescu, Effect of substitution of divalent ions on the electrical and magnetic properties of Ni-Zn-Me ferrites, IEEE Transactions of Magnetics, 36 (2000) 3962–3967. https://doi.org/10.1109/20.914348

[26] R. D. Waldron, Infrared Spectra of Ferrites, Physical Review, 99 (1955) 1725–1727. https://doi.org/10.1103/PhysRev.99.1727

[27] A. T. Raghavender. Sagar E. Shirsath, K. Vijaya Kumar, Synthesis and study of nanocrystalline Ni-Cu-Zn ferrites prepared by oxalate based precursor method, Journal Alloys and Compounds, 509 (2011) 7004–7008. https://doi.org/10.1016/j.jallcom.2011.03.127

[28] X. Sheena, C. Harry, P. J. Nimila, S. Thankachan, M. S. Rintu, E. M. Mohammed, Structural and antibacterial properties of silver substituted cobalt ferrite nanoparticles, Research Journal of Pharmaceutical, Biological and Chemical Sciences. 5(2014) 364–371.

Magnetic Oxides and Composites II   Materials Research Forum LLC
Materials Research Foundations 83 (2020) 134-156   https://doi.org/10.21741/9781644900970-6

Chapter 6

Applications of Metal/Metal Oxides Nanoparticles in Organic Transformations

Aniruddha Mondal[1*,a], S.K. Tarik Aziz[2,b], Ajay K. Potbhare[3,c], S. Mondal[3,d],
Trimurti L. Lambat[4,e], Ratiram Gomaji Chaudhary[3*,f], Ahmed A. Abdala[5*,g]

[1] Department of Chemical Engineering, Tatung University, Taipei 104, Taiwan, ROC

[2]Department of Chemistry, Bar Ilan Institute for Nanotechnology and Advanced Materials
(BINA), Bar Ilan University, Ramat-Gan 52900, Israel

[3]Post Graduate Department of Chemistry, Seth Kesarimal Porwal College of Arts, Science and
Commerce, Kamptee – 441 001, India

[4]Department of Chemistry, Manoharbhai Patel College of Arts, Commerce and Science, Deori -
441 901, India

[5]Chemical Engineering Program, Texas A&M University at Qatar, POB 23784, Doha, Qatar

[a]aniruddhacsmcri@gmail.com; [b]tarikazizchem@gmail.com; [c]ajaypotbhare2@gmail.com;
[d]sudipmondal5555@gmail.com; [e]lambatmbpc@gmail.com; [f]chaudhary_rati@yahoo.com;[g]
[g]ahmed.abdala@qatar.tamu.edu

Abstract

Catalysis plays a significant role in improving the catalytic processes to enhance
efficiency of chemical transformations by reducing wastes. The research on catalysis con-
temporarily emerged as one of the pioneered research in modern chemistry. In the last
few years, nanocatalysis has become an emerging area of science and technology owing
to their exceptional potential effects in catalysis and selectivity. On the nano-level, the
nanostructured materials possessed different size, which rendered different surface area,
that conveys the exclusive assets to nanocatalysts compared to their bulk counterpart. In
this chapter, we emphasise on the fundamental understanding of different nanocatalysts
and how they catalyse catalytic reactions.

Keywords

Metal Oxide Nanoparticles, Mesoporous Materials, Nanocatalysts, Heterogeneous
Catalysts, Organic Transformation

Contents

1.   Introduction

In recent times, catalysis dwell in a very critical role in the modern chemistry world and is frolicking a significant starring role in the different types of physico-chemical transformations of infinite chemicals from laboratory scale to industrial production [1,2]. Catalytic reagents have the capability to lower the reaction temperature, reduce the reagent based waste and enhance the corresponding selectivity of the individual desired products, leading to green technology by minimising the unwanted side reactions [3–5]. The majority of chemical manufacturing industries are based on catalytic processes and the catalytic reactions are at the core of making fine chemicals and materials for energy applications, polymer synthesis, oil refining, bio-fuel, pollution control, medical devices, food production, fertilizers and oil hydrogenation [6–8].

In earliest times, catalysts have been used for accelerating different classes of chemical reactions. In 1835, Berzelius was the first to introduce and advocate the term catalysis as used for simple reactions like ammonia synthesis by Haber process, different oxidation reactions, and methanol productions along with the increase of the catalytic processes [9,10, 2]. Though, in recent time, the chemical reaction systems became more complex to minimise the side reactions and reduce the undesired by-products [11,12]. Accelerating selective reaction pathway is very important for the manufacturing of industrially relevant chemicals. Reaction selectivity in catalysis is compulsory for promoting the desired reactions and catalysts are strictly modified for the improved selectivity. One way to improve the catalyst is by using nanosized catalyst, which is one of the most effective ways for further improvement of chemical reactions [13, 14].

Without any solid acid based nanocatalyst, assortment of important chemical based essential products for examples drugs, polymers, fibres, fuels, paints, fine chemicals,

Magnetic Oxides and Composites II   Materials Research Forum LLC
Materials Research Foundations **83** (2020) 134-156   https://doi.org/10.21741/9781644900970-6

lubricants and other value-added important materials would be not much achievable. Heterogeneous catalytic processes are supported and promoted the mechanisms by which different chemical conversions are possible. Therefore, significantly it has potential impact towards commercialisation of many chemical products. Therefore, chemical manufacturing processes that use nanostructured heterogeneous based catalysts end of additional financial benefits compared to their bulk counterparts making nanocatalyst based manufacturing processes more economic, green and sustainable [15-17].

Enzymatic catalysts are the most green and efficient catalyst compared to other catalytic processes, *i.e.* homogeneous and heterogeneous catalyst [18, 3]. In the cases of homogeneous and heterogeneous catalytic processes, there are a number of merits and demerits necessitating the need for new catalytic processes that combine the fast reaction kinetics of the homogeneous processes and the capability to recover and recycle the catalyst during heterogeneous processes [19–21]. Whereas, a potential nanocatalysts have advantages towards the heterogeneous catalytic system as they can accelerate the chemical transformations due to their enhanced specific surface area that boosts the contact amongst the catalyst and other reaction species leading to a high product yield and the ease of their separation and recovery, which is very important for the process of good economics (in terms of commercialisation) and product purity. One more important issue to point out is the separation of the solid based heterogeneous catalyst can easily separate from the reaction mixtures [22,23].

2. Chemistry of nanocatalysts

In general, synthesis of catalytic nanoparticles that fall into the 1-100 nm size range and exhibits different shapes were synthesised *via* either a top-down or bottom up approach (Fig. 1). These two methods were preferred for synthesis of nanomaterials, nanocatalysts are characterized by array of techniques to identify their shapes, size, morphology, chemical composition and oxidation state, surface area, band gap energy, absorption, scattering and diffraction properties through FTIR, transmission electron microscopy (TEM), scanning electron microscopy (SEM), atomic force microscopy (AFM), energy dispersive X-ray spectrometer (EDS), X-ray diffraction (XRD), X-ray photoelectron spectroscopy (XPS), ultra-violet photoelectron spectroscopy (UPS), extended X-ray adsorption fine structure spectroscopy (EXAFS), X-ray absorption near-edge spectroscopy (XANES), photoluminescence spectroscopy (PL), small angle X-ray scattering (SAXS) etc.

Magnetic Oxides and Composites II   Materials Research Forum LLC
Materials Research Foundations **83** (2020) 134-156   https://doi.org/10.21741/9781644900970-6

Nanostructured Materials

Top Down Method   **Bottom Up Method**

Top Down Method	Bottom Up Method
Metal vapour	Sol-Gel
Thermal Breakdown	Precipitation
Chemical Breakdown	Electrodeposition
Mechanical Grinding	Template assisted
Spontaneous Chemisorption	Micro-emulsion
Ball-milling	Microwave mediated
	Atomic layer deposition

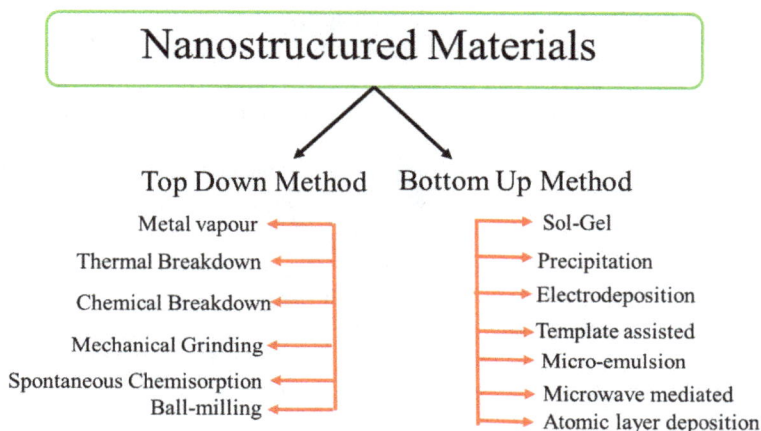

Fig. 1 Different approaches for nanoparticles synthesis.

Two most important factors of nanocatalyst materials are controllable size and shape which could be actively useful for the organic transformations [24,25]. Synthesis of nanocatalysts with well-defined size and shape is made *via* different classes of organic (soft) or inorganic (hard) templates and non-templates. Organic templates include variety of polymers such as (polyvinyl alcohols, polyvinyl pyrrolidone, various block copolymers), dendrimers (polyamidoamine), ligand, surfactant (thiols, phosphines, amines, and different salts). Inorganic templates are also used for regulating the shape, size and pore size distributions of nanocatalysts. In working towards the soft and hard templates for the synthesis of different size, shape and pore sizes of the nanostructured materials, self-templating methods are also capable of producing oriented nanostructured materials with different sizes and shapes [26,27]. The self-templating processes are two-step synthesis processes in which the corresponding templates are fully or partially incorporated into the individual architecture to the formation of diverse size and shapes of nanocatalysts. To get the large surface area properties of the synthesized nanocatalysts, selecting stabilising agents to tune the physicochemical properties of the nanostructured materials is a key parameter. In addition to the chemical composition and state of the nanocatalyst, the reactivity of the nanostructured materials depends primarily on two other factors, *i.e.* (i) tuning the nanocatalyst surface and, (ii) the nanocatalyst size and shape. Selectivity and reactivity of the nanocatalysts are greatly influenced by (a) the surface structure, (b) the crystallographic facets and, (c) the density of defects such as edges, corners and faces. Therefore, controlling the morphology of the nanostructured materials is an interesting and challenging endeavour globally [28,29, 19]. Synthesis of metal/metal oxide with different morphologies along with active surface properties, the selectivity of nanocatalysts can also be altered by controlling their inherent composition. Bimetallic nanostructured based nanocatalysts are emerging and a new field in nanotechnology. These systems exhibit new physicochemical and synergized catalytic properties due to the presence of two or more different metals. The properties of bimetallic nanocatalysts are also exclusively dependent on the surface of the nanocatalyst in nanochemistry. Moreover, incorporation of newly synthesized ligands and molecular atoms are ornamented to their surface, which can be tuned to yield new catalytic functionalities. Therefore, modification of the chemical properties of nanostructured materials *via* altering their different class of organic capping agents along with inorganic capping agents, with their own electronic and steric properties is of great interest [30,31].

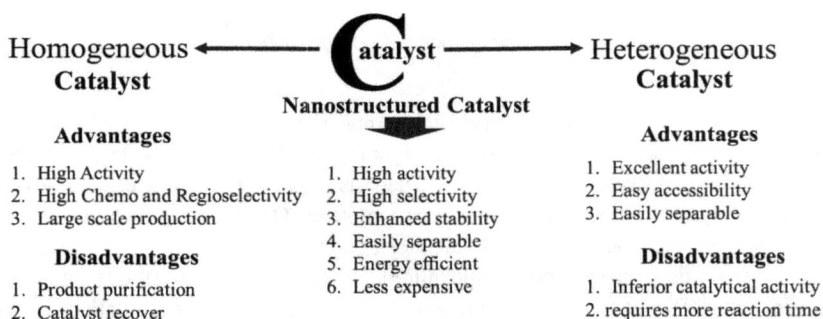

Homogeneous ◄——— **C**atalyst ———► Heterogeneous
    Catalyst   Catalyst
              Nanostructured Catalyst

Advantages   **Advantages**

Homogeneous Catalyst:
1. High Activity
2. High Chemo and Regioselectivity
3. Large scale production

Disadvantages
1. Product purification
2. Catalyst recover

Nanostructured Catalyst:
1. High activity
2. High selectivity
3. Enhanced stability
4. Easily separable
5. Energy efficient
6. Less expensive

Heterogeneous Catalyst:
1. Excellent activity
2. Easy accessibility
3. Easily separable

Disadvantages
1. Inferior catalytical activity
2. requires more reaction time

Fig. 2 Comparative efficiency of homogeneous, heterogeneous and nanocatalysts [13].

The key parameters for further improvement of nanocatalysts are the recyclability and recovery, where recyclability of nanocatalyst in solution-based processes remains a challenging task. Accordingly, deposition of identical tiny nanoparticles on to the surface of various porous supports (e.g. metal oxides, carbon, carbon nanotubes, membranes) for the easy to recover of heterogeneous systems has been explored in the last few years and magnetic supports have been used for the recovery of the heterogeneous nanocatalyst from the reaction media *via* magnetic forces, which is a robust, easily separable, efficient tool that has many advantages over different other physico-chemical process, like liquid-liquid extraction, liquid chromatography, distillation, filtration etc. Immobilization of nanocatalyst on the surface of magnetic nanostructured materials could be extracted from the corresponding reaction mixtures using simple external magnetic force and can be further re-dispersed in the non-appearance of the magnetic pitch [32-34]. In Fig. 2 are shown differences amid various reaction such as in homogeneous and heterogeneous [13,15].

In the last few decades, various metals and non-metals such as titanium, aluminium, clays and carbon-based nanostructured materials have been used in catalysis applications due to their excellent properties. The presence of the novel physicochemical properties of the nanocatalyst will empower the industrially relevant catalytic processes. Nanostructured materials as nanocatalyst was known as the most industrially relevant catalyst and possessing extensive applications in energy conversion and storage to chemical manufacturing industry compared to their bulk materials.

2.1   Theoretical aspects of nanocatalyst

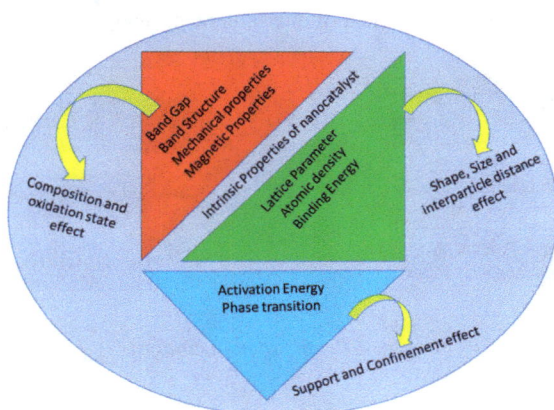

Fig. 3 Effect of intrinsic properties of nanomaterials on its catalytic activity.

Theoretical concepts of the nanocatalysts can be implicated discussed by accounting for the effects of the intrinsic properties of nanomaterials in catalysis (See Fig. 3). In nanostructured metal/metal oxide materials, the natural stuffs that encompass an energetic impact on their catalytic activity which can be thought-out as by following properties:

(a) The total quantity that can be reliably related to bond-length like spiteful framework constant, atomic density and binding energy.

(b) The residual atomic cohesive energy per discrete atom like self-organisation growth, thermal stability, coulomb barrier, critical temperature for the phase transition and the slow rate of evaporation in nano-solid.

(c) The intrinsic properties of nanocatalyst that differ with their binding energy density in the relaxed area such as Hamiltonian, determined the entire band energy and related properties like band gap, core level energy, photo absorption and photo emission.

(d) And the property nanocatalyst that can be tuned from the joint effect of the bonding energy density. The atomic cohesive energy such as the mechanical strength, stiffness, surface energy, surface stress, extensibility and compressibility of the nano-solids could lead to the development of even more fascinating and profitable materials [35, 36].

Presentation of solid materials or a bunch of atoms differ from that of an out-of-the-way atoms mostly due to the involvement of inter-atomic exchanges[37–39].

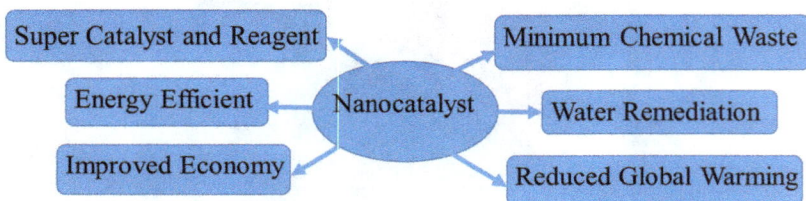

Fig. 4 Profits of nanocatalysts.

Modification of the comparative number of the under-coordinated surface atoms bears an additional autonomy that permits one to resist the properties of a nano-solid compared to their bulk counterpart. Hence, contributions from the under-coordinated atoms and the involvement of inter-atomic interactions could be the twitch to make a channel between an isolated atom and a bulk solid through their chemical and physical performances. The effects of diminution in the atomic level coordination (deviation of bond order, length, and angle) are considerable as they synergize the performance of a surface, a nanosolid, and a solid in amorphous state in terms of bond relaxation and the consequences on bond energy [40–42]. The unusual behaviour of a surface and a nano-solid has usually been understood and formulated as functions of reduction in atomic coordination and its derivatives on the atomic trapping potential, crystal binding intensity, and electron–phonon coupling. Specifically, controlling their size, shape, spatial distribution, surface alignment, surface electronic structure, thermal and chemical stability of the individual nanocomponents, it can be widely used in catalysis with newer properties and activity. Nanoscale based catalysts have been the area of interest in academic and industrial research consideration in recent times because of the plentiful potential profits that can accumulate through their use (Fig. 4).

2.2   Types of nanocatalysis and organic transformation

Nanocatalysts can be used in both homogeneous and heterogeneous catalytic processes. In homogeneous catalytic processes, a clear solution or suspension of nanocatalysts in different solvents are usually used and in this case prevention of nanocatalyst aggregation is key when designing a nanocatalyst system for use in homogenous catalytic processes [43,44]. Multifunctional groups-based polymer stabilization of nanoparticles to prevent their aggregation is recognized as the most operative way to become stable nanocatalysts in different reaction media. Long chain molecules (surface capping agents) make it difficult for the nanoparticles to aggregate or clump together forming large-size particles. However, such stabilization (surface capping) processes decrease the overall catalytic activity of nanocatalysts by plummeting the accessibility of their surface to the reacting molecules. Recovery of the highly colloidal nanocatalysts from the reaction mixture is another major concern for catalysis as it is disreputably problematic to eliminate nanoparticles from a solution and additional steps are required to achieve that, which may perhaps totally contradict the procedure over simplification associated with the use of catalyst in the first place. Non-recoverability of homogeneous colloidal nanocatalysts postures an environmental dangerous most nanoparticles cannot be degraded even by incineration and the property their accumulation in the ecosystems are mainly unknowns as well as threatening to the effectiveness and economics of the process.

Heterogeneous catalysis is always considered more environmentally friendly due to its high recoverability. The heterogeneous catalyst may be a solid or immobilized on an inert solid matrix. Research to explore the catalytic potential of numerous nanoparticles supports has gained significant interest from the scientific community [45,46]. Recent examples of nanocatalysts include copper, ruthenium; rhodium, silver, palladium, iron, gold, nickel and platinum nanoparticles supports on silica, clays, zeolite, alumina, bio-waste driven material, or carbon materials as shown in (Fig. 5). Nanoporous solids are another explored area of heterogeneous nanocatalysts. Nanoporous materials can be manufactured by growing the solid material around a molecular template and nanoscale features can also be etched onto the surface of a solid base catalyst using standard lithography methods - this can allow a degree of control over reactant flow on the catalyst surface, as well as increasing surface area [47, 30, 31]. One of the most motivating scientific and technological contests accompanying the expenditure of nanoparticles as catalysts is the indulgent of how the composition and atomic-scale structure of the nanocatalyst produce the most excellent catalytic features in term of enhancement of different reactions such as photocatalytic, electro-catalytic and thermal. The second encounter is to fabricate these nanoparticles with utmost management of the morphology, composition and microstructure. Current nanotechnology based synthetic methods undoubtedly provide a countless potential for future progresses in both characterization and synthesis of heterogeneous catalysts based on supported nanostructured materials [48–50]. Miao *et al.* has reported that catalytic activity of heterogeneous nanostructured MnO_2 catalyst for the synthesis of hydrogen gas (H_2) from formaldehyde at room temperature and proposed a conceivable reaction path for this transformation (Fig. 5) [50–51]. However, ultrathin MnO_2 nanosheets were required for improved production of hydrogen gas via formaldehyde reforming at room temperature with the use of simple de-ionised water.

Singha *et al.* [52] has reported that the catalyst exhibited for the photocatalytic activity towards amine formation in the absence of photoactivated azo compounds. Here, CdS and its nanocomposites have shown for the establishment of nitrene from side to side photoactivation of azide group. However, it also enhanced the decomposition of the azide molecule in certain extent which surpressed the azo compound formation.

Similarly, reduction of aromatic nitro compounds and their esterification reaction were reported by Galani *et al.* [53]. A modest and real-world procedure is proposed for the reduction of aromatic nitro compounds and the oxidative esterification of alcohols by using a recyclable 3D assembled porous rectangular ZnO nanoplatelets supported gold nanocatalysts (Au/ZnO) under mild conditions. The developed catalyst exhibited exceptional catalytic measurement for both reactions and led to good to excellent yield of up to 97% for nitro reduction and 90% yield for oxidative esterification. Moreover, the catalyst was recyclable for a minimum of five times lacking some noteworthy loss in catalytic property. The expenditure of the technologically advanced catalyst was advantageous because of the less expensive and accessibility of the ZnO support and beneficial with respect to catalyst recovered by simple filtration of the ZnO based nanostructured materials [53,54]. Giri *et al.* developed rectangular ZnO porous assembly with admirable catalytic activity. The ZnO based nanostructured material have shown outstanding heterogeneous catalytic activity for the synthesis of 5-substituted-1*H*-tetrazoles (yield = 94 %) [55, 56].

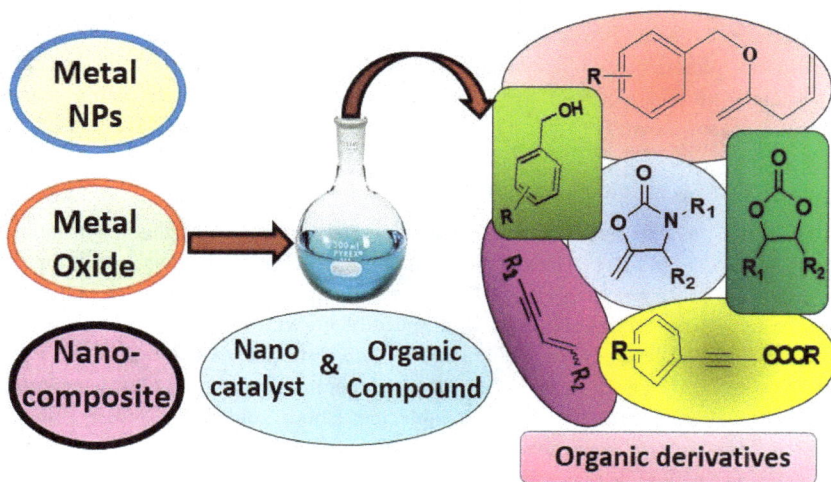

Fig. 5 Nanocatalysts and its organic transformation.

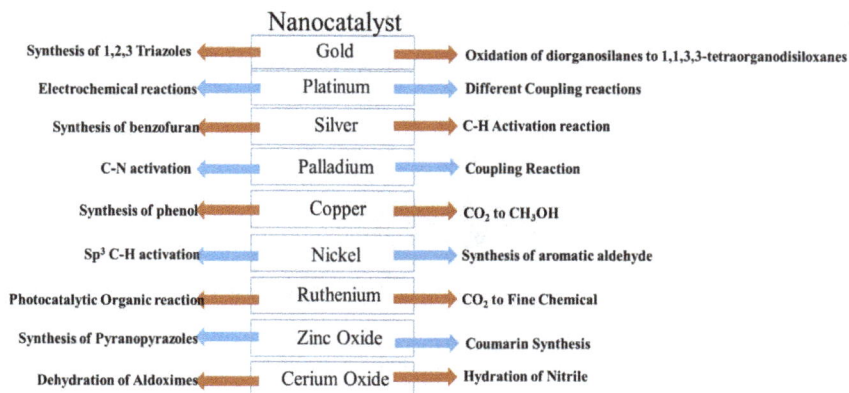

Scheme 1 Selective nanocatalysts and, their organic reactions.

Chaudhary *et al.* have reported the fabrication of spinel copper aluminates NCs for microbial assay and electrochemical performances via solvent-free synthesis of xanthanedione derivatives. Furthermore, NCs-catalyzed fabrication of xanthanedione to nanocatalyst that exhibited admirable yield and reusability with small reduction in effectiveness even after four successive sets [57, 58]. Few selective nanocatalyzed reactions are summarized in Scheme 1, which highlights the applications of nanocatalysts in organic synthesis. Furthermore, some organic derivatives prepared using selective nanocatalysts are shown in (Table 1).

3.   Future aspects of nanocatalysis

The nanocatalysis approach was acclaimed since 1950s before even the nanotechnology term was known [59-61]. The development of nanoscience and nanotechnology has unlocked new prospects in the area of catalyst and appliances of nanoscale materials as active catalysts or catalyst sustains is a speedily growing field of research that has reached several milestones. In 1986, Haruta *et al.* reported that, oxidation of carbon monoxide and hydrogen at near to the ground temperature using gold (Au) nanoparticles as a nanocatalyst. Ultimately a very reliable idea about gold for its lethargy towards chemical reaction has been completely transformed and a new access to interesting applications of nanoscale materials in catalysis is unlocked [62-67].

Nanocatalysis has the benefit of individually homogenous and heterogeneous catalysis *i.e.* high catalytic activity (turn over frequency (TOF)) and selectivity like homogeneous catalysis, regeneration and separation of catalyst after completion of reaction making it closer to the heterogeneous catalysis (Scheme 2).

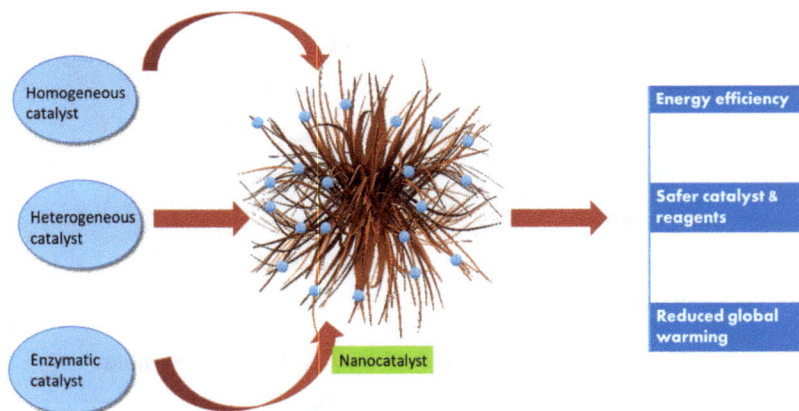

Scheme 2 Nanocatalysts for organic transformation.

Traditional heterogeneous catalyst systems have three main drawbacks match up to their homogeneous counterparts: (i) the minimum surface area which should be accessible to the reactant molecules. Thus, restraining their catalytic actions over the surface of concerned materials, (ii) expensive synthetic methods for the preparation of highly effective nanocatalyst and, (iii) most important is the expensive catalyst materials are highly spontaneous fuel. Nanoscale based catalytic materials can solve these problems by increasing the surface to volume ratio (S/V), which is base of the present status of nanocatalysis.

Table 1. A list of organic derivatives prepared from nanocatalyst

No	Nanocatalyst	Particles size (nm)	Organic derivatives	Ref.
1	Cu NPs	40	Chromenes	[71]
2	Ag NPs	14±2	Thiosemicarbazide	[72]
3	Pd NPs	2–6	1,3-enyne esters	[73]
4	CuO NPs	9 ± 16	Polyhydroquinoline	[74]
5	NiO NPs	20 ±60	Amidoalkylnapthol	[75]
6	PbO NPs	20.	Arylbenzodioxyxanthenedione	[76]
7	ZnO NPs	5-10	Spiro(indoline-pyranodioxine)	[77]
8	Al_2O_3 NPs	9± 15	Dihydropyrimidinones	[78]
9	Bi_2O_3 NPs	19-35	Dihydropyrano[3,2-b]chromenedionesmidazole	[79]
10	Fe_3O_4 NPs	80 ± 40.	2,4,6-triarylpyridines	[80]
11	$CuAl_2O_4$ NCs	10	Xanthanediones	[57]
12	$CuFe_2O_4$ NCs	35	Spirooxindoles	[81]
13	Silica-coated nickel oxide	30± 50	Malononitrile	[82]
14	Silica NPs	150–250	thioethers, thioesters, vinyl thioethers and thio-Michael	[83]
15	Copper NPs	100-150	Arylation of 4-Quinolone derivatives under ambient condition	[84]
16	ZnO NPs	50	2-Aryl benzimidazoles	[85]
17	Pd NPs	100-200	Isoxazole to 2-Azafluorenone	[86]
18	RuO_2/Cu_2O	20-40	Azo-coupling of anilines	[87]
19	Cu Nps	10-80	1H-Imidazoles	[88]
20	Fe_3O_4 NPs	40-50	Coumarin-3-carboxamide derivatives in aqueous ethanol	[89]
21	Mn_2O_3 nanoparticles	6-8	bifunctionalelectrocatalyst for oxygen reduction and oxygen evolution	[90]

Advances in nanocatalysis in the last few years comparing with present day's developments open a new vision for nanocatalysis regarding their economical synthesis process, stimulated design, industrially relevant processes, one catalyst multiple applications role and biologically important catalytic materials. Away from each other from these future aspects of nanocatalysis, its energy and environmental benefits will also be given responsiveness [68-70, 2].

Conclusions

In a conclusion, we have enlightened the catalytic performances of metal oxide nanocatalysts for high reaction activity, long-term stability and selectivity in commercial applications. The nanocatalysts illustrates an attractive paradigm but undeveloped concept for making corrosion resistive energy for conversion reactions. The significant development among the reported nanocatalysts was investigated comprehensively. Metal nanoparticles with controlled size, shape, microporosity and high surfaces were promising for greener heterogeneous catalysis-based processes. Moreover, focusing on industrial, environmental and social problems, researchers were able to synthesize catalytic nanoparticles supported on various materials to improve their durability for long cycles. Hence, the impact of the nanocatalyst supports clearly reflected by the increasing number of nanocatalyst related products in market, thus, with great anticipations, this chapter focuses on study of organic transformation and catalysis research development.

References

[1]   J. M. Thomas, R. Raja, Exploiting nanospace for asymmetric catalysis: confinement of immobilized, single-site chiral catalysts enhances enantioselectivity, Accounts of Chemical Research, 41 (2008) 708–720. https://doi.org/10.1021/ar700217y

[2]   S. Chatterjee, K. Sengupta, B. Mondal, S. Dey, A. Dey, Factors determining the rate and selectivity of $4e^-/4H^+$ electrocatalytic reduction of dioxygen by iron porphyrin complexes, Accounts of Chemical Research, 50 (2017) 1744–1753. https://doi.org/10.1021/acs.accounts.7b00192

[3]   D. Astruc, F. Lu, J.R. Aranzaes, Nanoparticles as recyclable catalysts: The frontier between homogeneous and heterogeneous catalysis, Angewandte Chemie International Edition, 44 (2005) 7852–7872. https://doi.org/10.1002/anie.200500766

[4]   Q. Sun, Z. Dai, X. Meng, L. Wang, F.S. Xiao, Task-specific design of porous polymer heterogeneous catalysts beyond homogeneous counterparts, ACS Catalysis,5 (2015) 4556–4567. https://doi.org/10.1021/acscatal.5b00757

[5]   E. Gross, F. Dean Toste, G.A. Somorjai, Polymer-encapsulated metallic
        nanoparticles as a bridge between homogeneous and heterogeneous catalysis,
        Catalysis Letters,145 (2015) 126–138. https://doi.org/10.1007/s10562-014-1436-9

[6]   M.R. Avhad, J.M. Marchetti, Innovation in solid heterogeneous catalysis for the
        generation of economically viable and ecofriendly biodiesel: A review, Catalysis
        Reviews-Science and Engineering, 58 (2016) 157–208.
        https://doi.org/10.1080/01614940.2015.1103594

[7]   R. Schlögl, Heterogeneous catalysis, Angewandte Chemie International Edition,
        54 (2015) 3465–3520. https://doi.org/10.1002/anie.201410738

[8]   P. Hu, M. Long, Cobalt-catalyzed sulfate radical-based advanced oxidation: A
        review on heterogeneous catalysts and applications, Applied Catalysis B:
        Environmental, 181 (2016) 103–117. https://doi.org/10.1016/j.apcatb.2015.07.024

[9] V. V. Ranade, S. S. Joshi, Industrial Catalytic Processes for Fine and Specialty
        Chemicals, Elsevier, Science-Direct, 2016, pp. 1-782. doi.org/10.1016/C2013-0-
        18518-2. https://doi.org/10.1016/B978-0-12-801457-8.00001-X

[10]   J. Beckman, The publication strategies of Jöns Jacob Berzelius (1779-1848):
        Negotiating national and linguistic boundaries in chemistry, Annals of Science, 73
        (2016) 195–207. https://doi.org/10.1080/00033790.2016.1138503

[11]   W. Yin, J. Yu, F. Lv, L. Yan, L.R. Zheng, Z. Gu, Y. Zhao, Functionalized nano-
        MoS_2 with peroxidase catalytic and near-infrared photothermal activities for safe
        and synergetic wound antibacterial applications, ACS Nano, 10 (2016)11000–
        11011. https://doi.org/10.1021/acsnano.6b05810

[12]   R.G. Chaudhary, J. Tanna, N. Gandhare, A. R. Rai, H. Juneja, Synthesis of nickel
        nanoparticles: Microscopic investigation, an efficient catalyst and effective
        antibacterial activity, Advanced Materials Letter, 6 (2015) 990–998.
        https://doi.org/10.5185/amlett.2015.5901

[13]   G. Zhao, X. Wu, R. Chai, Q. Zhang, X. Gong, J. Huang, Y. Lu, Tailoring nano-
        catalysts : turning gold nanoparticles on bulk metal oxides to inverse nano-metal
        oxides on large gold particles, Chemical Communication, 51 (2015) 5975–5978.
        https://doi.org/10.1039/C5CC00016E

[14]   D. Singappuli-arachchige, J.S. Manzano, L.M. Sherman, I. I. Slowing, Polarity
        control at interfaces : quantifying pseudo-solvent effects in nano-confined

systems,Chem Phys Chem, 17 (2016) 2982–2986.
https://doi.org/10.1002/cphc.201600740

[15]   Z. Xing, Q. Liu, A.M. Asiri, X. Sun, Closely interconnected network of
       molybdenum phosphide nanoparticles: A highly efficient electrocatalyst for
       generating hydrogen from water, Advanced Materials, 26 (2014) 5702–5707.
       https://doi.org/10.1002/adma.201401692

[16]   T. Ghosh, P. Ghosh, G. Maayan, A copper-peptoid as a highly stable, efficient,
       and reusable homogeneous water oxidation electrocatalyst, ACS Catalysis, 8
       (2018) 10631–10640. https://doi.org/10.1021/acscatal.8b03661

[17]   R.G. Chaudhary, G. S. Bhusari, A. Tiple, A. R. Rai, S. Somkuvar, A. K. Potbhare,
       T. Lambat, P. Ingle, A. Abdala, Metal/metal oxide nanoparticles: toxicity,
       applications, and future prospects, Current Pharmaceutical Design, 25 (2019)
       4013-4029. https://doi.org/10.2174/1381612825666191111091326

[18]   X. Cui, W. Li, P. Ryabchuk, K. Junge, M. Beller, Bridging homogeneous and
       heterogeneous catalysis by heterogeneous single-metal-site catalysts,Nature
       Catalysis, 1 (2018) 385–397. https://doi.org/10.1038/s41929-018-0090-9

[19]   Ä. Gallard, Catalytic decomposition of hydrogen peroxide by Fe (III) in
       homogeneous aqueous solution : mechanism and kinetic modeling, Environmental
       Science & Technology, 33 (1999) 2726–2732. https://doi.org/10.1021/es981171v

[20]   T. Lambat, A. Abdala, S. Mahmood, P. Ledade, R. Chaudhary, S. Banerjee,
       Sulfamic acid promoted one-pot multicomponent reaction: a facile synthesis of 4-
       oxo-tetrahydroindoles under ball milling conditions, RSC Advances, 9 (2019)
       39735-39742. https://doi.org/10.1039/C9RA08478A

[21]   J. Tanna, R. G. Chaudhary, V. Sonkusare, H. D. Juneja, CuO nanoparticles:
       synthesis, characterization and reusable catalyst for polyhydroquinoline
       derivatives under ultrasonication, Journal of the Chinese Advanced Materials
       Society, 4 (2016) 110-122. https://doi.org/10.1080/22243682.2016.1164618

[22]   Y. Yan, B.Y. Xia, X. Ge, Z. Liu, A. Fisher, X. Wang, A flexible electrode based
       on iron phosphide nanotubes for overall water splitting, Chemistry-A European
       Journal, 21 (2015) 18062–18067. https://doi.org/10.1002/chem.201503777

[23]   Z. Huang, Z. Chen, Z. Chen, C. Lv, H. Meng, C. Zhang, $Ni_{12}P_5$ nanoparticles as an
       efficient catalyst for hydrogen generation via electrolysis and photoelectrolysis,
       ACS Nano, 8 (2014) 8121–8129. https://doi.org/10.1021/nn5022204

Materials Research Forum LLC

https://doi.org/10.21741/9781644900970-6

[24] N.R. Elezovic, V.R. Radmilovic, N. V Krstajic, Platinum nanocatalysts on metal oxide based supports for low temperature fuel cell applications, RSC Advances, 6 (2016) 6788–6801. https://doi.org/10.1039/C5RA22403A

[25] J. Liu, Q. Ma, Z. Huang, G. Liu, H. Zhang, Recent progress in graphene-based noble-metal nanocomposites for electrocatalytic applications, Adavanced Materials, 31 (2019) 1–20. https://doi.org/10.1002/adma.201800696

[26] L. Liu, A. Corma, Metal catalysts for heterogeneous catalysis: from single atoms to nanoclusters and nanoparticles, Chemical Reviews, 118 (2018) 4981–5079. https://doi.org/10.1021/acs.chemrev.7b00776

[27] V.N. Sonkusare, R.G. Chaudhary, G.S. Bhusari, A. Mondal, A.K. Potbhare, R. Kumar Mishra, A.A. Abdala, H.D. Juneja, Mesoporous octahedron-shaped tricobalt tetroxide nanoparticles for photocatalytic degradation of toxic dyes, ACS Omega, 5 (2020) 7823-7835. https://doi.org/10.1021/acsomega.9b03998

[28] Y. Y. Huang, T. S. Zhao, G. Zhao, X. H. Yan, K. Xu, Manganese-tuned chemical etching of a platinum-copper nanocatalyst with platinum-rich surfaces, Journal of Power Sources, 304 (2016) 74–80. https://doi.org/10.1016/j.jpowsour.2015.11.038

[29] Z. C. Zhang, X. C. Tian, B. W. Zhang, L. Huang, F. C. Zhu, X. M. Qu, L. Liu, S. Liu, Y. X. Jiang, S.G. Sun, Engineering phase and surface composition of Pt_3Co nanocatalysts: A strategy for enhancing CO tolerance, Nano Energy, 34 (2017) 224–232. https://doi.org/10.1016/j.nanoen.2017.02.023

[30] H. R. Choi, H. Woo, S. Jang, J. Y. Cheon, C. Kim, J. Park, K.H. Park, S. H. Joo, Ordered mesoporous carbon supported colloidal Pd nanoparticle based Model catalysts for suzuki coupling reactions: impact of organic capping agents, Chem Cat Chem, 4 (2012) 1587–1594. https://doi.org/10.1002/cctc.201200220

[31] V. Sonkusare, R.G. Chaudhary, G.S. Bhusari, A.R. Rai, H.D. Juneja, Microwave-mediated synthesis, photocatalytic degradation and antibacterial activity of α-Bi_2O_3 microflowers/novel γ-Bi_2O_3 microspindles, Nano-Structure & Nano-Objects,13 (2018) 121-131. https://doi.org/10.1016/j.nanoso.2018.01.002

[32] N. Koukabi, E. Kolvari, M.A. Zolfigol, A. Khazaei, B.S. Shaghasemi, B. Fasahati, A magnetic particle-supported sulfonic acid catalyst: Tuning catalytic activity between homogeneous and heterogeneous catalysis, Advanced Synthesis and Catalysis,354 (2012) 2001–2008. https://doi.org/10.1002/adsc.201100352

[33] A. Schatz, O. Reiser, W.J. Stark, Nanoparticles as semi-heterogeneous catalyst supports, Chemistry—A European Journal, 16 (2010) 8950–8967. https://doi.org/10.1002/chem.200903462

[34] C. Ó. Dálaigh, S.A. Corr, Y. Gun'ko, S.J. Connon, A magnetic-nanoparticle-supported 4-N,N-dialkylaminopyridine catalyst: Excellent reactivity combined with facile catalyst recovery and recyclability, Angewandte Chemie International Edition, 46 (2007) 4329–4332. https://doi.org/10.1002/anie.200605216

[35] I. Ali, U. Kulsum, Z.A. AL-Othman, K. Saleem, Analyses of nonsteroidal anti-inflammatory drugs in human plasma using dispersive nano solid-phase extraction and high-performance liquid chromatography, Chromatographia, 79 (2016) 145–157. https://doi.org/10.1007/s10337-015-3020-x

[36] P. Wang, A. Kong, W. Wang, H. Zhu, Y. Shan, Facile preparation of ionic liquid functionalized magnetic nano-solid acid catalysts for acetalization reaction,Catalysis Letters,135 (2010) 159–164. https://doi.org/10.1007/s10562-010-0271-x

[37] A. Miyamoto, H. Himei, Y. Oka, E. Maruya, M. Katagiri, R. Vetrivel, M. Kubo, Computer-aided design of active catalysts for the removal of nitric oxide, Catalysis Today, 22 (1994) 87–96. https://doi.org/10.1016/0920-5861(94)80094-4

[38] G-J Li, T. Fujimoto, A. Fukuoka, M. Ichikawa, Ship-in-Bottle synthesis of Pt_9-Pt_{15} carbonyl clusters inside NaY and NaX zeolites,in-situ FTIR and EXAFS characterization and the catalytic behaviors in ^{13}CO exchange reaction and NO reduction by CO,Catalysis Letters,12 (1992) 171–185. https://doi.org/10.1007/BF00767199

[39] R. Lin, T. Zhao, M. Shang, J. Wang, W. Tang, V.E. Guterman, J. Ma, Effect of heat treatment on the activity and stability of PtCo/C catalyst and application of in-situ X-ray absorption near edge structure for proton exchange membrane fuel cell, Journal of Power Sources, 293 (2015) 274–282. https://doi.org/10.1016/j.jpowsour.2015.05.067

[40] E. C. Tyo, S. Vajda, Catalysis by clusters with precise numbers of atoms, Natural Publication Group, 10 (2015) 577–588. https://doi.org/10.1038/nnano.2015.140

[41] H. Abe, J. Liu, K. Ariga, Catalytic nanoarchitectonics for environmentally compatible energy generation, Biochemical Pharmacology, 19 (2016) 12–18. https://doi.org/10.1016/j.mattod.2015.08.021

[42]   C. Coutanceau, Electro-oxidation of glycerol at Pd based nano-catalysts for an
        application in alkaline fuel cells for chemicals and energy cogeneration, Applied
        Catalysis B: Environmental, 93 (2010) 354–362.
        https://doi.org/10.1016/j.apcatb.2009.10.008

[43]   S.W. Sheehan, J.M. Thomsen, U. Hintermair, R.H. Crabtree, G.W. Brudvig, C.A.
        Schmuttenmaer, A molecular catalyst for water oxidation that binds to metal oxide
        surfaces, Nature Communications, 11 (2015) 1–9.
        https://doi.org/10.1038/ncomms7469

[44]   M. L. Pegis, B. A. McKeown, N. Kumar, K. Lang, D.J. Wasylenko, X. P. Zhang,
        S. Raugei, J.M. Mayer, Homogenous electrocatalytic oxygen reduction rates
        correlate with reaction overpotential in acidic organic solutions, ACS Central
        Science, 2 (2016) 850–856. https://doi.org/10.1021/acscentsci.6b00261

[45]   E. Sorek, J. Ankri, G. Arbiv, R. Mol, I. Popov, H.-J. Freund, S. Shaikhutdinov, M.
        Asscher, Acetylene reactivity on Pd−Cu nanoparticles supported on thin silica
        films: the role of the underlying substrate, The Journal of Physical Chemistry C,
        123 (2019) 17425–17431. https://doi.org/10.1021/acs.jpcc.9b04722

[46]   S. Ganguly, P. Das, M. Bose, T.K. Das, S. Mondal, A.K. Das, N.C. Das,
        Sonochemical green reduction to prepare Ag nanoparticles decorated graphene
        sheets for catalytic performance and antibacterial application, Ultrasonics
        Sonochemistry,39 (2017) 577–588. https://doi.org/10.1016/j.ultsonch.2017.05.005

[47]   B. Zhang, X. Zheng, O. Voznyy, R. Comin, M. Bajdich, F.P.G. De Arquer, C.T.
        Dinh, F. Fan, M. Yuan, A. Janmohamed, H.L. Xin, H. Yang, Homogeneously
        dispersed multimetal oxygen-evolving catalysts, Science , 352 (2016) 333–338.
        https://doi.org/10.1126/science.aaf1525

[48]   C. Zhao, J. Guo, Q. Yang, L. Tong, J. Zhang, J. Zhang, C. Gong, J. Zhou, Z.
        Zhang, Preparation of magnetic Ni @ graphene nanocomposites and efficient
        removal organic dye under assistance of ultrasound, Applied Surface Science,357
        (2015) 22–30. https://doi.org/10.1016/j.apsusc.2015.08.031

[49]   A. Wong, Q. Liu, S. Griffin, A. Nicholls, J.R. Regalbuto, Synthesis of ultrasmall,
        homogeneously alloyed, bimetallic nanoparticles on silica supports, Science, 1430
        (2017) 1427–1430. https://doi.org/10.1126/science.aao6538

[50]   Q. Zhu, Q. Xu, Immobilization of ultrafine metal nanoparticles to high-surface-area materials and their catalytic applications, Chemistry, 1 (2016) 220–245. https://doi.org/10.1016/j.chempr.2016.07.005

[51]   L. Miao, Q. Nie, J. Wang, G. Zhang, P. Zhang, Applied catalysis B : Environmental ultrathin MnO_2 nanosheets for optimized hydrogen evolution via formaldehyde reforming in water at room temperature, Applied Catalysis B: Environmental, 248 (2019) 466–476. https://doi.org/10.1016/j.apcatb.2019.02.047

[52]   K. Singha, A. Mondal, S.C. Ghosh, A. Baran, Visible-light-driven efficient photocatalytic reduction of organic azides to amines over cds sheet – rgo nanocomposite, Chemistry: An Asian Journal, 15 (2018) 255–260. https://doi.org/10.1002/asia.201701614

[53]   S. M. Galani, A. K. Giri, S.C. Ghosh, A.B. Panda, Development of easily separable ZnO-supported Au nanocatalyst for the oxidative esterification of alcohols and reduction of nitroarenes, ChemistrySelect, 4 (2018) 9414–9421. https://doi.org/10.1002/slct.201801730

[54]   A. K. Potbhare, R.G. Chaudhary, P. Chouke, S. Yerpude, A. Mondal, V. Sonkusare, A. Rai, H. Juneja. Phytosynthesis of nearly monodisperse CuO nanospheres using *Phyllanthus reticulatus/Conyza bonariensis* and its antioxidant/antibacterial assays, Material Science and Engineering C, 99 (2019) 783-793. https://doi.org/10.1016/j.msec.2019.02.010

[55]   A. K. Giri, A. Saha, A. Mondal, S.C. Ghosh, S. Kundu, A.B. Panda,Rectangular ZnO porous nano-plate assembly with excellent acetone sensing performance and catalytic activity, RSC Advances, 5 (2015) 102134–102142. https://doi.org/10.1039/C5RA19828C

[56]   P. Chouke, A. Potbhare, G. Bhusari, S. Somkuwar, Dadamia PMD Shaik, R. Mishra, R.G. Chaudhary, Green fabrication of zinc oxide nanospheres by *Aspidopterys cordata* for effective antioxidant and antibacterial activity, Advanced Materials Letter,10 (2019) 355–360. https://doi.org/10.5185/amlett.2019.2235

[57]   R. G. Chaudhary, V.N. Sonkusare, G.S. Bhusari, A. Mondal, D.P.M.D. Shaik, H.D. Juneja, Microwave-mediated synthesis of spinel $CuAl_2O_4$ nanocomposites for enhanced electrochemical and catalytic performance, Research on Chemical Intermediates, 44 (2018) 2039–2060. https://doi.org/10.1007/s11164-017-3213-z

[58]   A. Potbhare, P. Chauke, S. Zahra, V. Sonkusare, R. Bagade, M. Umekar, R. Chaudhary, Microwave-mediated fabrication of mesoporous Bi-doped $CuAl_2O_4$ nanocomposites for antioxidant and antibacterial performances, Materials Today: Proceedings, 15 (2019) 454–463. https://doi.org/10.1016/j.matpr.2019.04.107

[59]   J. Jiang, J. R. Swierk, K. L. Materna, S. Hedström, S. H. Lee, R. H. Crabtree, C. A. Schmuttenmaer, V. S. Batista, G. W. Brudvig, High-potential porphyrins supported on SnO_2 and TiO_2 surfaces for photoelectrochemical applications, The Journal of Physical Chemistry C, 2 120 (2016) 28971–28982. https://doi.org/10.1021/acs.jpcc.6b10350

[60]   M. J. Kahlich, H.A. Gasteiger, R.J. Behm, Kinetics of the selective low-temperature oxidation of CO in H_2 -rich gas over Au / α -Fe_2O_3, Journal of Catalysis, 440 (1999) 430–440. https://doi.org/10.1006/jcat.1998.2333

[61]   B. K. Min, C.M. Friend, Heterogeneous gold-based catalysis for green chemistry : low-temperature CO oxidation and propene oxidation, Chemical Reviews,107 (2007) 2709–2724. https://doi.org/10.1021/cr050954d

[62]   N. Goswami, S. Chaudhuri, A. Giri, P. Lemmens, S. K. Pal, Surface engineering for controlled nanocatalysis : key dynamical events from ultrafast electronic spectroscopy,The Journal of Physical Chemistry C, 118 (2014)23434-23442. https://doi.org/10.1021/jp507456n

[63]   S. Zhang, L. Nguyen, Y. Zhu, S. Zhan, In-situ studies of nanocatalysis, Accounts of Chemical Research, 46 (2013) 1731–1739. https://doi.org/10.1021/ar300245g

[64]   L. L. Chng, N. Erathodiyil, J. Y. Ying, Nanostructured catalysts for organic transformations, Accounts of Chemical Research, 46 (2013) 1825–1837. https://doi.org/10.1021/ar300197s

[65]   A. Zuliani, F. Ivars, R. Luque, Advances in nanocatalyst design for biofuel production, Chem Cat Chem, 10 (2018) 1968–1981. https://doi.org/10.1002/cctc.201701712

[66]   Y. Shi, B. Zhang, Recent advances in transition metal phosphide nanomaterials: Synthesis and applications in hydrogen evolution reaction, Chemical Society Reviews, 45 (2016) 1529–1541. https://doi.org/10.1039/C5CS00434A

[67]   M. Haruta , Size- and support-dependency in the catalysis of gold, Catalysis Today, 36 (1997) 153–166. https://doi.org/10.1016/S0920-5861(96)00208-8

[68] D. Wang, L. Bin Kong, M.C. Liu, Y.C. Luo, L. Kang, An approach to preparing Ni-P with different phases for use as supercapacitor electrode materials, Chemistry-A European Journal, 21 (2015) 17897–17903. https://doi.org/10.1002/chem.201502269

[69] Z. Li, M. Li, Z. Bian, Y. Kathiraser, S. Kawi, Design of highly stable and selective core / yolk – shell nanocatalysts-A review, Applied Catalysis B: Environmental, 188 (2016) 324–341.

[70] Y. Yin, R.M. Rioux, C.K. Erdonmez, S. Hughes, G.A. Somorjai, A.P. Alivisatos, Formation of hollow nanocrystals through the nanoscale Kirkendall Effect, Science, 304 (2004) 711–714. https://doi.org/10.1126/science.1096566

[71] J. Tanna, R.G. Chaudhary, N. Gandhare, A. Rai, S. Yerpude, H.D. Juneja, Copper nanoparticles catalysed an efficient one-pot multicomponents synthesis of chromenes derivatives and its antibacterial activity, Journal of Experimental Nanoscience, 11(2016) 884–900. https://doi.org/10.1080/17458080.2016.1177216

[72] S. Chandra, A. Kumar, Green approach for the synthesis of silver-nanoparticles and efficient use in oxidative cyclization, International Journal of Applied Biology and Pharmaceutical, 2 (2011) 78–85.

[73] B. Ranu, K. Chattopadhyay, L. Adak, A. Saha, S. Bhadra, R.Dey, and D. Saha, Metal nanoparticles as efficient catalysts for organic reactions, Pure and Applied Chemistry, 81 (2009) 2337–2354. https://doi.org/10.1351/PAC-CON-08-11-19

[74] J.A. Tanna, R.G. Chaudhary, V.N. Sonkusare, H.D Juneja, CuO nanoparticles: synthesis, characterization and reusable catalyst for polyhydroquinoline derivatives under ultrasonication, Journal of the Chinese Advanced Materials Society, 4 (2016) 110–122. https://doi.org/10.1080/22243682.2016.1164618

[75] J. Tanna, R. G. Chaudhary , N. V. Gandhare, A. R. Rai, H. D. Juneja, Nickel oxide nanoparticles: synthesis, characterization and recyclable catalyst, International Journal of Scientific and Engineering Research, 6 (2015) 93–98.

[76] T.Lambat, R.G. Chaudhary, A. Abdala, R. Mishra, S. Mahmood, S. Banerjee, Mesoporous PbO nanoparticle-catalyzed synthesis of arylbenzodioxy xanthenedione scaffolds under solvent-free conditions in a ball mill, RSC Advances, 9 (2019) 31683–31690. https://doi.org/10.1039/C9RA05903B

[77]   H. Sachdeva, R. Saroj and, D. Dwivedi, Nano-ZnO catalyzed multicomponent
        one-pot synthesis of novel spiro(indoline-pyranodioxine) derivatives, The
        Scientific World Journal, 18 (2014) 427195. https://doi.org/10.1155/2014/427195

[78]   J. A. Tanna, R. G. Chaudhary, N. V Gandhare, H. D. Juneja, Alumina
        nanoparticles :A new and reusable catalyst for synthesis of dihydropyrimidinones
        derivatives, Advanced Materials Letter, 7 (2016) 933–938.
        https://doi.org/10.5185/amlett.2016.6245

[79]   M. Ziraka, M. Azinfara and M. Khalili, Three-component reactions of kojic acid:
        Efficient synthesis of Dihydropyrano[3,2-b]chromenediones and
        aminopyranopyrans catalyzed with Nano-Bi$_2$O$_3$-ZnO and Nano-ZnO, Current
        Chemistry Letters, 6 (2017) 105–116. https://doi.org/10.5267/j.ccl.2017.4.001

[80]   A. Maleki and R. Firouzi-Haji, L-Proline functionalized magnetic nanoparticles:
        A novel magnetically reusable nanocatalyst for one-pot synthesis of 2,4,6-
        triarylpyridines, Scientific Reports, 8 (2018) 17303–17909.
        https://doi.org/10.1038/s41598-018-35676-x

[81]   A. Bazgir, G. Hosseini, R. Ghahremanzadeh, Copper ferrite nanoparticles: an
        efficient and reusable nanocatalyst for a green one-pot, three-component synthesis
        of spirooxindoles in water,ACS Combinatorial Science, 15 (2013) 530–534.
        https://doi.org/10.1021/co400057h

[82]   R.G. Chaudhary, J. Tanna, A. Mondal, N. Gandhare, H. D. Juneja, Silica-coated
        nickel oxide a core-shell nanostructure: synthesis, characterization and its catalytic
        property in one-pot synthesis of malononitrile derivative, Journal of the Chinese
        Advanced Materials Society, 5 (2017) 103–117.
        https://doi.org/10.1080/22243682.2017.1296371

[83]   S. Banerjee, J. Das, R. P. Alvareza, S. Santra, Silica nanoparticles as a reusable
        catalyst: a straightforward route for the synthesis of thioethers, thioesters, vinyl
        thioethers and thio-Michael adducts under neutral reaction conditions, New
        Journal of Chemistry, 34 (2010) 302–306. https://doi.org/10.1039/B9NJ00399A

[84]   P. Ghosh, S. Das, Ligand free approach for the copper (II)-mediated C-NH$_2$
        Arylation of 4-Quinolone derivatives under ambient condition, ChemistrySelect, 3
        (2018) 8624–8627. https://doi.org/10.1002/slct.201801756

Magnetic Oxides and Composites II   Materials Research Forum LLC
Materials Research Foundations **83** (2020) 134-156   https://doi.org/10.21741/9781644900970-6

[85]   S. K. Banjare, S. Payra, A. Saha, S. Banerjee, Efficient room temperature synthesis of 2-Aryl benzimidazoles using ZnO nanoparticles as reusable catalyst, Organic & Medicinal Chemistry International Journal, 1 (2017) 555–568.

[86]   S. Das, D. Hong, Z. Chen, Z. She, W. H. Hersh, G. Subramaniam, Y. Chen, Auto-tandem palladium catalysis: from Isoxazole to 2-Azafluorenone, Organic Letters, 47 (2016) 5578–5581. https://doi.org/10.1021/acs.orglett.5b02731

[87]   A. Saha, S. Payra, B. Selvaratnam, S. Bhattacharya, S. Pal, R. T. Koodali, S. Banerjee, Hierarchical mesoporous RuO_2/Cu_2O nanoparticle-catalyzed oxidative homo/hetero Azo-coupling of anilines, ACS Sustainable Chemistry and Engineering, 6 (2018) 11345−11352. https://doi.org/10.1021/acssuschemeng.8b01179

[88]   R. Pagadala, N. V. Gandhare,U. Kusampally, V. Jetti,J. Meshram, H. D. Juneja, Synthesis of 1H-Imidazoles catalyzed by Cu-nanoparticle and its physicochemical properties, Journal of Heterocyclic Chemistry, 45 (2014) 116–122. https://doi.org/10.1002/chin.201428111

[89]   S. Payra, A. Saha, S. Benerjee, Magnetically Recoverable Fe_3O_4 Nanoparticles for the One-Pot Synthesis of coumarin-3-carboxamide derivatives in aqueous ethanol, ChemistrySelect, 3 (2018) 7535–7540. https://doi.org/10.1002/slct.201800523

[90]   K. K. Hazarika, C. Goswami, H. Saikia, B. J. Borah, P. Bharali, Cubic Mn_2O_3 nanoparticles on carbon as bifunctional electrocatalyst for oxygen reduction and oxygen evolution reactions, Molecular Catalysis, 451 (2018) 153–160. https://doi.org/10.1016/j.mcat.2017.12.012

Magnetic Oxides and Composites II
Materials Research Foundations **83** (2020) 157-179

Materials Research Forum LLC
https://doi.org/10.21741/9781644900970-7

Chapter 7

Ameliorating Physical Properties of $Co_{1-x}Ca_xFe_2O_4$ Nano Ferrites for Technological Applications

Ebtesam E. Ateia[1,a] and M. Farag Shokry[1,b]

[1]Physics Department, Faculty of Science, Cairo University, Giza, Egypt

[a]ebtesam@sci.cu.eg, [a]drebtesam2000@yahoo.com

[b]mohamedfaragshokry@gmail.com

Abstract

In the present chapter, we report physical properties of $Co_{1-x}Ca_xFe_2O_4$ ($0.0 \leq x \leq 1.0$) nano ferrites, prepared by a citrate auto combustion method for technological applications. The evaluation of XRD patterns and HRTEM images indicated fine particle nature. The prepared calcium sample showed orthorhombic phase structure, while the other investigated samples showed cubic spinel structure. The magnetic hysteresis loops at different temperatures (77, 100, and 300 K) were recorded using a vibrating sample magnetometer. The obtained data shows that, the shape of investigated nano ferrite particles can be used as a powerful tool for adapting magnetic properties.

Keywords

Spinel Ferrite, Orthorhombic, Magnetic Moment, Physical Properties, Maximum Energy Density

Contents

1.   Introduction

A ferrite is a ceramic material made by mixing and firing large proportions iron(III) oxide (Fe_2O_3) blended with small proportions of one or more additional metallic elements, such as cobalt, calcium , nickel, and zinc [1]. Recently, ferrite nanoparticles are of great interest due to their extensive use in the chemical and technological applications. One of the fascinating properties of ferrites is the possibility to synthesize different compositions and thereby modify their properties.

Depending upon the crystal structure, ferrites are divided into four types.

1. Spinel

2. Garnet

3. Perovskite

4. Hexagonal or hexaferrites

Both, spinel (AB_2O_4) and perovskite (ABO_3) are promising magnetic materials for the applications in microelectronics.

An important property of the spinel ferrites is the distribution of the metal ions (Me) over the tetrahedral (A) and octahedral (B) sites as shown in Fig. 1. The distribution is called normal if Me^{2+} ions occupy on A-sites and Me^{3+} ions on B-sites. Zinc ferrite can be classified as normal spinel, which can be written as $Zn^{2+}[Fe^{3+}]O_4$. In an inverse spinel, Me^{2+} ions occupy on B-sites where as Me^{3+} ions are equally distributed between A- and B-sites. Cobalt ferrite is an example of an inverse spinel and its ionic distribution can be written as $Fe^{3+}[Co^{2+}Fe^{3+}]O_4$. The distribution can also be intermediate between normal

Materials Research Forum LLC
https://doi.org/10.21741/9781644900970-7

and inverse, e.g. manganese ferrite [2,3], which distribution is given by $Mn^{2+}_{0.8}Fe^{3+}_{0.2}$ $[Mn^{2+}_{0.2}. Fe^{3+}_{10.8}]O_4$.

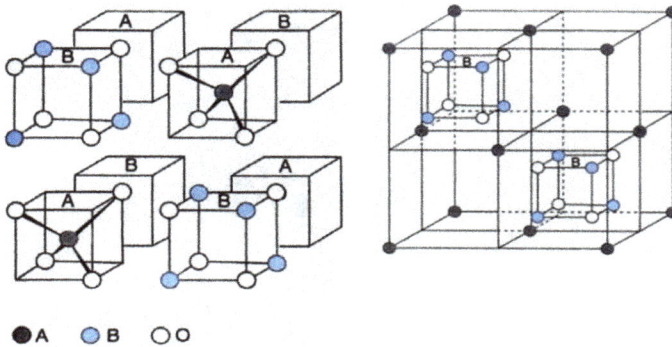

●A ◐B ○O

Fig. 1 The unit cell of AB$_2$O$_4$ spinel [4].

The spinel ferrites have remarkable physical properties. Among them, $CoFe_2O_4$ has attracted more attention due to its interesting magnetic properties like high coercivity, moderate saturation magnetization, chemical stability, good mechanical hardness and large energy products (i.e., MH_{max}) [5]. It also exhibits high cubic magneto crystalline anisotropy, which is the common feature of spinel ferrite [6]. Due to these features, cobalt nano ferrite can be used for numerous technological and permanent magnet applications like high density magnetic storage device [7]. In comparison with $CoFe_2O_4$, calcium ferrite has a significant advantage; it is biocompatible and eco-friendly due to the presence of Ca^{2+} instead of heavy metals [8]. However, it reveals very weak ferromagnetism at room temperature.

A lot of efforts have been carried out for the synthesis of Co/Ca Fe_2O_4 with improved magnetic and physical properties [9-10].The electrical, structural and magnetic properties of cobalt can be tailored either by changing the microstructure or by incorporation of different metal ions [11-15].

Perovskite structured ceramics has become one of the worldwide materials due to their typical properties. The cubic perovskite is called the ideal perovskite. This class of materials has great potential for a variety of applications due to their simple crystal structures and unique ferroelectric and ferromagnetic properties. [16, 17]. The perovskite structure (ABO_3) is shown schematically in Fig. 2. Large divalent or trivalent ions (A) occupy the corners of a cube and small trivalent or tetravalent metal ions (B) occupy the centre of the cube. The oxygen ions are situated centrally on the faces of the cube.

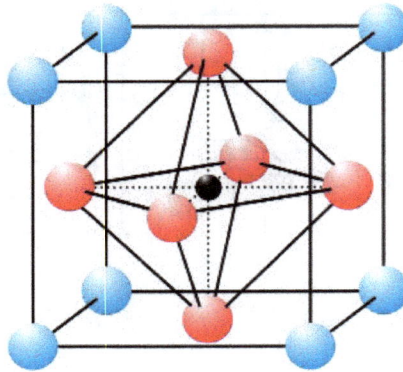

Fig. 2 A schematic of a perovskite crystal structure (blue spheres represent the 'A' cations, black spheres represent the 'B' cations and red spheres represent the O^{2-} anions).

In pure perovskites there are two extreme structural variants, expressed by the tolerance factor T as given in Eq.1 [18].

$$T = \frac{r_A + r_O}{\sqrt{2}(r_B + r_O)}$$

(1)

Where r_A, r_B and r_O are the ionic radii of the A, B cations and the oxygen anion respectively.

Several chemical and physical techniques are available for the synthesis of ferrite nano particles. Wet methods for the preparation of fine particles include citrate auto combustion, sol-gel processing, spray-drying, co-precipitation, flash, hot-spraying, laser-induced vapor phase reactions and oxalate precursor [19-22]. Among the previous methods, the citrate auto-combustion method is the effective in providing smallest particles size with good chemical homogeneity, narrow particle size distribution and it is easy to control the stoichiometry especially for substituted ferrites. Additionally, it is less complicated and comparatively inexpensive method than the others.

In the present study, an attempt has been made to prepare nano ferrite samples with chemical composition $Co_{1-x}Ca_xFe_2O_4$ ($0.0 \leq x \leq 1.0$) by a citrate auto combustion method. The effect of substitution of nonmagnetic ion (Ca^{2+}) on the structural, morphological and magnetic behavior was investigated.

2.   Experimental procedure

2.1   Synthesis

The chemicals used for the synthesis of the samples were of high purity and some of their properties are reported in Table 1.

Table 1. The elements used in the present work.

Elements[23]	Electronic configuration	Ionic radius[24] (Å)	Chemical formula	Magnetic type[23]
$^{55.845}Fe_{26}$	$[Ar]\,4s^2 3d^6$	$(0.645)^{VI}$ $(0.490)^{IV}$	$FeN_3O_9 \cdot 9H_2O$	Ferromagnetic
$^{58.933}Co_{27}$	$[Ar]\,4s^2 3d^7$	$(0.58)^{IV}$ $(0.745)^{VI}$	$CoN_2O_6 \cdot 6H_2O$	Ferromagnetic
$^{40.078}Ca_{20}$	$[Ar]4s^2$	$(1.34)^{XII}$	$CaN_2O_6 \cdot 4H_2O$	Paramagnetic

Ionic radius that in six-fold symmetry [VI]

Ionic radius of ions that in four fold symmetry [IV]

Ionic radius of ions that in twelve fold symmetry [XII]

2.2   Samples preparation

Nano ferrite samples of $Co_{1-x}Ca_xFe_2O_4$ ($x = 0.1, 0.5, 1.0$) were synthesized by citrate auto combustion method. Initial ingredients of metal nitrate salts of high purity (BDH) have been used. The initial ingredients are cobalt nitrate, calcium nitrate, and iron (III) nitrate. The citric acid ($C(OH)(COOH)(CH_2 \cdot COOH)_2 \cdot H_2O$) was used because it is a weak acid. It helps in the homogenous distribution and segregation of the metal ions. During water dehydration, it suppresses the precipitation of metal nitrates. Therefore, at a relative low temperature the precursors can form a homogenous single phase. The flow chart for preparing nano ferrite powder is shown in Fig. 3.

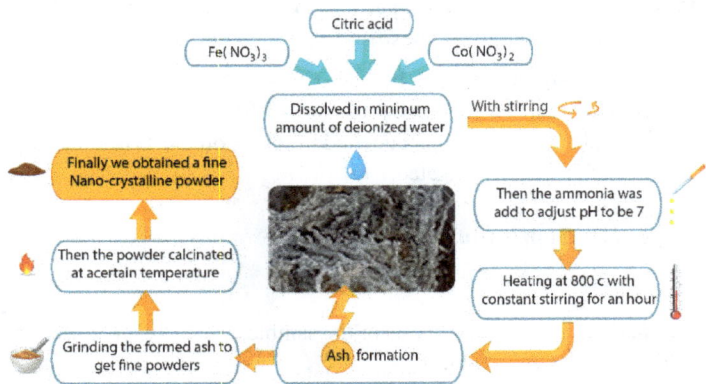

Fig. 3 The flow chart for preparing ferrite nanoparticles

2.3   Characterization techniques

The X-ray diffraction patterns of the heated samples were recorded at room temperature using DIANO X-ray diffraction system (the 2100E X-ray diffractometer) of target Cu-K$_\alpha$ (λ=1.5418 Å) to check phase purity and crystal structure of heated samples.

The lattice parameter (*a*) was calculated from X-ray data using the following relations [25]:

For cubic: $\dfrac{1}{d^2} = \dfrac{h^2+k^2+l^2}{a^2}$ (2)

For orthorhombic: $\dfrac{1}{d^2} = \dfrac{h^2}{a^2} + \dfrac{k^2}{b^2} + \dfrac{l^2}{c^2}$ (3)

Where *d* is the interplaner distance for a certain series (*hkl*).

The surface morphology of the samples was examined by field emission scanning electron microscopy (FESEM, using the SEM model Quanta 250 FEG) attached with EDAX unit (energy dispersive X-ray analysis) and high transmission electron microscope (HRTEM, model: Tecnai G20, Super twin, double tilt).

In the development of nano-ferrite materials the VSM has been widely used to measure the saturation magnetization, remanence magnetization, coercivity, anisotropy fields, etc., and also to measure the temperature dependent parameters of interest. The room temperature and low temperature magnetic hysteresis loops of heated samples were recorded using Vibrating Sample Magnetometer (VSM, Model Lakeshore7410) with a maximum applied magnetic field of $\pm$ 20 kOe.

3.   Results and discussion

3.1   Structural analysis

3.1.1 FT-IR analysis

Fig. 4 shows FT-IR spectra of the $Co_{1-x}Ca_xFe_2O_4$ nano particles recorded in wave number range of 450–4000 cm^{-1} at room temperature. One noticeable absorption band v_1 is identified in between the range 566–513 cm^{-1}; while the band v_2 is not seen. These bands can be attributed to the stretching vibrations due to the interactions between the oxygen atom and the cations in tetrahedral (A) and octahedral [B] sites, respectively [26]. From Fig.4, it is clear that the value of v_1 band (corresponding to A-sites is shifted towards lower wavelengths (higher frequencies). This is because Ca ions preferably occupied on the B and A-sites. Also the difference in the ionic radii of A-site cations (Fe^{3+}, Ca^{2+}) are much smaller than that of B-site cations (Fe^{3+}, Ca^{2+} and Co^{2+}). As shown in Fig. 4 that the bands in the range of 3397 – 3370 cm^{-1} correspond to O–H stretching vibration. In addition, the bands near 1600 cm^{-1}– 1580 cm^{-1} correspond to the asymmetrical stretching of C-O [27].

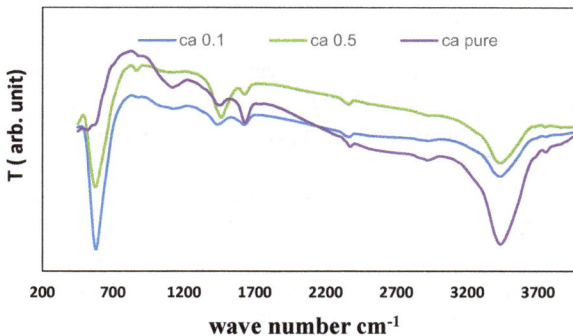

Fig. 4 FT-IR spectra of $Co_{1-x}Ca_x Fe_2O_4$ (x = 0.1, 0.5 and 1.0).

3.1.2 X-ray diffrection analysis

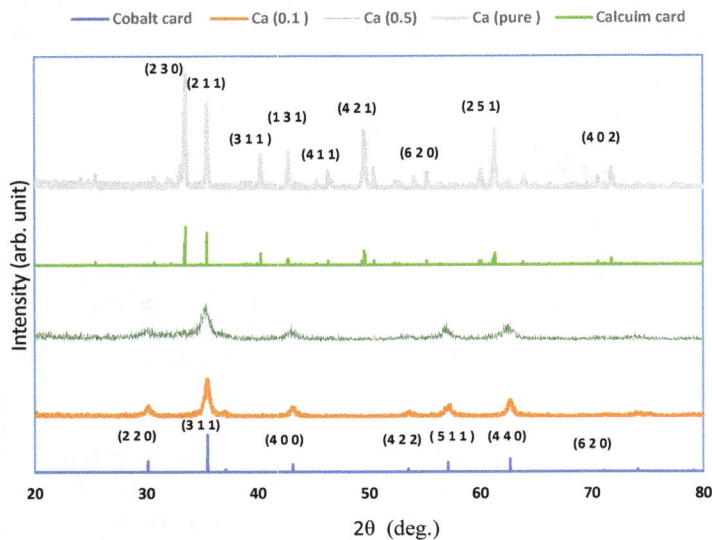

Fig. 5 *The X-ray diffraction patterns of* $Co_{1-x}Ca_xFe_2O_4$ *(x = 0.0, 0.1, 0.5 and 1.0) ferrite powder.*

The structural properties of heated nano ferrite samples $Co_{1-x}Ca_xFe_2O_4$ ($0.0 \leq x \leq 1.0$) were investigated using X ray diffraction (XRD) technique and obtained XRD patterns are shown in Fig. 5. The XRD patterns were compared and indexed using ICDD card no. 04-014-8388, 04-010-1893. X-ray analysis of samples ($x < 1.0$) corresponds to spinel lattice with a cubic structure and space group *Fd-3m*. The diffraction peaks corresponding to planes (220), (311), (222) (400), (422), (511), and (440) provide a clear indication for the formation of spinel structure of the samples. For pure calcium sample $x = 1$ the diffraction peak indicates the orthorhombic phase whose space group is *Pbnm* [28]. In the obtained orthorhombic phase, the distorted BO_6 octahedra sharing edges and corners generate a framework, with the Ca^{2+} ions. It is clear from the figure 5 that the broader diffraction peaks designate the nano crystalline nature of prepared samples [29].

X-ray density (D_{xrd}) was calculated from the relation [30]:

$$D_{xrd} = \frac{ZM}{NV} \tag{4}$$

Where Z is the number of molecules per unit cell, M is the molecular weight, N is the Avogadro's number and V is the unit cell volume. The experimental lattice parameter "a", the theoretical density "Dx," and tolerance factor T [18] of the investigated samples are also summarized in Table 2. As shown from the table 2, the calculated tolerance factor designates that the structure of the investigated samples depends on the calcium ion content.

The crystallite size are determined from the FWHM of the most intense peaks corresponding to (002) (311) planes for calcium and cobalt nano ferrite samples respectively [31]. The obtained data is tabulated in the Table 2.

Table 2   The crystallite size from XRD, particle size from HRTEM, Exp. lattice parameter, X-ray density (D_{xrd}) and tolerance factor for $Co_{1-x}Ca_xFe_2O_4$ (x=0.0, 0.1, 0.5, 0.1)

Samples	Crystallite Size (nm)	Particle size (nm)	Exp. Lattice Parameter ($a_{Exp.}$) (Å)	D_x (g/cm^3)	Tolerance Factor (T)
CoFe$_2$O$_4$	37.73	38.9	a = 8.381	5.295	0.991
Co$_{0.9}$Ca$_{0.1}$Fe$_2$O$_4$	14.31	14.44	a = 8.386	5.241	0.995
Co$_{0.5}$Ca$_{0.5}$Fe$_2$O$_4$	7.98	8.14	a = 8.582	4.726	0.961
CaFe$_2$O$_4$	37.61	37.88	a = 10.703 b=9.229 c= 3.024	9.592	0.890

3.1.3 Thermal anaylsis

In order to study the formation of the spinel, orthorhombic phases, thermo gravimetric (TGA) and differential scanning calorimetry/DTG analysis for the samples are carried out in the temperature range 50 to 1000 °C as shown in Fig.6 (a-c). The TGA curves for each of the compositions can be divided into three regions based on different processes taking place in each of them, first for CaFe$_2$O$_4$ (i) 50–250 °C, (ii) 250–550 °C, and (iii) 600–750 °C then for CoFe$_2$O$_4$ (i)50- 350 °C, (ii) 350-500 °C, and (iii) 550–650 °C. In region (i) the first step of weight loss (~3 %) corresponds to the volatilization of the organic solvents. For pure calcium at 650 °C a sharp mass loss is observed on TG–DTG curves. The DTG curves show presence of exothermic peaks at ~ 600 °C for pure and doped samples (at third region) owing to the loss of water molecules from the samples. At T ~ 650 °C, an intense exothermic effect appears on the DTG plots, which are accompanied with the combustion reaction. This process is associated with a sharp mass loss on the thermogravimetry (TG) curve. Above about 650, 800 °C, no mass loss can be detected on TG–DTG curves for cobalt.

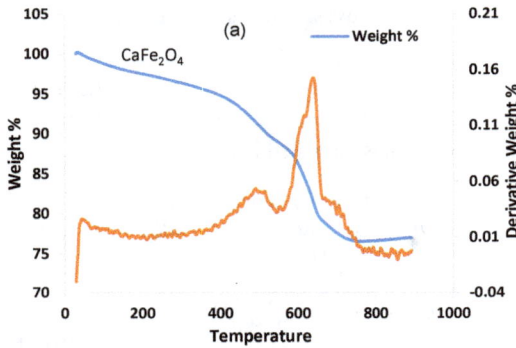

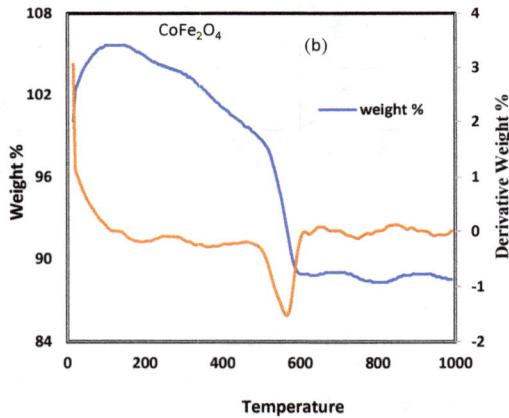

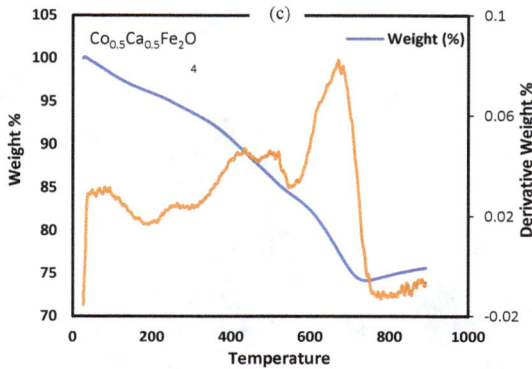

Fig. 6 (a-c) (TG) and (DTG) curves for $Co_{1-x}Ca_xFe_2O_4$, (x = 0.0, 0.5, 1.0) ferrites.

Both the exothermic peaks on the DTG curve and the rapid and significant mass loss on the TG curve are characteristic for combustion reactions. The high exothermic peaks are obtained for Ca-doped cobalt nano ferrite, which has an important role in obstructing the grain growth as mentioned in the previous work [32].

3.1.4 Energy dispersive x-ray analysis (EDAX)

Furthermore, the energy dispersive X-ray analysis (EDAX) spectra and elemental mapping data of heated samples are shown in Fig. 7 (a-d). It is clear from Fig. 7 (a-d) that the studied samples are composed of Fe, Co, O and Ca elements. The atomic ratios of Co/Fe and Ca/(Fe) are very close to ½ and 1 respectively. One of the benefits of mapping is that, it provides the analyst the ability to grasp what elements are present and where they are found. Significant benefit of mapping is the ability to see how the elements in a sample occur in combination with each other. Element maps show the cobalt, calcium, and iron contributions. One can select three elements of interest for comparison and then ascribes a primary color (green, blue, or red) to each element. In our case, if blue is assigned to cobalt, red is assigned to iron and green is assigned to calcium, where as Co-Ca–Fe areas will be a blend of blue, red and green).

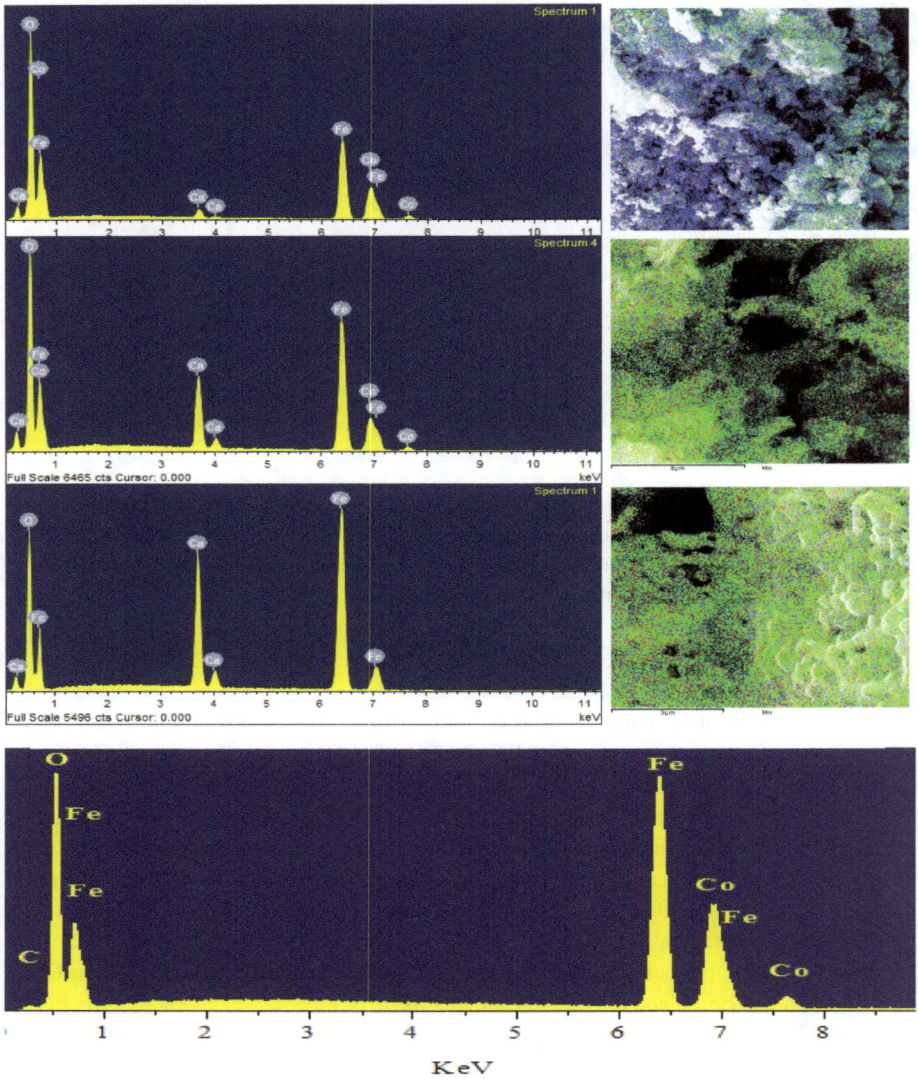

Fig. 7(a-d) Elemental mapping micrographs and EDAX of $Co_{1-x}Ca_xFe_2O_4$ (x = 0.0, 0.1, 0.5, 1.0) ferrite.

3.1.5 HRTEM analysis 3

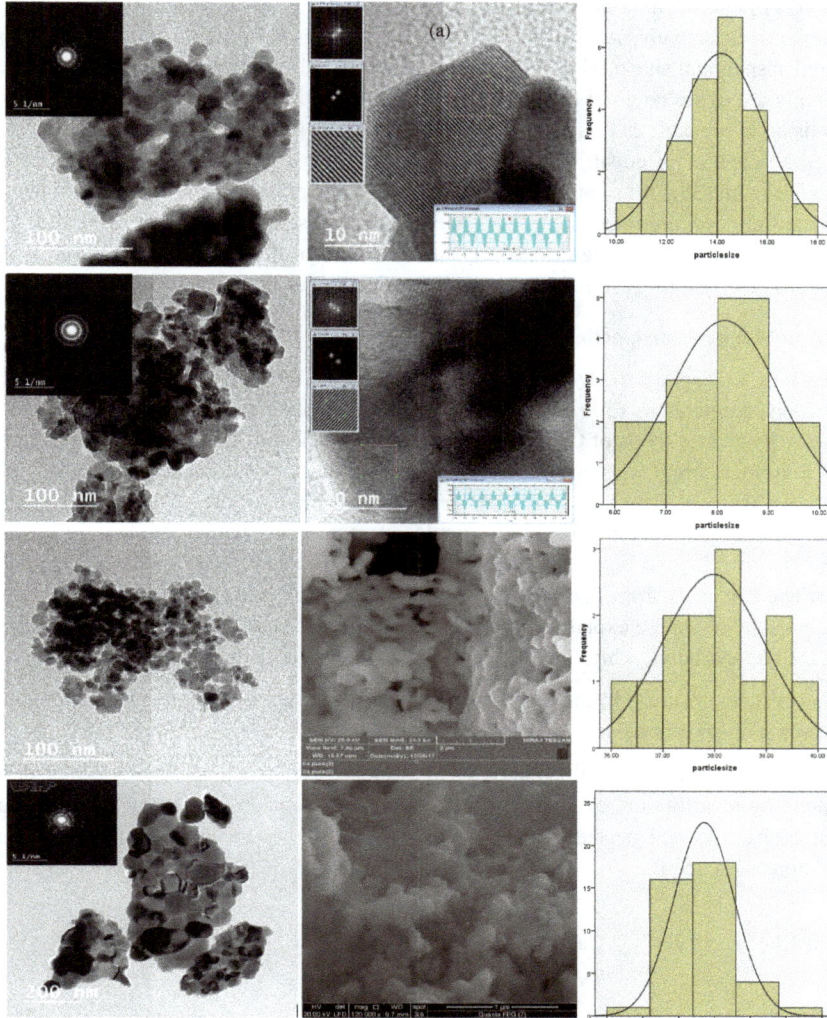

Fig. 8 (a-d) HRTEM micrographs, the selected-area electron diffraction SAED) patterns, the size distribution, of $Co_{1-x}Ca_xFe_2O_4$ (x = 0.0, 0.1, 0.5,1.0) ferrite powder.

Fig. 8 (a-d) shows the HRTEM micrographs, the selected-area electron diffraction (SAED) patterns, and the size distribution of $Co_{1-x}Ca_xFe_2O_4$ (x = 0.0, 0.1, 0.5, 1.0) and calcium respectively. As shown from the figure (Fig. 8 (a-d)) all the samples exhibit regular spherical shape with uniform size except x = 1. The mean particle sizes of each sample are in the range of 8.140–38.910 nm, which is in good agreement with XRD data as listed in the table 2. It is clear from table 2 that, the Ca^{2+} substitution greatly decreases the particle size of cobalt nano ferrites, which agrees well with the results obtained by Saafan et al. [33]. As seen from the figure that the perfect lattice fringes verify the high crystallinity of prepared samples. Moreover, the selected area electron diffraction (SAED) pattern indicating the polycrystalline nature of the nanoparticles (inset of Fig.8 (a-d). The bright rings signify different diffraction planes in the single unit cell.

3.2   Magnetic properties

Fig. 9 (a–c) displays the hysteresis loops. Fig. 9(a) shows hysteresis loops of Co sample at room temperature (300 K) recorded under an applied field of ±30 kOe. Fig. 9 (b) shows hysteresis loops of Ca sample recorded at 100 K and 300 K under an applied field of ± 40 kOe. Fig.9 (c) shows hysteresis loops for calcium substituted cobalt nanoferrite samples (Ca 0.1, Ca 0.5) recorded under an applied field of ± 30 kOe , which designates the soft ferrimagnetic nature of Ca samples at room temperature. While the Co sample reveals very high coercivity value indicating that cobalt belongs to the group of hard ferrites. It is clear from Fig.9 (b) that for pure Ca sample's magnetization is unsaturated. This behavior can be explained on the basis of disordered and canted spins at the surface of the nanoparticles, which are difficult to align along the field direction [34].

However, an increase in the magnetization is detected up to x = 0.5 and then it starts to decreases with increasing Ca contents.

This behavior can be explained in terms of the super exchange interactions, which depend upon the distribution of magnetic and non-magnetic ions in the spinel system at tetrahedral (A) and octahedral [B] sites [35]. The suggested cation distribution can be written as Eq. (5)

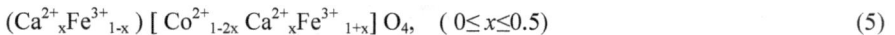

$$(Ca^{2+}_xFe^{3+}_{1-x}) [Co^{2+}_{1-2x} Ca^{2+}_x Fe^{3+}_{1+x}] O_4, \quad (0 \leq x \leq 0.5) \tag{5}$$

From the hysteresis loop the saturation magnetization M_s, the remanence magnetization M_r, the coercivity H_c, energy loss, squareness ratio (M_r/M_s) anisotropy constant (K), are obtained and listed in Table 3. It is clear from the table 3 that, the replacement of Co^{+2}

(the higher magnetic moment~ 3 BM) by Ca^{+2} (the zero magnetic moment) decreases the saturation magnetization and coercivity of the doped samples.

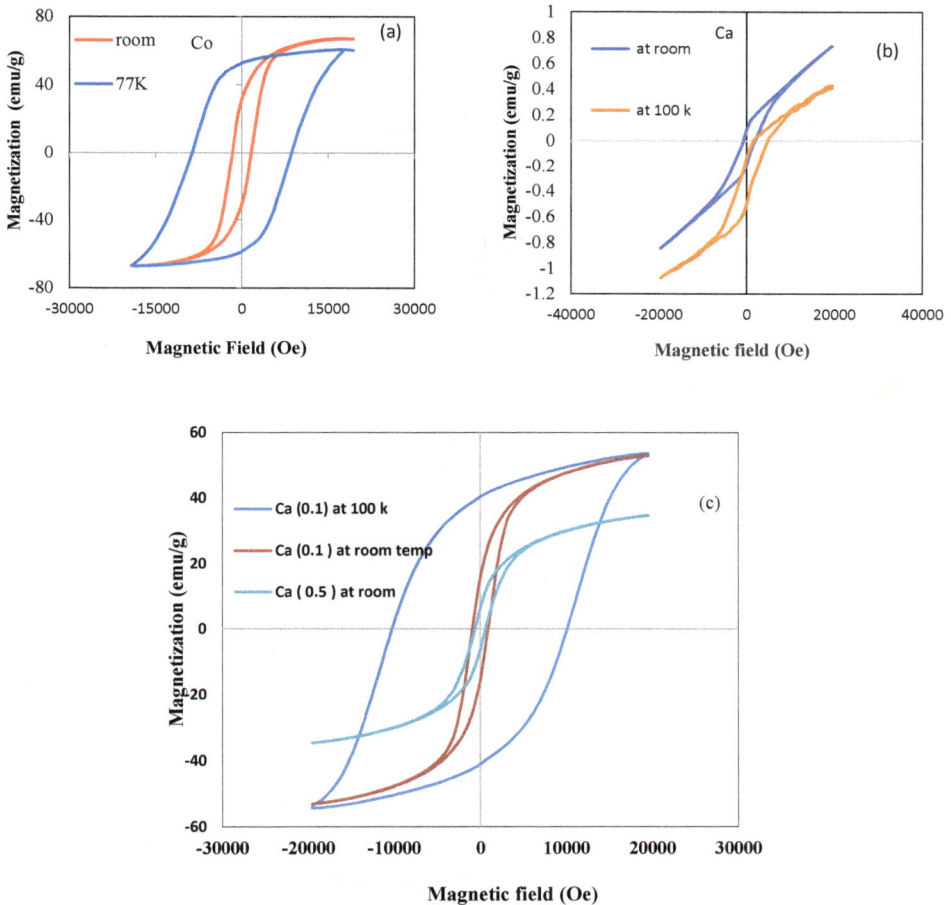

Fig. 9 (a–c) Magnetic hysteresis loops for $Co_{1-x}Ca_xFe_2O_4$

Table 3. Values of saturation magnetization (M_s), remanent magnetization (Mr), coercive field (H_c) energy loss, squareness ratio (Mr/Ms), anisotropy constant (K), for the investigated samples at 300 K.

sample	M_s (emu/g)	M_r remanant (emu/g)	H_c (Oe)	Squareness ratio M_r/M_s	Energy loss (j/g)	Anistropy constant (K) (emu Oe/g)
300 K						
$CoFe_2O_4$	66.847	31.114	1641	0.465	3.37 E-02	114290
$Co_{0.9}Ca_{0.1}Fe_2O_4$	53.053	15.554	920	0.293	1.22E-02	50872
$Co_{0.5}Ca_{0.5}Fe_2O_4$	34.655	6.303	608	0.182	5.72E-03	21954
$CaFe_2O_4$	0.78882	0.134	1295	0.169	2.07E-04	1064
77 K						
$CoFe_2O_4$	63.708	52.529	8656	0.825		574450
100 K						
$Co_{0.9}Ca_{0.1}Fe_2O_4$	53.988	40.663	10000	0.753	1.81E-01	571261
$Co_{0.7}Ca_{0.3}Fe_2O_4$	62.608	46.261	76	0.739	1.60E-01	4967
$CaFe_2O_4$	0.751	0.185	1450	0.247	3.08E-04	1135

In addition to the nature of the cation itself, its relative distribution in the crystal structure is equally important, particularly in the case of spinel structures, where the distribution of cations in both sites describes the type of magnetic behavior.

Squareness ratio (SQR) is a specific magnetic factor that depends on anisotropy. It provides an indication about the ease of the magnetization direction to be reoriented to the nearest easy axis of magnetization direction after the magnetic field is removed [36]. In our case it has been found to be 0.465 for the pure cobalt ferrite and lower than this value for the Ca-doped samples. This value suggests that $CoFe_2O_4$ nano particles have a mixed cubic/uniaxial anisotropy. Though, the value of $1 > [M_r/M_s] > 0.5$ confirms the existence of exchange coupling particles [37].

The squareness ratio values of the investigated samples at 300 K are less than 0.5. The particles in this case interact by magneto static interaction. On the other hand, cobalt particles interaction changes from magneto static at 300K to exchange coupling at 100 K. The obtained data indicates that substitution of Ca^{2+} and the consequent decrease of Co^{2+} at 100 K led to a decrease of the anisotropy compared to the non-substituted sample. The main change of the magnetic parameters is due to the decrease in temperature from 300 to 100 K, can be attributed to continuous reduction of thermal effects that acting on the nucleation process. The area within a hysteresis loop signifies a magnetic energy loss per unit volume of samples per magnetization-demagnetization cycle; this energy is revealed as heat. It is most desirable to minimize the energy losses in the investigated samples. From the table 3 it is clear that, the replacement of cobalt by calcium decreases the magnetic energy loss. However, for this reason special attention is given to calcium nano ferrites, which are suitable candidates to be used in transformer cores and high density recording media.

The maximum energy density of a permanent magnet $(MH)_{max}$ is determined by the point on the second-quadrant side of the $M–H$ hysteresis (Fig. 9 (a-c)). This gives the largest area for an enclosed rectangle. We achieved a significant decrease of the $(MH)_{max}$ for samples with $x = 0.1$ (1.114 kJ/m^2) compared with pure cobalt ferrite sample (1.592 kJ/m^2) due to their high coercivity and remnant magnetization. According to the obtained values of MH_{max}, one can classify the investigated ferrite samples as conventional magnetic materials.

Another important magnetic parameter is the switching field distribution (SFD) as shown in Fig. 10 (a-c). The SFD is quantified in two ways: direct hysteresis loop differentiation and the so-called $\Delta H(M, \Delta M)$ method [38, 39]. It is of particular importance in characterizing the magnetic properties of magnetic media, and can be determined by using the Eq. (6).

$$SFD = \Delta H / H_c \tag{6}$$

Where, ΔH is the full width at half maxima of the differentiated curve (dM/dH) as shown in Fig.10 (a-c). This amount measures the energy barrier distribution that, in a nano particle system, is accompanying with a distribution of particle coercivity. Generally, the SFD is associated with recording factors such as optimal bias current and noise. System with small SFD and high H_c are appropriate for high density recording as mentioned before [40].

Furthermore, the derivative curves provide valuable data concerning the magnetic properties of the system. However, the appearance of double peaks in SFD reveals the competition between strong dipolar interactions and exchange coupling [41].

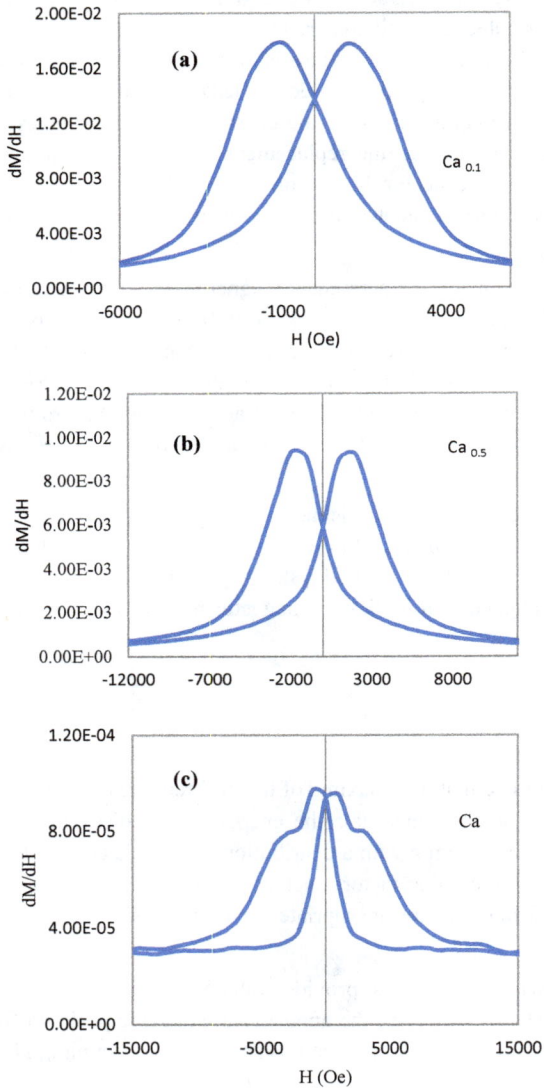

Fig. 10 (a-c) Switching field distribution for $Co_{1-x}Ca_xFe_2O_4$.

Magnetic Oxides and Composites II   Materials Research Forum LLC
Materials Research Foundations 83 (2020) 157-179   https://doi.org/10.21741/9781644900970-7

It is clear from Table 4 that Ca substitution has the advantage of reducing the SFD value for the investigated samples.

Table 4 The switching field distribution SFD for $Co_{1-x}Ca_xFe_2O_4$.

Calcium concentration	H_c (Oe)	SFD
300 K		
0.1	920	21.28
		5.39
0.5	608	6.17
		18.98
1.0	1295	1.50
		7.91
100 K		
1.0	1450	1.63
		6.99

Conclusions

$Co_{1-x}Ca_xFe_2O_4$ ($x = 0, 0.1, 0.5, 1.0$) ferrites are successfully prepared in single phase cubic/ orthorhombic structure. The crystallite sizes of the investigated samples are in the range of 7.98 nm to 37.73 nm.

For the studied samples, the minor elements (calcium) added to and substituted within the cobalt are definitely identified and located in the elemental maps, and their spectra are accurately quantified. This analytical technique is useful in a wide range of industrial labs and manufacturing.

The replacement of cobalt by calcium decreases the magnetic energy loss, crystallite sizes and the switching field distribution (SFD).

The investigated samples with small SFD and high Hc are suitable for high density recording media applications.

The appearance of double peaks in switching field distribution (SFD) reveals the competition between strong dipolar interactions and exchange coupling.

References

[1]   Carter, C. Barry; Norton, M. Grant Ceramic Materials: Science and Engineering. Springer. (2007) pp. 212–15. ISBN 0-387-46270-8.

[2]   M Vadivel, R Ramesh Babu, M. Arivanandhan, K Ramamurthi. Structural, spectral, morphological, dielectric, magnetic,and optical properties of La-Ni ions co-substituted $CoFe_2O_4$ nanoparticles, Journal of Superconductivity Novel Magnetism, 30 (2) (2017) 441-453. https://doi.org/10.1007/s10948-016-3760-3

[3]   A. T. Raghavender, R. G. Kulkarni, and K. M. Jadhav, Magnetic properties of mixed Cobalt-Aluminum ferrite nanoparticles, Chinese Journal of physics, 48 (4) (2012) 512-522.

[4]   L. Smart, E. Moore, Solid state Chemistry: an Introduction, 3rd ed. Boca Raton: CRC Press, 2005.

[5]   Lawrence Kumar, Pawan Kumar, Manoranjan Kar, Influence of Mn substitution on crystal structure and magnetocrystalline anisotropy of nanocrystalline $Co_{1-x}Mn_xFe_{2-2x}Mn_{2x}O_4,$Applied Nano sciences, 3 (2013)75-82. https://doi.org/10.1007/s13204-012-0071-2

[6]   R. Valenzuela, Magnetic Ceramics, Cambridge University Press, New York (NY), USA, (1994).

[7]   A. Goldman, U K. Modern Ferrite Technology, Springer, New York, (2006).

[8]   U. Yenial, F Pagnenelli, Calcium ferrite nanoparticle production from mining wastes marble dust and pyrite ash. Conference paper, https://www. researchgate.net/publication/ 324485362 , (2017) 587-594 .

[9]   Sobhi Hcini, Aref Omri, Michel Boudard, Microstructural, magnetic and electrical properties of $Zn_{0.4}M_{0.3}Co_{0.3}Fe_2O_4$ (M = Ni and Cu) ferrites synthesized by sol–gel method, Journal of Materials Science: Materials in Electronics, 29(8) (2018) 6879-6891. https://doi.org/10.1007/s10854-018-8674-3

[10] H.F. Abosheiasha, S.T. Assar, Effects of sintering process on the structural, magnetic and thermal properties of $Ni_{0.92}Ca_{0.08}Fe_2O_4$ nanoferrite, Journal of Magnetism Magnetic Materials, 370 (2014) 54–61. https://doi.org/10.1016/j.jmmm.2014.06.054

[11] E Ebtesam Ateia, S. Fatma Soliman, Modification of Co/Cu nanoferrites properties via $Gd^{3+}Er^{3+}$doping. Applied physics A, Materials science processing, 123 (2017) 312. https://doi.org/10.1007/s00339-017-0948-8

[12]   Ebtesam Ateia, Asmaa A. H., El-Bassuony, Fascinating improvement in physical
        properties of Cd/Co nanoferrites using different rare earth ions, Journal of
        Materials Science: Materials in Electronics, 28 (2017) 11482–11490.
        https://doi.org/10.1007/s10854-017-6944-0

[13]   Ebtesam E. Ateia, Galila Abdelatif, Fatma S. Soliman, Optimizing the physical
        properties of calcium nanoferrites to be suitable in many applications, 28 (2017)
        5846 –5851. https://doi.org/10.1007/s10854-016-6256-9

[14]   D. H. Kim, D. E. Nikles, D. T. Johnson and C. S. Brazel, Heat generation of
        aqueously dispersed CoFeO$_4$ nanoparticles as heating agent for magnetically
        activated drug delivery and hyperthermia, Journal of Magnetism and Magnetic
        Materials, 320 (2008) 2390–2396. https://doi.org/10.1016/j.jmmm.2008.05.023

[15]   M. Veverka, P. Veverka, Z. Jirák, O. Kaman, K. Knížek, M. Maryško, E. Pollert
        and K. Závěta, Synthesis and magnetic properties of Co$_{1-x}$Zn$_x$Fe$_2$O$_{4+\gamma}$
        nanoparticles as materials for magnetic fluid hyperthermia, Journal of Magnetism
        and Magnetic Materials, 322 (2010) 2386–2389.
        https://doi.org/10.1016/j.jmmm.2010.02.042

[16]   L. Jiang, J. Guo, H. Liu, M. Zhu, X. Zhou, P. Wu, C. Li, Prediction of lattice
        constant in cubic perovskites, Journal of Physics and Chemistry of Solids, 67
        (2006)1531-1536. https://doi.org/10.1016/j.jpcs.2006.02.004

[17]   A. Bokov, Z.-G. Ye, Recent progress in relaxor ferroelectrics with perovskite
        structure, in: Frontiers of Ferroelectricity, Springer, 2006, pp. 31-52.
        https://doi.org/10.1007/978-0-387-38039-1_4

[18]   Parkin, editors-in-chief, Helmut Kronmller, Stuart; Mats Johnsson; Peter
        Lemmens Handbook of magnetism and advanced magnetic materials . Hoboken,
        NJ: John Wiley & Sons. ISBN 978-0-470-02217-7. Retrieved 2012.

[19]   V. Pillai, D.O. Shah, Synthesis of high-coercivity cobalt ferrite particles using
        water-in-oil microemulsions, Journal of Magnetism and Magnetic Materials 163
        (1996) 243–248. https://doi.org/10.1016/S0304-8853(96)00280-6

[20]   K. Maaz, Arif Mumtaz, S.K. Hasanain, Abdullah Ceylan, Synthesis and magnetic
        properties of cobalt ferrite (CoFe$_2$O$_4$) nanoparticles prepared by wet chemical
        route, Journal of Magnetism and Magnetic Materials 308 (2007) 289–295.
        https://doi.org/10.1016/j.jmmm.2006.06.003

[21]   E. Ebtesam Ateia, M. K. Abdelamksoud, M. A. Rizk, Improvement of the
        physical properties of novel (1−x)CoFe$_2$O$_4$+(x)LaFeO$_3$ nanocomposites for

technological   applications journal of Materials Science Materials in Electronics 28(2017)16547–16553. https://doi.org/10.1007/s10854-017-7567-1

[22]   M Sagrario. L. A.Montemayor, Garcı´a-Cerda, J.R. Torres-Lubia´n, Preparation and characterization of cobalt ferrite by the polymerized complex method, Materials Letters, 59 (2005) 1056– 1060. https://doi.org/10.1016/j.matlet.2004.12.004

[23]   Periodic Table, SARGENT-WELCH, Scientific Company, 7300 Linder Avenue, Skokie, Illinois 60076, Catalog Number 5-18806.

[24]   R. D. Shannon, Acta Crystallographica, 32A (1976) 751–767. https://doi.org/10.1107/S0567739476001551

[25]   B. D. Cullity, Elements of X-ray Diffraction, Adison-Wesley Publ. Co., London (1967).

[26]   R.D. Waldron , Infrared Spectra of Ferrites. Physical Review 99 (1955)1727– 1735. https://doi.org/10.1103/PhysRev.99.1727

[27]   http://scholar.lib.vt.edu/theses/available/etd-04262006-181958/unrestricted/ Appendix A. pdf.

[28]   L. Khanna, N.K. Verma, Size-dependent magnetic properties of calcium ferrite nanoparticles, Journal of Magnetism and Magnetic Materials, 336(2013) 1–7. https://doi.org/10.1016/j.jmmm.2013.02.016

[29]   S. Manouchehrei, S.T.M. Benehi, M.H. Yousefi, Effect of aluminum doping on the structural and magnetic properties of Mg-Mn ferrite nanoparticles prepared by coprecipitation method, Journal of Superconductivity and Novel Magnism 29 (2016) 2179–2188. https://doi.org/10.1007/s10948-016-3546-7

[30]   A. M. El-Sayed, Influence of zinc content on some properties of Ni–Zn ferrites Ceramics International, 28 (4) (2002) 363–367. https://doi.org/10.1016/S0272-8842(01)00103-1

[31]   Ahmad Monshi, Mohammad Reza Foroughi, Mohammad Reza Monshi, Modified Scherrer equation to estimate more accurately nano-crystallite size using XRD, World Journal of Nano Science and Engineering , 2 (3) (2012) 2161–4954. https://doi.org/10.4236/wjnse.2012.23020

[32]   Ebtesam E. Ateia, Amira T. Mohamed, M. Morsy, Humidity sensor applications based on mesopores $LaCoO_3$ Journal of Materials Science: Materials in Electronics 30 (21) (2019) 19254–19261. https://doi.org/10.1007/s10854-019-02284-y

[33]   S.A. Saafan, S.T Assar, S.F. Mansour, Magnetic and electrical properties of $Co_{1-x}Ca_x Fe_2O_4$ nanoparticles synthesized by the auto combustion method, Journal of Alloys and Compounds 542 (2012) 192–198. https://doi.org/10.1016/j.jallcom.2012.07.050

[34]   O scar Iglesias, Amilcar Labarta , Role of surface disorder on the magnetic properties and hysteresis of nanoparticles, Physica B 343 (2004) 286–292. https://doi.org/10.1016/j.physb.2003.08.109

[35]   E. Ebtesam Ateia, A. Asmaa El-Bassuony, Galila Abdelatif, S. Fatma Soliman, Novelty characterization and enhancement of magnetic properties of Co and Cu nanoferrites, journal of materials science Materials in Electronics 28 (2017) 241–249. https://doi.org/10.1007/s10854-016-5517-y

[36]   S.T. Assar, H.F. Abosheiasha, S.A. Saafan, M.K. EL Nimr, Preparation, characterization   and magnetization of nano and bulk $Ni_{0.5}Co_{0.5-2x}Fe_{2+x}O_4$ samples, J. Molecular Structure 1084 (2015) 128–134. https://doi.org/10.1016/j.molstruc.2014.12.031

[37]   Saulo Gregory Carneiro Fonsecaa, Laédna Souto Neivab, Maria Aparecida Ribeiro Bonifácioc Tunable magnetic and electrical properties of cobalt and zinc ferrites $Co_{1-x}Zn_xFe_2O_4$ Synthesized by Combustion Route, Materials Research, 21(3) (2018) e20170861. https://doi.org/10.1590/1980-5373-mr-2017-0861

[38]   A.Berger, Y.H.Xu, B. Lengsfield, Y.Ikeda, and E.E.Fullerton, spl Delta/H(M,/spl Delta/M) method for the determination of intrinsic switching field distributions in perpendicular media, IEEE Transations on Magnetics, 41(2005) 3178. https://doi.org/10.1109/TMAG.2005.855285

[39]   A. Berger,B. Lengsfield, andY. Ikeda, Determination of intrinsic switching field distributions in perpendicular recording media Journal Applied Physics, 99 (2006) 08E705. https://doi.org/10.1063/1.2164416

[40]   J.C.Lodder, Handbook of Magnetic Materials , chapter 2, Magnetic recording hard disk thin film media, in Handbook of Magnetic Materials, 11(1998) 291-405. https://doi.org/10.1016/S1567-2719(98)11006-5

[41]   A. Berger and H. Hopster, Magnetization reversal properties near the reorientation phase transition of ultrathin Fe/Ag(100) films J. Applied Physics 79 (1996) 5619. https://doi.org/10.1063/1.362261

Magnetic Oxides and Composites II
Materials Research Foundations **83** (2020) 180-192

Materials Research Forum LLC
https://doi.org/10.21741/9781644900970-8

Chapter 8

Lithium Ferrites Prepared Differently and its Magnetic Properties

S. Soreto Teixeira[1,a], M.P.F. Graça[1,b], L.C. Costa[1,c], M.A. Valente[1,d]

[1]i3N and Physics Department, University of Aveiro, 3810-193 Aveiro, Portugal

[a]silvia.soreto@ua.pt, [b]mpfg@ua.pt, [c]kady@ua.pt, [d]mav@ua.pt

Abstract

The purpose of this chapter is to compare the structural, morphologic and magnetic properties of lithium ferrite nanoparticles obtained by solid state reaction (SSR) and sol-gel (SG) methods using nitrates as raw materials. Samples structure was studied by X-ray diffraction and Raman spectroscopy and their morphology by scanning electron microscopy. SG method favored the appearance of the single lithium ferrite crystal phase at lower heat treatment temperature. The more suitable samples to be applied in electronic devices are the ones treated at 1200 °C, which present the highest levels of saturation magnetization ($\approx$68 emu/g at 300 K).

Keywords

Ferrites, X-Ray Diffraction, Magnetic Properties, Sol Gel, Solid State Reaction

Contents

1.   Introduction

Ferrites, especially the lithium ferrite, due to its interesting properties is one of the materials with much interest in the electronic field. This material is sought after because it can be a potential substitute for others used in microwave devices [1], in enhancing magnetic resonance, magnetically guided drug delivery, magnetic fluids, ferrofluid technology, refrigeration, magnetocaloric and high magnetic density recording. Structurally, it can crystallize as spinel, $MO \cdot Fe_2O_3$, where M represents cations with double positive charge such as Mn^{2+}, Ni^{2+}, Zn^{2+}, Co^{2+}, Fe^{2+}, Cu^{2+}, Mg^{2+}; hexagonal, $MO \cdot 6Fe_2O_3$ where M represents Ba^{2+}, Ca^{2+}, Sr^{2+} and as garnet type, $3M_2O_3 \cdot 5Fe_2O_3$ where M represents triple positive charge cations, such as Y or elements of rare earth group [2].

Lithium ferrite is one of the most interesting and known cubic ferrites belonging to the magnetic soft materials. Due to its characteristics such as high Curie temperature (620 °C), high saturation magnetization and a square hysteresis loop it can be used as ferrimagnet in magnetic storage devices [3] and also is a probable candidate in rechargeable lithium ion batteries as a cathode material [4]. Both, ordered (α-$LiFe_5O_8$) and disordered (β-$LiFe_5O_8$) crystal phases of lithium ferrite can be identified [1], [5], [6] (Fig.1).

Magnetic Oxides and Composites II
Materials Research Foundations **83** (2020) 180-192

Materials Research Forum LLC
https://doi.org/10.21741/9781644900970-8

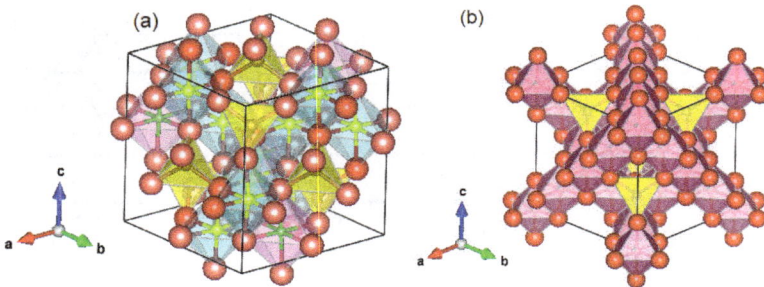

Fig. 1 Cubic crystalline system of LiFe₅O₈: (a) ordered phase, α, with special group P4₁32 (213), built by tetraethers (yellow) with iron ions (brown) in the centre and octahedra's (blue and rose) and lithium ions (green), respectively; (b) disordered phase, β, (special group Fd3m) built by tetrahedra's (yellow) with iron ions (brown) in the centre and octahedra's (rose) with lithium ions (green).

The ordered phase can be achieved by rapid cooling from temperatures above 800 °C to room temperature, with slow cooling and below 750 °C [7]. An order-to-disorder transition can occur between 735 and 755 °C in temperature. The opposite transition is reflected in the diffraction plates by extra reflections with an enlarged profile [8].

To prepare LiFe₅O₈ by solid-state reaction method, the big issue is due to the volatility of lithium above 1000 °C on high temperatures (≈1200 °C) are required. Consequently, the obtained powders could have low quality because the sintering process usually primes to low specific surface areas [9], affecting its electrical and magnetic properties. For synthesised lithium ferrite at low temperatures, several chemical methods could be used, such as glass crystallization, co-precipitation, mechanical alloying, hydrothermal and sol-gel methods.

Therefore, the preparation of LiFe₅O₈ at low temperatures is a subject of interest. To improve its magnetic properties, the comparison of results for both preparation methods, such as solid state reaction (SSR) and sol-gel (SG), is the main goal of this work.

2.   Experimental description

$LiFe_5O_8$ powders were prepared by solid state reaction and sol-gel methods. The raw materials used in both processes were iron (III) nitrate (Fe $(NO_3)_3 \cdot 9H_2O$) and lithium nitrate ($LiNO_3$) (Merck KGaA, Darmstadt, Germany).

2.1   Solid state reaction method

Lithium iron nitrates were weighed with 1:5 molar ratio of Li and Fe ions and homogenized in a planetary ball mill system (Fritsch - Pulverisette 7.0) at 250 rpm for 1 h. An equal volume of powders and balls of 10 mm diameter was used. After this first mixture, 10 mL of ethanol was added. The wet mixture was again placed in the ball mill system for 3 h at 500 rpm. After this process, the mixing vessel was placed in an oven at 80 °C for 24 h to evaporate the ethanol.

2.2   Sol-gel method

In order to obtain lithium ferrite crystal phase, the Pechini route [10] was used. Citric acid (CA) was the chelating agent and ethylene glycol (EG), which is the responsible for the polymerization process, respectively. The first step was to prepare the suspensions of each nitrate, first with the excess of CA (1:3 molar ratio between metal precursor and CA) and then with EG (2:3 molar ratio between CA and EG). To promote solubility, each suspension was stirred for 30 minutes at room temperature and after both suspensions was mixed for 24 h. The obtained gel was dried in a muffle at 250 °C for 60 h. Also, the temperature and time were chosen to ensure the maximum solvent evaporation.

For both methods, the two powders were sintered between 400 and 1200 °C, for 4 h, with a heating rate of 5 °C/min. The temperature of the heat treatments was defined taking into account the results of the thermal analysis by DTA, using a Linseis equipment, in a temperature range that goes from room temperature to 1200 °C, with a heating rate of 20 °C/min. Then, its structural properties were studied by X-ray diffraction (XRD), using a X'Pert MPD Philips diffractometer (Cu-Kα radiation, λ = 1.5406 Å) and for Raman spectroscopy, using HR-800-UV Jobin Yvon Horiba spectrometer (532 nm laser line) using a microscope objective (50×). After identifying the samples with lithium ferrite as single crystal phase, the morphology of its surface was analysed, by scanning electron microscopy (SEM) Hitachi S4100-1, and magnetic measurements were performed using the vibrating sample magnetometer (VSM) technique.

3. Results and discussion

3.1 Thermal analysis

The results of the thermal analysis were analysed to choose the heat-treatment temperatures. In the DTA thermograms (Fig. 2) the exothermic bands, indicating significant structural changes such as crystallisation, are centered around 280, 550 and 850 °C for the SSR method and at 500 and 750 °C for the samples prepared by the SG. In cause of that, the performed heat treatments were at 400, 1000, 1050, 1100, 1150 and 1200 °C for SSR and at 400, 600, 800 and 1200 °C for SG methods, respectively.

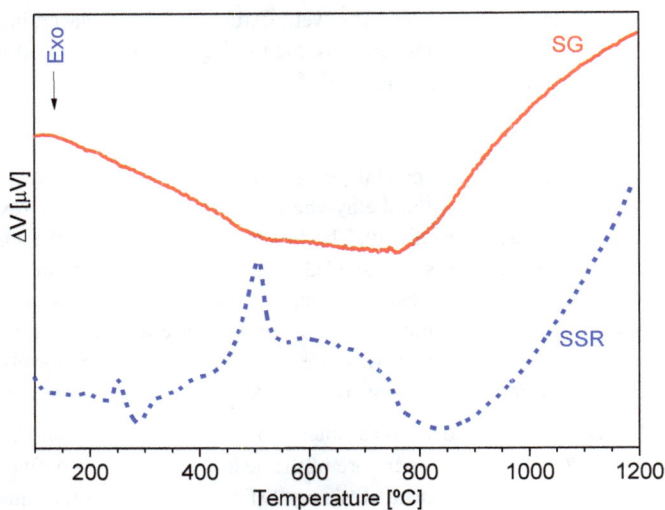

Fig. 2 Thermograms of powders prepared by SSR (dash line) and SG (solid line) methods showing the endothermic and exothermic phenomena, resulting in structural changes.

3.2 XRD analysis

X-rays diffraction patterns of the powders heat treated at temperatures between 400 and 1200 °C by SSR and SG methods are shown in Fig. 3(a) and 3(b). According to XRD diffractograms of the samples, prepared with both methods, the $LiFe_5O_8$ crystal phase is present for samples treated between 400 °C and 1200 °C for both methods.

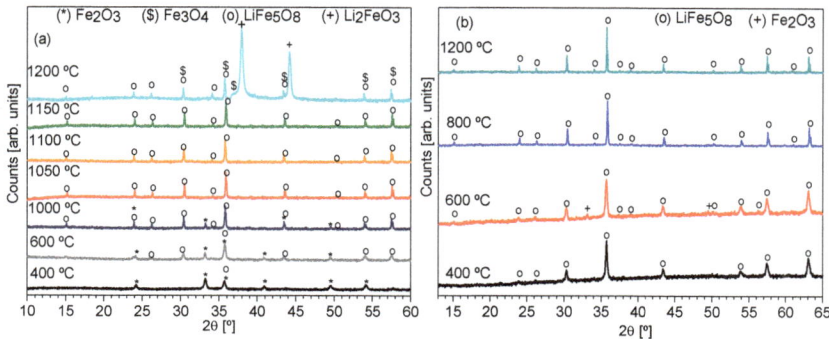

Fig. 3 X-rays diffraction patterns of the powders heat treated at temperatures between 400 and 1200 °C by (a) SSR and (b) SG methods. The LiFe₅O₈ crystal phase is detected, as single phase, at 800 °C (SG method) and at 1050 °C (SSR method).

Whereas, using SSR method this crystal phase is detected, as single phase, only at 1050 °C continued until 1150 °C of temperature, using the SG it is detected at a lower temperature, namely at 800 °C maintained until 1200 °C. Additionally, for the SSR method above the temperature of 1200 °C, the lithium ferrite is transformed in lithium ferrate (Li_2FeO_3) and magnetite (Fe_3O_4), according to the equation (1):

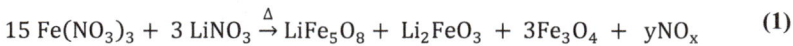

$$15\ Fe(NO_3)_3 + 3\ LiNO_3 \xrightarrow{\Delta} LiFe_5O_8 + Li_2FeO_3 + 3Fe_3O_4 + yNO_x \qquad \textbf{(1)}$$

For both preparation methods, lithium ferrite and hematite crystal phases were detected in samples treated at temperatures below 800 °C. By the SG method, the hematite phase was not detected at the sample treated at 400 °C, as had occurred by SSR, indicating a possible presence of an amorphous state.

The $LiFe_5O_8$ crystal phase presents in both diffractograms (Fig. 3 (a), (b)) is attributed to the disordered phase of lithium ferrite (β-$LiFe_5O_8$) [11].

3.3 Raman spectroscopy

Raman spectra for samples treated at 1100 °C (SSR) and at 800 °C (SG) are shown in Fig. 4.

Magnetic Oxides and Composites II   Materials Research Forum LLC
Materials Research Foundations **83** (2020) 180-192   https://doi.org/10.21741/9781644900970-8

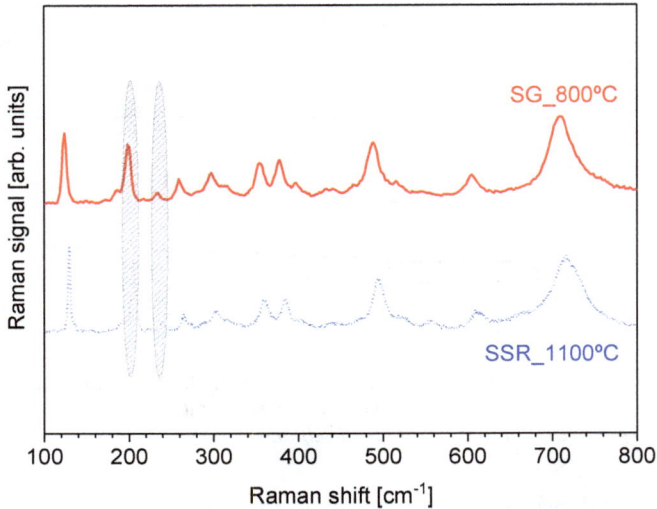

Fig. 4 Raman spectra for samples treated at 1100 °C (SSR) and at 800 °C (SG) showing the vibration bands characteristics for both, ordered and disordered lithium ferrite crystal phases, and also, the highlight indicating the presence of the ordered phase, α-LiFe$_5$O$_8$.

However, the Raman spectroscopy results showed vibrations; which are characteristics of the lithium ferrite ordered phase (α-LiFe$_5$O$_8$) centred at 202 and 236 cm^{-1} and the others centred around 263, 300, 320, 357, 385, 407, 439, 475, 492, 558, 612 and 719 cm^{-1} which are characteristics of both, ordered and disordered, phases [11], [12] (Fig. 4). Consequently it was confirmed that besides the disordered phase of lithium ferrite (β-LiFe$_5$O$_8$), which had been identified by previous XRD results, it is also present the ordered one (α-LiFe$_5$O$_8$).

3.4   Surface morphology

Fig. 5 represents SEM images of prepared samples. The samples surface microscopy shows that, for both the methods, the particles size increases with the temperature of heat treatment.

The lithium ferrite grains habit is prismatic. It is also evident that at lower temperatures of heat treatment, the samples prepared by the SG method exhibits higher grain sizes than for SSR preparation method. After the structural and morphological characterizations of the samples which only the lithium ferrite crystal phase was present, it was performed the magnetic measurements using the VSM.

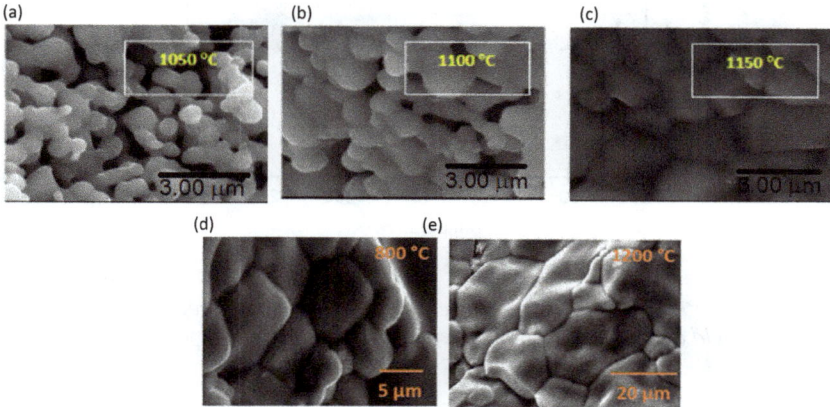

Fig. 5 *SEM micrographs of the samples with LiFe$_5$O$_8$ as single crystal phase: (a), (b) and (c) obtained by SSR method, (d) and (e) by SG method.*

3.5 Magnetic properties

3.5.1 Magnetic susceptibility

Fig. 6 shows the magnetic susceptibility, at 5 and 300 K, with an applied magnetic field of $H = 1$ kOe. The samples prepared by the SG method have the highest magnetic susceptibility, being the sample treated at 1200 °C with the maximum value (≈ 580 emu.g^{-1}.T^{-1}). On the other hand, the sample prepared by SSR and heat-treated at 1050 °C is the one that has lower values (≈ 470 emu.g^{-1}.T^{-1}). According to the SEM micrographs of samples prepared by the sol-gel method (Fig. 5(d) and (e)) an increase in mean grain size, from approximately 5 μm, in the sample treated at 800 °C, to above 20 μm, for the sample treated at 1200 °C is observed. This phenomenon promotes an increase in the probability of random distribution of magnetic moments and thus an increase in the characteristic response of soft magnetic materials.

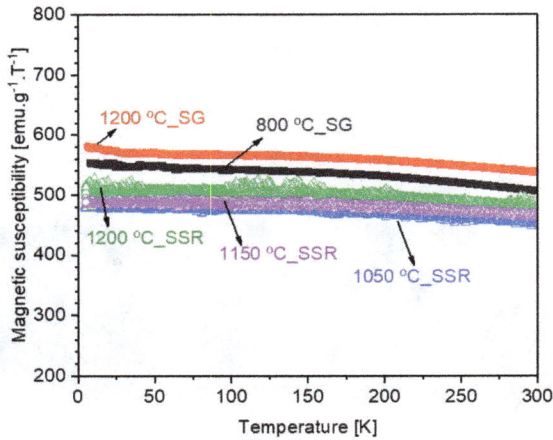

Fig. 6 Magnetic susceptibility vs. temperature, recorded at FC regime under a magnetic field of H=1 kOe, of lithium ferrite samples, obtained by SSR and SG methods, and treated at different temperatures.

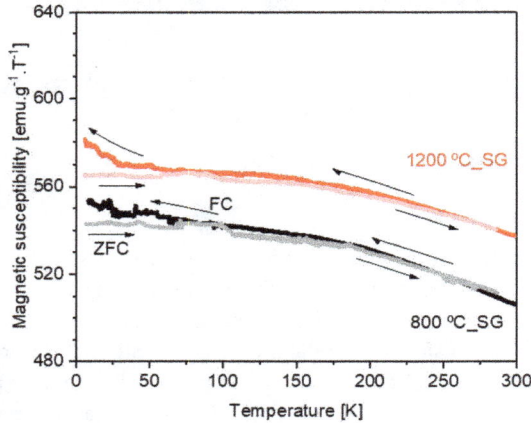

Fig. 7 Magnetic susceptibility vs. temperature, recorded at ZFC and FC regimes, under a magnetic field of H=1 kOe, of lithium ferrite samples, heat treated at 800 and 1200 °C, obtained by the SG method.

Analysing the magnetic susceptibility for the sol-gel preparing method samples (Fig. 7), it is shown a slow increase of the blocking temperature, T_B, with the heat treatment temperature, being ≈137 K (sample 800 °C_SG) and ≈ 230 K (sample 1200 °C_SG). This behavior can be justified by the increases of the grain coalescence with the heat treatment temperature, having a consequence of an improvement of the grain size, causing also an increase of T_B [13].

3.5.2 Hysteresis loops

Figure 8 shows the hysteresis curve of all samples at 5 K of temperature. Both samples obtained by sol-gel method, treated at 800 and 1200 °C, have a magnetization of ≈87 emu/g for a magnetic field of 20 kOe. This value is very close to that obtained in the literature for this material prepared by the aerosol method [14, 15].

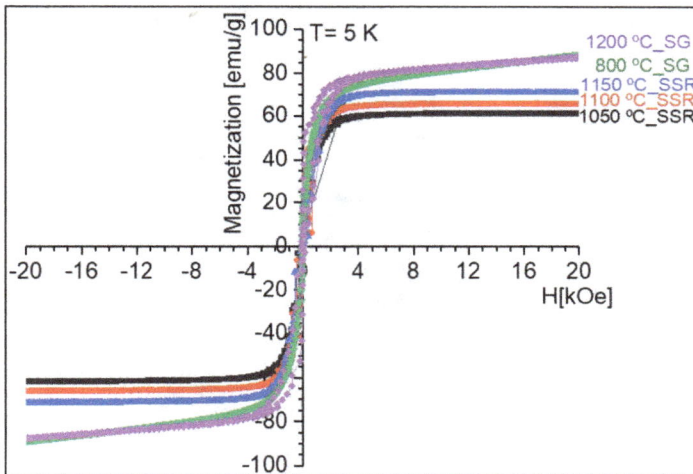

Fig. 8 Hysteresis curves at T = 5 K, of lithium ferrite samples, obtained by SSR and SG methods, and treated at different temperatures.

For others authors, the magnetization of lithium ferrite is about 60 emu/g [1], [16], which is close to that obtained in the treated samples between 1050 °C and 1150 °C, by SSR method (Fig. 8). The variation of magnetization is connected to the XRD (Fig. 2(a)) and

Raman (Fig. 3) results showing that samples heat treated at temperatures between 1000 and 1200 ° C have the ordered phase, α-LiFe$_5$O$_8$, as main crystal phase. Most likely, this ordered ferrite phase was developed from the thermally activated reaction between α-Fe$_2$O$_3$ and the "free" lithium ions present in the network, suggesting the simultaneous existence of an amorphous phase, which decreases with the increase of the treatment temperature.

Comparing the magnetization (M) as a function of the magnetic field (H) applied at different temperatures, 5 and 300 K, for the sample treated at 1200 °C (Fig. 9), it is lower at 300 K, saturating at ≈ 68 emu/g, with a magnetic field of 2.7 kOe, while at 5 K, this sample did not saturate for 50 kOe. According to the literature [17], samples of LiFe$_5$O$_8$ prepared by the sol-gel method, show that the saturation magnetization values ranging from 34 to 52 emu/g at 50 kOe, being maximum for higher size grain particles (45 nm). This results reported in the literature are lower than the obtained (≈ 68 emu/g) however, it should be noted that the average particle size is larger (Fig.4).

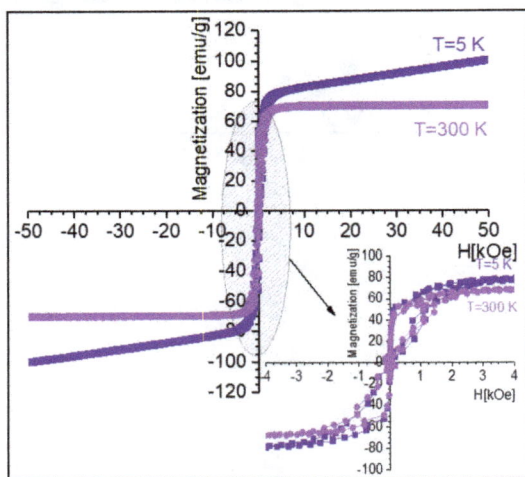

Fig. 9 Hysteresis curves, at 5 and 300 K, of the sample prepared by SG method and heat treated at 1200 °C.

Magnetic Oxides and Composites II   Materials Research Forum LLC
Materials Research Foundations **83** (2020) 180-192   https://doi.org/10.21741/9781644900970-8

Conclusions

The preparation of lithium ferrite crystals can be obtained either by solid state reaction and sol-gel methods using iron and lithium nitrates as starters. It was found that by SG method the lithium ferrite crystal phase is detected at earlier temperatures, being a single phase in a large range of temperatures ($\Delta T \approx 400$ °C) of heat treatment than for SSR method ($\Delta T \approx 1000$ °C). The presence of both $LiFe_5O_8$ crystal forms, ordered and disordered, was confirmed by Raman spectroscopy. The magnetic properties are influenced by structural and morphologic characteristics being the sol-gel samples the ones with highest values of magnetic susceptibility and magnetization.

References

[1]   N. G. Jović, Jović, N. G., Masadeh, A. S., Kremenovic, A. S., Antic, B. V., Blanusa, J. L., Cvjeticanin, N. D., Goya, G. F., Antisari, M. V., and Bozin, E. S., Effects of thermal annealing on structural and magnetic properties of lithium ferrite nanoparticles, Journal of Physical Chemistry C, 113 (48) (2009) 20559–20567. https://doi.org/10.1021/jp907559y

[2]   D. Gingasu, I. Mindru, L. Patron, and S. Stoleriu, Synthesis of lithium ferrites from polymetallic carboxylates, Journal of the Serbian Chemical Society, 73 (10) (2008) 979–988. https://doi.org/10.2298/JSC0810979G

[3]   A. Goldman, Modern ferrite technology, 2nd Ed. Pittsburgh, PA, USA: Springer US, (2006).

[4]   S. S. Teixeira, M. P. F. Graça, and L. C. Costa, Dielectric and Structural Properties of Lithium Ferrites, Journal of Physical Chemistry C, 47 (5), (2014) 356–362. https://doi.org/10.1080/00387010.2013.840316

[5]   S. Dey, A. Roy, D. Das, and J. Ghose, Preparation and characterization of nanocrystalline disordered lithium ferrite by citrate precursor method, Journal of Magnetism and Magnetic Materials, 270 (1–2) (2004) 224–229. https://doi.org/10.1016/j.jmmm.2003.08.024

[6]   S. S. Teixeira, M. P. F. Graça, and L. C. Costa, Dielectric, morphological and structural properties of lithium ferrite powders prepared by solid state method, Journal of Non-Crystalline Solids, 358 (16) (2012) 1924–1929. https://doi.org/10.1016/j.jnoncrysol.2012.06.003

[7]   A. Ahniyaz, T. Fujiwara, S. Song, and S. Yoshimura, Low temperature preparation of β-$LiFe_5O_8$ fine particles by hydrothermal ball milling, Solid State Ionics, 151 (2002) 419–423. https://doi.org/10.1016/S0167-2738(02)00548-9

[8]   M. N. Iliev, V. G. Ivanov, N. D. Todorov, V. Marinova, M. V. Abrashev, R.
      Petrova, Y. Q. Wang, and A. P. Litvinchuk, Lattice dynamics of the α and β phases
      of $LiFe_5O_8$, Physical Review B: Condensed Matter, 83 (2011) 174111-1-174111–7.
      https://doi.org/10.1103/PhysRevB.83.174111

[9]   H. M. Widatallah and F. J. Berry, The influence of mechanical milling and
      subsequent calcination on the formation of lithium ferrites, Journal of Solid State
      Chemistry, 164 (2002). 230–236. https://doi.org/10.1006/jssc.2001.9466

[10] M. P. Pechini, Method of preparing lead and alkaline earth titanates and niobates
      and coating method using the same to form a capacitor, US Pat. 3330697 (1967).

[11] E. Wolska, P. Piszora, W. Nowicki, and J. Darul, Vibrational spectra of lithium
      ferrites: infrared spectroscopic studies of Mn-substituted $LiFe_5O_8$, International
      Journal of Inorganic Materials, 3 (2001) 503–507. https://doi.org/10.1016/S1466-
      6049(01)00069-1

[12] W. Cook and M. Manley, Raman characterization of α and β-$LiFe_5O_8$ prepared
      through a solid-state reaction pathway, Journal of Solid State Chemistry, 183 (2)
      (2010) 322–326. https://doi.org/10.1016/j.jssc.2009.11.011

[13] R. K. Zheng, H. Gu, B. Xu, and X. X. Zhang, The origin of the non-monotonic
      field dependence of the blocking temperature in magnetic nanoparticles, Journal of
      Physics Condensed Matter, 18 (26) (2006) 5905–5910.
      https://doi.org/10.1088/0953-8984/18/26/010

[14] S. Singhal and K. Chandra, Cation distribution in lithium ferrite ($LiFe_5O_8$) prepared
      via aerosol route, Journal of Electromagnetic Analysis and Applications, 2 (01)
      (2010) 51–55. https://doi.org/10.4236/jemaa.2010.21008

[15] A. K. Singh, T. C. Goel, R. G. Mendiratta, O. P. Thakur, and C. Prakash, Dielectric
      properties of Mn-substituted Ni–Zn ferrites, Journal of Applied Physics, 91 (2002)
      6626–6630. https://doi.org/10.1063/1.1470256

[16] S. Verma, J. Karande, A. Patidar, and P. A. Joy, Low-temperature synthesis of
      nanocrystalline powders of lithium ferrite by an autocombustion method using
      citric acid and glycine, Materials Letters, 59 (2005) 2630–2633.
      https://doi.org/10.1016/j.matlet.2005.04.005

[17] S. E. Shirsath, R. H. Kadam, A. S. Gaikwad, A. Ghasemi, and A. Morisako, Effect
      of sintering temperature and the particle size on the structural and magnetic
      properties of nanocrystalline $Li_{0.5}Fe_{2.5}O_4$, Journal of Magnetism and Magnetic
      Materials, 323 (2011) 3104–3108. https://doi.org/10.1016/j.jmmm.2011.06.065

Magnetic Oxides and Composites II
Materials Research Foundations **83** (2020) 193-232

Materials Research Forum LLC
https://doi.org/10.21741/9781644900970-9

Chapter 9

Multifunctional Ferrites: Synthesis, Behavior and Biomedical Applications

Amar K. Nandanwar[1,a], Kishor G. Rewatkar[1,b]

[1]Department of Physics, Dr. Ambedkar College, Deeksha Bhoomi, Nagpur – 440 010, MS, India

[a]amarkn13@gmail.com, [b]kgrewatkar@gmail.com

Abstract

Ferrites are a family of oxides with outstanding magnetic properties. Superparamagnetic iron oxide nanoparticles (SPIONs) are widely used in experimentation for various *in vivo* applications. All biomedical applications require high value of magnetization and the average size of particles should be less than 100 nm. To prepare multifunctional ferrite nanoparticles, sol-gel auto combustion synthesis technique was adopted. It was found that Cd-Ni substitution results in improving the saturation magnetization and reducing the coercivity values favorable for hyperthermia treatment. In this review, we have discussed magnetic parameters with increasing heating potential of nanoparticles.

Keywords

Magnetization, Multifunctional Ferrites, Magnetic Hyperthermia, SPION, Specific Absorption Ratio

Contents

1.   Introduction

Material science encompasses numerous disciplines, namely, Physics, Chemistry, Biology and Engineering and, is a real knowledge domain in nature [1]. The evolution of material science is usually an indicator of man's progress. Their urge to enhance the existing and replace the obsolete with novel materials often results in fewer resources. The emergence of nanoscience and engineering science as a leading technology of the 21st century has not solely accelerated the growth of material science, but also opened a new window to technology [2]. Today, nanoscience and engineering science become more substitutable with materials technology. Magnetism and magnetic materials are playing a crucial role in one's life. The magnetic industry is prepared to surpass the semiconductor industry with the proliferation of the latest gadgets based on magnetic materials and new innovations within the scope of nano-magnetism [3].

Nano-magnetic materials already exist in the applications of data storage, sensing element and device technologies; however more and more emphasis is given on exploring these in the life sciences and drug delivery [4]. So, it's solely stated that magnetism and magnetic materials at the nano regime attract the attention of researchers worldwide. It is used in the latest areas of research like spintronic devices; giant magneto resistance (GMR) based sensors, magnetic random access memories (RAMs) and alternative novel gadgets based on nano-magnetism [5].

Nano-science and engineering are providing us a new perceptive and control of matter at the atomic and molecular scale. Especially, nanoparticles have attracted more attention of researchers and techno habitual people due to their outstanding magnetic, optical and electronic properties [6]. The surprisingly small size range of these particles builds them ideal candidature for nano-engineering of surfaces, a large range of nano-sensors, textile industries, especially defense and security purposes and also the production of functional nanostructures. On the basis of nano-engineering size modifications in nanoparticles (NPs) are intrinsically accessible to use in the field of bio-medicinal applications, such as color contrast agents, for magnetic resonance imaging (MRI) and for targeted drug delivery in tumor therapy [7]. New materials are being designed and developed forever so new characterization methods are endlessly developed. In the development of new advance biomedical devices, the dynamic mechanism of the material magnetization has to be showing enough importance to permit any operations required for the device to work [6]. Slow relaxation of the magnetic nanoparticle is therefore essential.

The superparamagnetic nature of iron oxide nanoparticles (SPIONs) plays an important role in magnetic targeting [8]. The identification of the dynamic mechanism of magnetic relaxation method is the crucial step for the improvement in particle design with the essential in which blood flow exerts a force on it. External magnetic field targets SPIONs to the specific area to settle down over it [9]. The effect of magnetic field theory depends on various parameters like field strength, magnetic field gradient and latent volume of the SPIONs. Often, the gradient of magnetic field can be generated externally using AC-magnetic field [10].

Iron oxide magnetic nanoparticles are the subject of innovative research because of their versatile applications in an area such as biomedical and diagnostics. In the last three decades, the magnificent area of *in vitro* diagnostics has been increased by advanced applications of nano particles [11]. The magnetic nanoparticles need alteration to overcome on bacterial activity and increase the biocompatibility before the period of implementation for drug delivery applications.

The objectives of this chapter are four folds. The first objective is to synthesize transition metal substituted multifunctional magnetic iron-based $Cd_xNi_{1-x}Fe_2O_4$ (x = 0.0, 0.2, 0.4, 0.6, 0.8, 1.0) spinel ferrites by the sol-gel process using urea as a catalyst. The second objective is to study the basic principle of magnetic heating mechanism and magnetic parameters to increase heating potential. The third one is to characterize SPIONs. The fourth objective is to explore future applications of magnetic nanoparticles in the medical field

2.   Synthesis of superparamgnetic iron oxide nanoparticles (SPIONs)

Advance in synthesis of functional magnetic nanostructured materials has attracted tremendous interest in recent years due to their exciting properties and potential technological application in many fields [12]. The sol-gel technique is a chemical route; it offers good control over molecular homogeneity, elemental composition, powder morphology and uniformly nano-sized metal clusters, which is crucial for enhancing the saturation magnetization, coercivity, electrical conductivity and optical reflectance phenomenon of the nanoparticles. These advantages make the sol-gel method a favorable alternative over other chemical methods for the synthesis of ceramics nanoparticles [13].

When the metallic chemicals or ceramic compounds are dissolved in relevant solvent then this mixture is termed as a *sol.* When this *sol* of the metallic compounds is heated, it converts into viscous semi-liquid termed *gel*, from which conclusively the final powder can be obtained. It is very important to note that the mechanism of getting the final product depends on colloidal discipline in which the metallic cations are suspended in the final viscous liquid solution [14].

2.1   Sol-gel auto combustion technique

The sol-gel auto combustion method is widely used for the synthesis of nanoparticles over the other conventional methods. This technique, however, has low processing temperature, fast reaction time, better control over the grain size and better molecular homogeneity. To synthesize nanoparticles, the following steps are required;

Step (i) – Appropriate amounts of nickel nitrate ($Ni(NO_3)_2 \cdot 6H_2O$), iron nitrate ($Fe(NO_3)_3 \cdot 9H_2O$), and cadmium nitrate ($Cd(NO_3)_2 \cdot 6H_2O$) with Fe : Ni : Cd molar ratios of ($2 : 1\text{-}x : x$) are dissolved into 20 ml. de-ionized water under constant stirring. Urea powder is used as fuel, added in mixed solution of metal nitrates under constant stirring to prepare homogeneous compositions. This solution is often referred to as "precursor" since it is the basis of the following steps that lead to the final powder compound.

Step (ii) - This is an essential step to remove most of the solvent. The aqueous solution of metallic ions is gradually heated at 80 °C for 2 h on a hot plate under constant stirring to convert the solution into rigid brownish color gel. The gel is highly viscous in nature.

Step (iii) - The dry ash was obtained after the gel was heated in the microwave oven at power 800 watts for three minutes. This burnt ash forming compounds were crushed using a mortar and pestle for 6 h to obtain the final product.

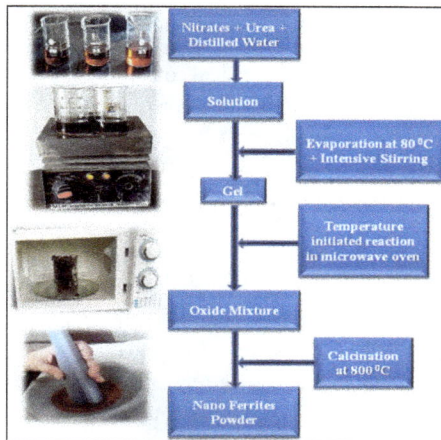

Fig. 1 Flow chart to prepare $Cd_xNi_{1-x}Fe_2O_4$ ferrite powder.

Step (iv) - This burnt powder was calcined in a muffle furnace at 800 °C for 4 h. After calcinations, again the powder was crushed for 3 h. The resultant powder is the required nano crystalline ferrite sample which is ready to use for characterization.

3. Theoretical aspects of magnetic relaxation

Magnetic relaxation is a process to generate heat for the treatment of tumors cells. Magnetic particle relaxation adoption was first studied by R Medal *et al.* in the year 1957 [14] and still researchers have been engaged in the same area. Recently, it has been found that magnetic nanoparticles are more effective in the field of cancer treatment as compared to other radiation therapy [22-24]. A minimum clinical trial required that magnetic nanoclusters should be biodegradable, chemically stable and well dispersed in ionic cellular cytoplasm. Most significant, in biomedical environment they must exhibit excellent superparamagnetic properties [15]; therefore, magnetic behavior of nanoparticles has become the first crucial step in feasible design for *in vivo* applications. It was found that the magnetic characterizations of magnetic nano-composites are strongly correlated with their thermal dissipative relaxation behavior. The dynamics mechanism of magnetic particles, dispersed in liquid medium in general, involves two fundamental relaxation mechanisms regarding the alignment of the magnetization vector with an applied external magnetic field: Brownian and Néel relaxation [16].

3.1 Brownian relaxation mechanism

The Brownian relaxation occurs by the physical rotation of particles suspended in a fluid in response to an applied magnetic field. The corresponding Brownian relaxation time (τ_B) is given by the following equation,

$$\tau_B = \frac{3V_{hyd}}{k_B T}\eta \qquad (1)$$

A complex magnetic nature expressed by the dynamic response of ferromagnetic nanoparticles suspended in a liquid. This complex magnetic nature varies with a small variable magnetic field with frequency (υ) [17]. Nevertheless, the dynamical susceptibility $\chi(\upsilon)$ depends upon the particle relaxation mechanism and relaxation times of magnetization [18]. Therefore, the relaxation time τ_B is experimentally accessible from the frequency (υ) dependant complex magnetic susceptibility of the poly-dispersed ferrofluids system of non- interacting rigid dipoles suspended in a liquid carrier. This term define as Debye terms with specific relaxation time:

Magnetic Oxides and Composites II
Materials Research Foundations **83** (2020) 193-232

Materials Research Forum LLC
https://doi.org/10.21741/9781644900970-9

$$\chi(\upsilon) = \frac{1}{3V\mu_0 k_B T} \sum_{i=1}^{N} \frac{m^2}{1-2\pi i \upsilon \tau_B} \tag{2}$$

Here, 'V' is the average volume of the spherical particles, and 'm' is the giant magnetic moment of single domain particle.

3.2 Néel relaxation mechanism

The Néel relaxation of particle is defined as the magnetic vector that may rotate within the particle with regard to relaxation time about crystal axis in the liquid [19].

Néel relaxation time (τ_N) is given by the following equation,

$$\tau_N = \tau_o \exp\left(\frac{\Delta E}{k_B T}\right) \tag{3}$$

Where, τ_o is a constant usually approximated as 10^{-9} s, and the Néel relaxation time (τ_N) varies exponentially with(ΔE), the potential energy barrier assuming uniaxial identical non interacting particles.

$$\Delta E = KV(1-h)^2 \tag{4}$$

Here, $h = (H/H_k)$ is the reduced field, H is the external magnetic field and H_k is the internal field due to anisotropy. In real ferro-fluids there is always a distribution of energy barriers due to a distribution of the magnetic anisotropy constant K as well as of the volumes of the magnetic particle cores V.

On the other hand magnetic anisotropy energy displays an influence on thermal energy. If the magnetic anisotropy energy is more with respect to thermal energy ($K > \Delta E$) then magnetic dipoles are no longer allowed to perform oscillations. Finally, the Néel relaxation becomes inconsequential. In this limit the MNPs can be considered as rigid particles having constant magnetic moment firmly attach to it. In this rigid dipole model, the torque is created by the magnetic moment due to the external driving magnetic field, which is transferred to the rigid particle. In an oscillating magnetic field, for instance, the MNPs have initiated oscillations with their Néel relaxation and finally will rotate as Brownian [20].

On the basis of relaxation time model, nanoparticles with relaxation time much faster than the measured time are called 'superparamagnetic nanoparticles' [21]. Some nanoparticles that do not relax by the Néel relaxation mechanism within the measurement time period are called 'blocked particles' and they do not contribute to the magnetization change within the considered time window.

In the hyperthermia process, when an external magnetic field is applied to the serum then heat energy is generated due to work done by the MNPs with relaxation time. Simultaneously, the generated energy dissipated locally by the magnetic nanoparticles. A high dissipation rate is achieved if both Néel and Brownian relaxations equally contribute [22]. To address the contribution of both relaxations, magnetic moment and particle rotations have to be considered. For that reason, the total effective relaxation time constant (τ) is given by the following equation:

$$\tau = \frac{(\tau_N \tau_B)}{(\tau_N + \tau_B)} \tag{5}$$

However, such a construction of the arithmetical equation for the fast mechanism appears unbelievable, particularly for colloidal magnetic nanoclusters. An external applied magnetic field, the inter-particle dipole-dipole interactions within a single nanocluster may outcome in an effective moment to physically oscillate the nanocluster. Even though the Néel relaxation time constant (τ_N) of the cores must be shorter than the Brownian relaxation time constant (τ_B) of the respective nanocluster, it is observed that the Brownian dynamics is always dominant at low frequency region [23].

According to the practical approach of magnetic particles in a fluid, the rotation of the magnetic dipole moment between different easy axes in Néel relaxation and Brownian relaxation have described rotation of entire particles. Fig. 2 shows the Néel rotation Vs. Brownian rotation of magnetic fluid components of the magnetic relaxation mechanism.

In recent years, few researchers discussed several practical and theoretical issues related to important aspects of relaxation time. It is very important to study temperature variation of the nanoparticles suspension during an applied external AC magnetic field, and the different models of fitting the experimental data to extract the nanoparticles heating power. From the study of Rudolf Hergt et al. [24], for particle diameter of 50 nm one may estimate experimental data, calculated anisotropy is about 1.5 J/kg, at frequency of 300 kHz and finally measured heating power of 450 W/g for the full hysteresis loop (100 kA/m). According to O'Grady et al. [25], the time-dependent magnetization of cobalt ferrofluids at 77 K using zero field cooling VSM and observed logarithmic time dependence of magnetization. Determination of relaxation time distribution of the colloidally dispersed nano-owes by Philip B et al. [26], produces distinct peaks in experimentally accessible time duration successively found. The Brownian relaxation is indicated by the peak of large time range, while the other peak of small time range can be attributed to Néel relaxation. Liquid dispersive solution measurements were carried out on the SPIONs offers fair hydrodynamic size distributions.

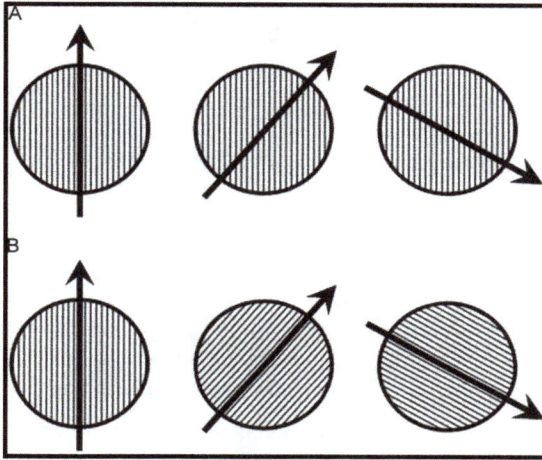

Fig. 2 *Néel rotation Vs. Brownian rotation mechanism of a magnetic fluid. (A) Néel rotation, (B) Brownian rotation.*

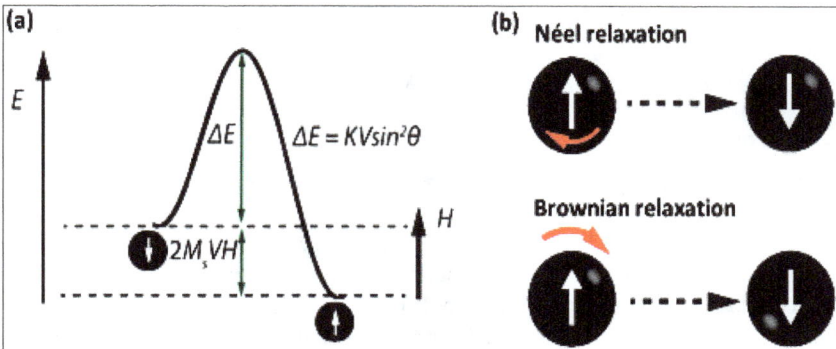

Fig. 3(a) *Representation of energy barrier for the single domain particles, (b). Relaxation processes that influence the heating properties of magnetic nanoparticles.*

Cayless *et al.* [27] reported Néel relaxation measurement in fine particle system at 4.2 K using RF SQUID magnetometer in a time window 60 s and 3600 s. The magnetic relaxation study prevails that *ln(t)* is collinearly dependent on relaxation for the compounds freeze prior while in samples frozen in zero fields did not show linearity [28].

Bogardus *et al.* [29] studied the relaxation of ferro-fluid liquid within the time range using a special measurement system.

4.   Key role of magnetic parameters to affect heating potential

In magnetism, there is a fundamental dissimilarity between intrinsic and extrinsic properties. Intrinsic properties such as the particle size effect, the magnetocrystalline anisotropy (K) and Curie temperature (T_c) are realized on atomic length and time scale. They can, in general, be considered as equilibrium properties. While, extrinsic magnetic properties, such as the coercivity (H_c), remanence (M_r), saturation magnetization (M_s) etc. reflect real structural morphology of the magnet. Therefore, for the influence of heating potential, intrinsic and extrinsic parameters are very essential to study.

Here we describe for better understanding and possible improvements in intrinsic as well as extrinsic parameters models on the basis of various magnetic behaviors.

4.1   Intrinsic parameters

In order to develop a relaxation model for the magnetic heating mechanism in the SPIONs, it is a prerequisite to know accurately the intrinsic magnetic parameters of the magnetic nanomaterials such as the particle size effect, magnetocrystalline anisotropy constant and Curie temperature, etc.

4.1.1   Particle size effect

In general study, magnetic hysteresis loss and relaxation loss are two mechanisms that are responsible to dissipated thermal energy in ferro fluid. In primary level, these mechanisms are studied in bulk materials of particles size $\approx 1\mu m$. In practice, particles in bulk form or larger in size affect many biomedical applications and they do not form a stable colloidal as well as very difficult to dispersed within tumors [16, 19]. In addition, inside the body bacterial defense mechanisms may activate due to larger particle size [20].

Hysteresis loss factor increases its utility in particles greater than their critical grain size, which forms strong magnetic behavior within particles, and it shows ferromagnetism. Nano particles of less than 20 nm size show superparamagnetic nature with single domain structure and so hysteresis losses become quickly negligible. Therefore, in this range, iron oxides particles show both Néel relaxation as well as Brownian relaxation mechanism.

Generally, for Brownian relaxation the most suitable particles range between 10 nm to 100 nm. Brownian relaxation prefers larger size range particles. The Néel relaxation occurs due to magnetic orientation within the magnetic particles therefore it is most significant in the smaller size particles. Nevertheless, without consideration of magnetic anisotropy constant this division cannot occur [30].

The suitable value of the total magnetic anisotropy constant (K) for magnetic nano particles in the range of about 20 kJ/m^3, then the MNPs below 7 nm diameter fail to satisfy this condition and can't generate significant heating. Therefore for Néel relaxation, the dominant mechanism, which is directly, depends upon particle diameter between 7 to 15 nm. Within this range, Néel relaxation carries an exponential dependence on the particles volume (Eq. 1). To generate large thermal energy using the Specific Absorption Ratio (SAR) is extremely sensitive to the dimension of (spherical) nanoparticles.

4.1.2 Anisotropy

The magnetic anisotropy is a directional dependent. If the MNPs possess magnetically anisotropic nature then it is very complicated to find magnetization equilibrium. In an anisotropic condition, an easy axis plays a vital role in saturation magnetization. In agreement with the nature of magnetic moments easy axis behavior, one can classify the various anisotropies like magnetically crystalline anisotropy, the shape anisotropy, stress anisotropy, and the exchange anisotropy.

According to anisotropic conditions, when an external magnetic field is applied to the magnetic materials, magnetic dipoles align to the direction of an easy axis within the particles [8]. Its dynamics have anticipated on the basis of classical approximation. It is observed that alignment of micro-spin is greatly influenced by the thermal fluctuations originating from the couples of the spin [5, 9–16]. MNPs dissolved in a viscous liquid. Fig 3(a) shows that the transition takes place between magnetic dipole moment to the easy axis of the MNPs to a state with anisotropic energy dependant free rotation.

The uniaxial anisotropy energy $E\,(\theta)$ is a simple type of magnetic anisotropy, that helps to identify the combined effects of both surface and volume anisotropy (Eq. (6)):

$$E(\theta) = (K_V V + K_S S)\,Sin^2\theta \qquad (6)$$

Where K_V is the volume anisotropic constant, V is volume the particle, Ks is the surface anisotropy constant, S is the particle surface and θ is the angle between the vector of the particles magnetic moments 'm' and an anisotropic axis.

In absence of surface anisotropy, magnetic anisotropy energy for volume of the material is given by eq. (7):

$$E(\theta) = (K_V V) \, Sin^2\theta \qquad (7)$$

In general, an external magnetic field is present, rotation of the individual particle's magnetic moment has been reached to the orientation corresponding to stability having a minimum energy, requires overcoming an energy barrier, $\Delta E \sim K_V V$. On the basis of magnetic anisotropy, derived from the above equations, one conclude that high oscillations performed by the MNPs that provide large energy dissipation in the given liquid. Fig. 3 (a) shows the dependency of magnetic energy of magnetic particles upon the direction of its magnetization vector [18].

4.1.3 Curie temperature

The Curie temperature (T_c) is the temperature above which ferromagnetic materials behave as paramagnetic materials with losing permanent magnetic order and the magnetism completely disappears. In hyperthermia treatment, Curie temperature shows a crucial role in the generation of heat within a tumor cell. It is observed that when tumor cells are exposed to a temperature between 42 °C – 47 °C (315 K– 319 K), the cells are destroyed [13-15]. Hence, it is favorable temperature for the nanoparticles which does not exceed this level. In actual practice, it is very difficult to measure the temperature of cells correctly. To overcome this difficulty, we can use the type of nano particles those depend upon Curie temperature. Thus when nanoparticles reached to Curie temperature, they automatically lose their magnetic property itself. Hence, without removing oscillating external magnetic field, heating get stop. Therefore it is called self regulated temperature. With the aim to improve self regulated hyperthermia treatment, the researchers have responsibility to synthesize new materials and structures of MNPs. Here we have discussed some of the reported research work on controlling (T_c) on MNPs. Under the title of 'self control hyperthermia' Martirosyan *et al.* [32] studied theoretically the role of structural parameters and compositions on (T_c) of superparamagnetic nano particles. It becomes clear that reducing the particle size might result in a decrease of (T_c). It was suggested by Apostova *et al.* [33] that introducing magnetic nanoparticles in the ferromagnetic materials result in lower (T_c). It witnessed because of reduction of the exchange magnetic interactions between the magnetic ions in the NPs. [33].

4.2   Extrinsic parameters

The most important extrinsic properties *i.e.* properties not only depend on the compositions, inter-metallic phases and crystal structures, but also depend on production parameters as *e.g.* heat treatment, particle size etc., can be evaluated from the full hysteresis loop as a function of temperature. Typical data deduce from a hysteresis loop are not only the saturation magnetization, the remanence magnetization, the coercivity but also the initial magnetization curve after thermal demagnetization.

4.2.1   Saturation magnetization

To carry out systematic information about the magnetic properties of sample, the magnetic hysteresis curve is often recorded till saturation of magnetic moment region [34]. The magnetic saturation (M_s) value is a characteristic of magnetic materials. Therefore large (M_s) value gives more authority to control magnetic moment of the MNPs. This valuable property of MNPs becomes applicable for treatment of cancer cells using an external magnetic field. On the other hand, (M_s) also depends on temperature. It was recorded that, magnetic moment shows fluctuation with change in temperature. For high temperature or above (T_c) the magnetic nature of MNPs disappears and shows non magnetic behavior. To overcome this problem Bloch's suggested the formula known as Bloch's law. He examines that on the basis of Boss - Einstein temperature condensation the variation in (M_s) value as the function of temperature is given by [35]:

$$M_S(T) = M(0) \left[1 - \left(\tfrac{T}{T_0} \right)^\alpha \right] \tag{8}$$

Where, T_0 is the temperature for zero magnetization ($M_s = 0$) and M (0) is the (M_s) at zero Kelvin. Several studies discussed the temperature dependence of the saturation magnetization in MNPs and reported deviations from Bloch's law at low temperatures [36].

The Bloch's law could be due to the magnetic particles interaction, variation of particle size, core cladding interactions and surface disordered that influences the surface anisotropy and asymmetric anisotropy of the particles. Chantrell *et al.* derived a relation for time dependant magnetization of system of nanoparticles after magnetization in the external magnetic field *H* for a time *t*. It was found that saturation magnetization is directly proportioning to log ratio between the sum of T_c and *T* to an absolute temperature

(T), Where (T_c) is the Curie temperate [37]. Finally we conclude that saturation magnetization is one of essential parameter for hyperthermia treatment.

4.2.2 Magnetic coercivity and remanence magnetization

Fig. 4 represents magnetic hysteresis loops of paramagnetic, superparamagnetic and ferromagnetic materials. The nanoparticles with superparamagnetic state having negligible remanence and coercivity possess a huge magnetic moment and appear like a giant paramagnetic atom which responds very quickly to magnetic fields oscillation. The essential condition for hyperthermia application, MNPs possesses high value of saturation magnetization, which results in improving thermal energy dissipation by the MNPs to the tumor cells. The various simulations were conducted by E. Obaidat *et al.* [38] to establish that the coercivity of given magnetic materials depends on the frequency of external oscillating magnetic field. When the external magnetic field is directed towards anisotropic easy direction with $\phi = 0$, then the results convey good agreement for the coercivity parameter [30]. These results are obtained by the following equation:

$$\mu_0 H_c = \mu_0 H_K \left(1 - K^{1/2}\right) \tag{9}$$

For random orientation case,

$$\mu_0 H_c = 0.48 \, \mu_0 H_k (1 - k^n) \tag{10}$$

Where, $n = 0.8$ and magnetic coercivity dimensionless parameter (k) that includes temperature and takes in to account the comprehensive rate of magnetic field. The various simulations methods studied by T. Nattermann *et al.* [39] to investigate the dynamic hysteresis area with various temperatures and frequency range [39]. The researcher also deduces suitable formula to calculate the area of the major loops using Stoner-Wolfforth based model [16]. It is analytically estimated that the oscillation of whole MNPs with an external applied magnetic field the easy axis was assigned to be fixed [40, 41]. Finally they conclude that, the result is in fair agreement with the Néel relaxation theory.

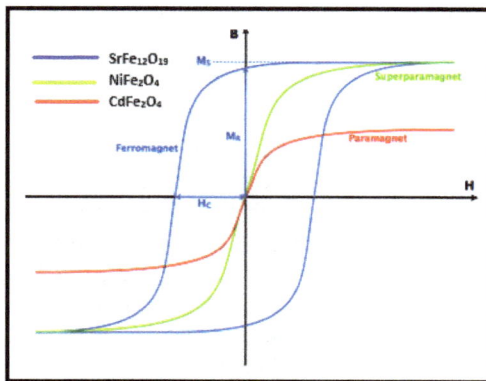

Fig. 4 Hysteresis loops of paramagnetic, superparamagnetic and ferromagnetic materials.

5.   Characterization

The structural analysis of the synthesized samples was carried out using X-ray diffraction (XRD) technique. X-ray diffraction patterns of all samples were recorded at room temperature using X-ray diffractometer (Bruker AXS D8 Advance XRD System, Cu-kα radiation, λ=1.5406 Å). The scanning 2θ range was between 20-80° (scan step of 0.02°). The unit cell parameter was determined using the Cellref software [8] and, the crystallite size was calculated from full width half maxima (FWHM) value of the reflection peak broadening using the Debye-Scherer's formula, [9]. The IR spectra of all samples were recorded at room temperature in wavenumber range of 400-4000 cm^{-1} using a FTIR spectrometer (Thermo Nicolet, Avatar 370). Surface morphology was examined using SEM micrographs and an energy dispersive spectroscopy technique (EDX, SEM- JEOL-6100) was used for elemental analysis. The recorded SEM micrographs assist to find out average particle size distribution using histogram curves. The average size of particles was calculated through a statistical analysis performed by counting about 250 particles by assuming spherical geometry. To identify the morphology of materials and to investigate their internal structure, transmission electron pictures were recorded using a TEM (JEOL-100), working at 100 kV and equipped with a high resolution sensitive camera. A vibrating sample magnetometer (Lakeshore VSM 7410) was used to record magnetic hysteresis loops of prepared samples under an applied magnetic field of up to ± 15 kilo Gauss and magnetic parameters were obtained from magnetic hysteresis loops. For hyperthermia treatment, magnetic particles were tested heating potential using a magnetic induction heating instrument.

5.1 Structural analysis

5.1.1 XRD analysis

Phase and structure of $Cd_xNi_{1-x}Fe_2O_4$ ($0.0 \leq x \leq 1.0$) ferrites were elucidated by X-ray diffraction technique and the obtained XRD patterns are shown in Fig. 5. XRD patterns of $NiFe_2O_4$ and $CdFe_2O_4$ were well matched with standard JCPDS file numbers 10-0325 and 22-1063 of ferrites. It can be seen that the observed peaks are sharp, reflecting highly crystalline character of the prepared samples. The reflection peaks were indexed as (220), (311), (222), (400), (422), (511), (440), (620) and (533), which are characteristics of single phase cubic spinel structure. XRD analysis confirms formation of cubic spinel phase with space group Fd_3m (227). It is interesting to note that the XRD-peak shifts towards small diffraction angle as shown in Fig. 6. Basically we have just focused on the average shifting of individual peaks of various spectra. We can conclude that (311) prominent peak shows larger shift towards lower diffraction angle. On the basis of Bragg's Law we conclude that shifting of diffraction peaks with an increase in Cd content (x) may be attributed to larger ionic radii of Cd^{+2}(0.97 Å) compared to Ni^{2+} cations (0.69 Å) [33, 34]

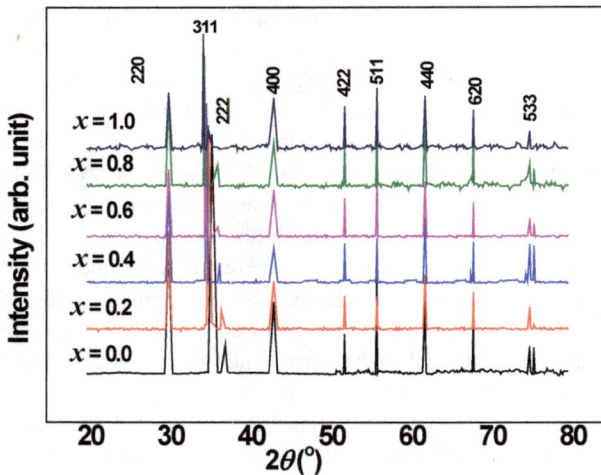

Fig. 5 XRD patterns of $Cd_xNi_{1-x}Fe_2O_4$ ($0.0 \leq x \leq 1.0$) spinel ferrites

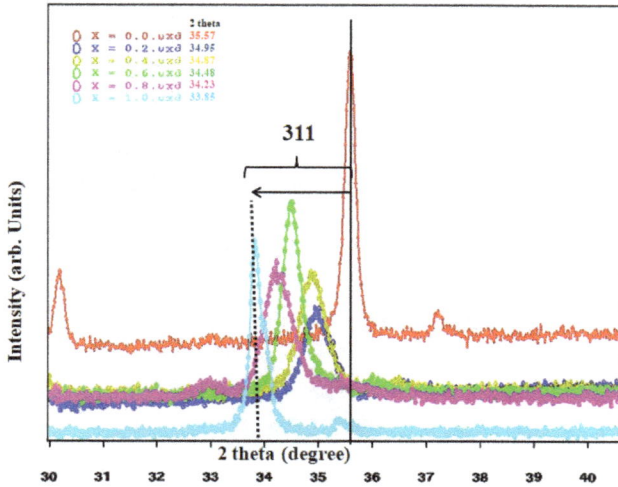

Fig. 6 Shifting of (311) peak towards lower 2θ value.

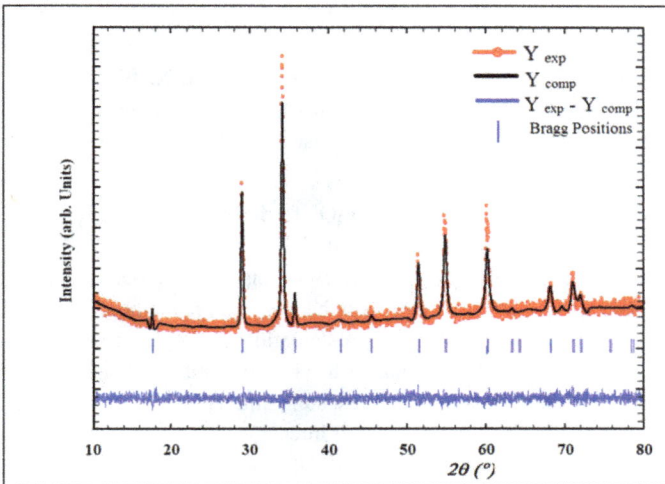

Fig. 7 Reitveld refinement pattern of $NiFe_2O_4$ ferrite sample.

Table 1 Structural parameters of $Cd_xNi_{1-x}Fe_2O_4$ (0.0 $\leq$ x $\leq$ 1.0) ferrite sample.

Formula	Molecular weight	Lattice parameter	Bulk density	X-Ray density	Porosity	Particle Size
	g/mol	Å	g/cm^3	g/cm^3	%	nm
$NiFe_2O_4$	1162.00	8.3501	3.445	5.349	35.00	12.81
$Cd_{0.2}Ni_{0.8}Fe_2O_4$	1166.34	8.4811	3.280	5.330	33.40	14.10
$Cd_{0.4}Ni_{0.6}Fe_2O_4$	1169.87	8.5183	5.500	5.500	32.10	14.47
$Cd_{0.6}Ni_{0.4}Fe_2O_4$	1173.40	8.6125	3.452	5.550	31.80	19.48
$Cd_{0.8}Ni_{0.2}Fe_2O_4$	1176.94	8.6630	3.375	5.668	30.45	21.66
$CdFe_2O_4$	1180.47	8.7382	3.592	5.733	27.34	23.19

The full width at half-maximum (FWHM) of the strongest XRD peak (311) value was estimated to calculate average crystallite size of this synthesized powder using the Debye-Scherer's formula,

$$D = \frac{0.9\,\lambda}{\beta cos\theta} \tag{11}$$

Where (D_{xrd}) is crystalline size, (λ) is denoted as wavelength of Cu-kα radiation; (β) is the FWHM and θ is the diffraction angle of strongest characteristic peak. Crystalline size (D_{xrd}), Lattice constant (a), X-ray density (D_x) and surface area (S) were determined from the XRD patterns and are depicted in Table 1. The typical values of crystalline size were found to be 12.81 – 23.19 nm, which are in good conformity with the reported value by M. Arshad et al. [42]. This indicates that impurity free Cd-Ni ferrites were obtained at 800 °C. Similar results have revered that, in case of Cd-Zn ferrites prepared by soft chemical synthesis method [26], here an increase in the lattice parameter directly affected by the cadmium concentration.

The Rietveld's refinement pattern of $NiFe_2O_4$ ferrite nanoparticles was done using the Full Proof Suit software and obtained pattern is shown in Fig 7. Here, Successive refinement of the structural analysis such as cation distribution, occupancies, particle size and lattice parameter was estimated through the least square method. The peak shape was set as to be pseudo-Voigt with range for calculation of single reflection in the unit of FWHM (full width half maxima) equal to 8. Here the red color lines with dots circles represent an experimental data and a black solid line shows measured intensities. The difference between measured and calculated intensities represented at the bottom by the blue line. The fitness of pattern has been checked for given XRD data. The goodness of fit (GoF) is defined as:

Materials Research Forum LLC
https://doi.org/10.21741/9781644900970-9

$$GoF = \frac{R_{wp}}{R_{exp}} \qquad (12)$$

Where *(Rwp)* is the weighted residual error and *(Rexp)* is the expected error in the pattern. Refinement has been continually refined until convergence was reached for a low value of *(GoF)*, which was confirmed by the goodness of refinement. We have observed low values of *GoF* in between 4.36 to 5.13, which justify the goodness of refinement [43]. The obtain lattice parameter values from the refinement are listed in Table 1. It is observed that the lattice parameter increases with increasing concentration of Cd^{2+} ions. As pointed out by F S. Tehrani *et al.* [44], that variation in ionic radii could be attributed due to discrepancy between ionic radii Cd^{2+} (0.97 Å) and Ni^{2+} (0.69 Å). In addition to the shift of prominent peak (311) towards small diffraction angle with Cd^{2+} content (*x*) may be credited with larger lattice parameter. It is observed that that the XRD analysis of the studied compounds are in good agreement with the Rietveld refinements for the cation distribution reported by S.V.A. Prasad *et al.* [45].

5.1.2 FTIR analysis

The recorded FTIR spectra at room temperature in the wave number range of 4000-380 cm^{-1} for the Cd-Ni ferrite samples are shown in Fig. 8. In the IR band, variation in spectrum is usually assigned to vibrations of ions in the crystal [46, 47]. Two main broad cation-oxygen bands are seen in the IR spectra of $Cd_xNi_{1-x}Fe_2O_4$ (*x* = 0.4 and 0.8) spinel ferrites between 540-600 (v_1) and 390-450 (v_2). In the present case, the peaks appear at 410 cm^{-1} and 420.1 cm^{-1} (v_2) and 579.14 cm^{-1} and 559.25 cm^{-1} (v_1) in *x* = 0.4 and 0.8 compositions, respectively [48].

The broad band near 3450 cm^{-1} corresponds to O-H stretching vibrations ascribed to water molecules present in a sample. Sabale *et al.* [49] reported an IR spectroscopy study for the application of hyperthermia treatment and agreements with the above mentioned respected band assigned values. The shift in the band positions indicates that the Cd^{2+} ions have been incorporated in the spinel lattice, which helps the formation of nanocrystalline $Cd_xNi_{1-x}Fe_2O_4$ spinel ferrites and shifting of metallic cation from tetrahedral to the octahedral site of Cd^{2+}. The distribution of metallic cations generates a compressive strain between the two sites (A and B sites). This transformation of cation distribution depends on the variation of lattice constant and bond lengths [50].

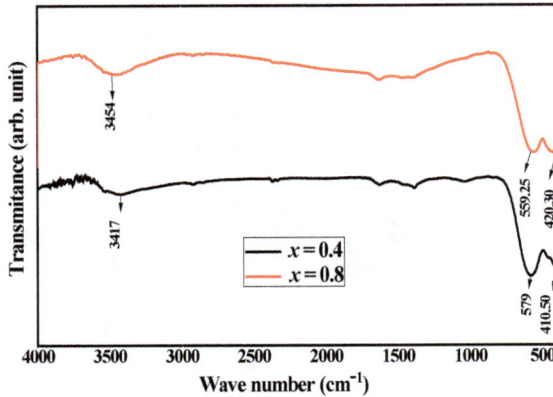

Fig. 8 FTIR spectra of $Cd_xNi_{1-x}Fe_2O_4$ (x = 0.4 and 0.8) spinel ferrites.

The variation observed in both v_1 and v_2 which caused by the perturbation to take place in the $Fe^{3+} - O^{2-}$ bonds by the replacement of ions as shown in Table 3. It is observed that the free energy play a vital role in transformation of cations. Unless their free energy has a minimum rate the migration of ions between the two sites are possible.

5.2   SEM micrographs

Typical FE-SEM images of $Cd_xNi_{1-x}Fe_2O_4$ (x = 0.4 and 0.8) samples annealed at 800 °C are shown in Fig.9. It is clear from Fig.9 that formed grains have non homogeneous distribution and an almost non uniform shape in the range of 12 to 40 nm as shown in histogram plot in Fig.10. The FE-SEM micrographs shows that formed grains size are affected by the concentration with increasing grain size listed in Table 1. M. G. Naseri *et al.* [51] showed that the inter cooperation between magnetic nano particles showed particle clusters area in images. It is also observed that function of calcinations temperature has favorable for grain size agglomeration.

Fig. 9 Scanning Electron Micrographs of $Cd_xNi_{1-x}Fe_2O_4$ (x = 0.4 and 0.8) spinel ferrites.

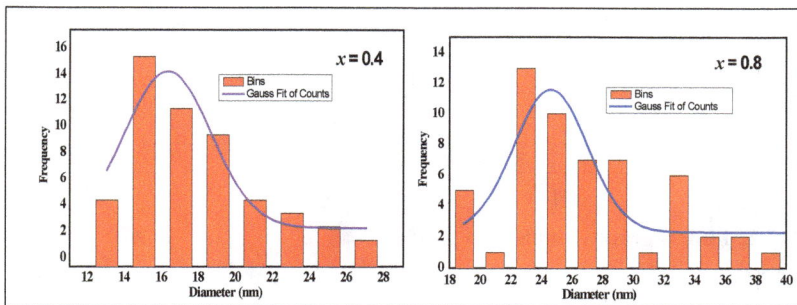

Fig. 10 Histogram curves for grain size distribution of $Cd_xNi_{1-x}Fe_2O_4$ (x = 0.4 and 0.8) spinel ferrites.

5.3 Elemental analysis by EDX

The element composition of the synthesized $Cd_xNi_{1-x}Fe_2O_4$ (x = 0.4 and 0.8) spinel ferrites has been investigated by the energy dispersive X-ray analysis (EDX) and obtained spectra are shown in Fig. 11. The EDX spectra of $Cd_xNi_{1-x}Fe_2O_4$ (x = 0.4 and 0.8) confirmed the presence of Cd, Ni, Fe, and O in the sample. For both (x = 0.4 and 0.8) samples, the EDX result shows almost the same ratio of chemicals taken in stoichiometry. The atomic weight percentage of obtained elements are listed in Table 2, for both (x = 0.4 and 0.8) samples found values matched to the expected composition ratio. Actually the measured Fe/Ni atomic ratio was found to be very close to the $Cd_xNi_{1-x}Fe_2O_4$ (x = 0.4 and 0.8) compositions [49].

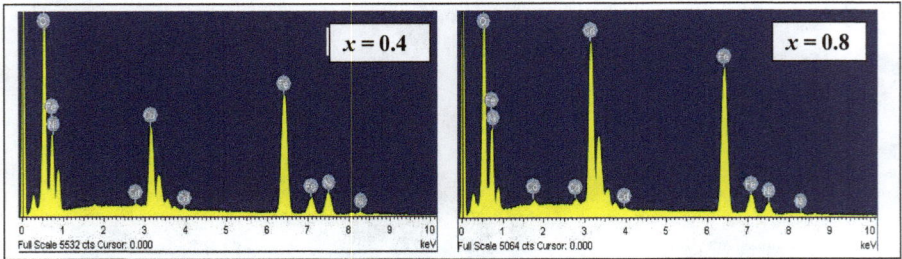

Fig. 11 EDX spectra of $Cd_xNi_{1-x}Fe_2O_4$ (x = 0.4 and 0.8) spinel ferrites.

Table 2 Atomic and weight percentages.

Elements	Weight %		Atomic %	
$Cd_xNi_{1-x}Fe_2O_4$	x = 0.4	x = 0.8	x = 0.4	x = 0.8
O K	25.55	25.35	57.86	59.35
Fe K	43.34	41.34	28.12	27.73
Ni L	13.57	05.32	08.37	03.40
Cd L	17.55	27.98	05.66	09.52
Total	100.00	100.00	100.00	100.00

5.3.1 TEM micrographs and SAED patterns

The TEM micrographs of $Cd_xNi_{1-x}Fe_2O_4$ (x = 0.4 and 0.8) ferrites are shown in Fig. 12. The TEM images of the samples show agglomerated nanocrystallites of quasi-spherical shape dispersed uniformly over the surface having size ranges approximately between 20 nm to 60 nm. N. Sharma *et al.* [47] reported that magnetic nanoparticles of ferrites are generally found agglomerated due to the attraction force among magnetic nanoparticles [48]. The observed particle size data from TEM micrographs of mixed Cd-Ni (x = 0.4 and 0.8) ferrites are in fair agreement with the particle size determined from XRD data. For more clarity, the high-resolution electron microscopy (HR-TEM) was used to check crystalline quality by observation. It is clearly observed from the HR-TEM micrographs (Inset of Fig. 12) that a set of parallel fringes across the crystal structure. These uniformity distributed fringes prominently show the symmetric arrangement of lattice places. Finally, we conclude that good formation of spinel polycrystalline structure with proper chemical compositions, without Frenkel defects, such as plane dislocations, point defects, crystalline stacking faults, grain boundary defects such as amorphous outer-shell [48]. The Selected Area Electron Diffraction (SAED) patterns of both (x = 0.4 and 0.8) samples are shown in Fig. 13, indicating that formed Cd-Ni nanocomposites are well crystallized. No other rings except for the preferred planes are found in SAED pattern. No other rings except for the preferred planes are found in the pattern.

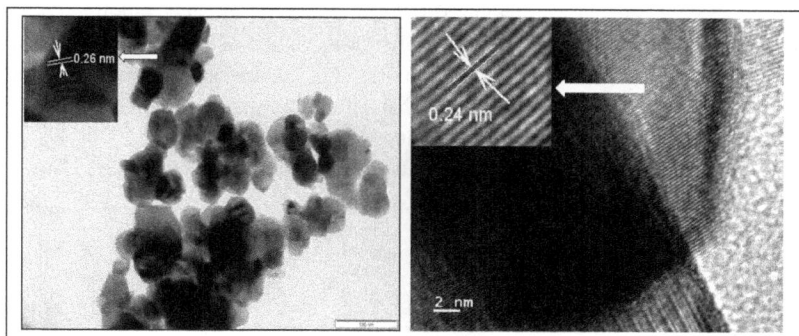

Fig. 12 Transmission Electron Micrographs of $Cd_xNi_{1-x}Fe_2O_4$ (x = 0.4 and 0.8) spinel ferrites.

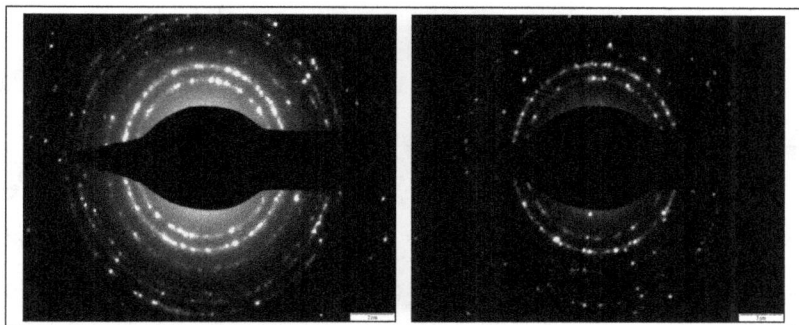

Fig. 13 Selected area electron diffraction (SAED) patterns of $Cd_xNi_{1-x}Fe_2O_4$ (x = 0.4 and 0.8) spinel ferrites.

5.4 Magnetic properties

The magnetic hysteresis loops of $Cd_xNi_{1-x}Fe_2O_4$ (x = 0.2, 0.4, 0.6 and 0.8) nanoparticles were recorded using a vibration sample magnetometer (VSM, Lakeshore 7410) at 300 K under an applied magnetic field up to ± 15 kG. Fig. 14 (a) shows relation between magnetization and applied magnetic field (*M-H* loops). It can be attributed to the competition of ferromagnetic ions such as Fe^{3+} and nonmagnetic ions such as Cd^{2+} ions in the occupancy of the tetrahedral and octahedral sites. Fig. 14 (b) shows the hysteresis loops up to $H = \pm 1$ kG. As expected, they exhibited superparamagnetic behavior. The variation in the value of the (H_C) with Ni^{2+} depends on the concentration and particle size. It can be explained on the basis of domain structure, cation distributions and the anisotropy of the crystals [53]. The narrow loop area of the magnetization curve shows the soft magnetic nature of prepared samples. The values of saturation magnetization (M_s), coercivity (H_c) were calculated from hysteresis loops and magnetic parameters listed in Table 3. It is clear from Table 3 that saturation magnetization (M_s) decreasing with Cd^{2+} contents i.e., the maximum value of (M_s) is found in the $Cd_{0.2}Ni_{0.8}Fe_2O_4$ sample. Akash et al. also found similar results in Cd-Zn doped nickel ferrite [53].

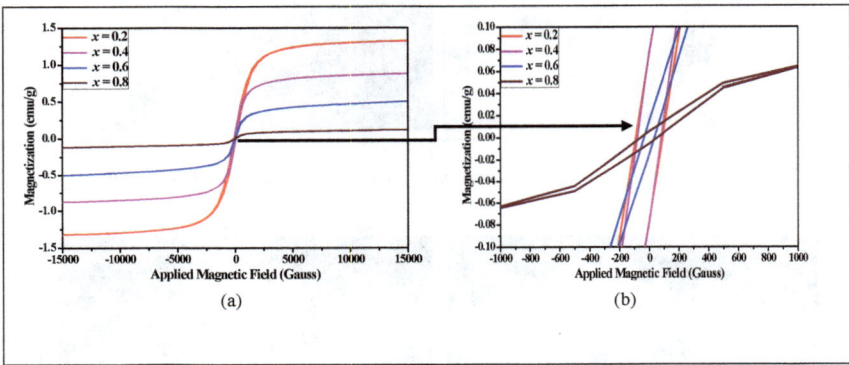

Fig. 14 Magnetic hysteresis loops of $Cd_xNi_{1-x}Fe_2O_4$ $(0.2 \leq x \leq 0.8)$ spinel ferrites.

Table 3 Cationic distributions and magnetic parameters of $Cd_xNi_{1-x}Fe_2O_4$ (x = 0.2, 0.4, 0.6 and 0.8) spinel ferrites.

Cd Content	Cation distributions	Crystallite size	Lattice parameter	Saturation magnetization	Coercivity	Magnetic moment	SQR (M_r/M_s)
(x)		(nm)	(Å)	(emu/g)	(Gauss)	(Am²)	
0.2	$(Cd_{0.2}Fe_{0.8})_{tet}$ $[Ni_{0.4}Fe_{1.6}]_{oct}$	14.10	8.4811	51.70	90.93	2.27	0.0015
0.4	$(Cd_{0.4}Fe_{0.6})_{tet}$ $[Ni_{0.3}Fe_{1.7}]_{oct}$	14.47	8.5183	48.99	80.62	2.24	0.0016
0.6	$(Cd_{0.6}Fe_{0.4})_{tet}$ $[Ni_{0.2}Fe_{1.8}]_{oct}$	19.48	8.6125	22.86	56.36	1.089	0.0007
0.8	$(Cd_{0.8}Fe_{0.2})_{tet}$ $[Ni_{0.1}Fe_{1.9}]_{oct}$	21.66	8.6630	4.90	35.02	0.243	0.0010

The saturation magnetization (M_s) of Cd-Ni nanoparticles (at 300K) calculated from magnetic hysteresis loops for x = 0.2, 0.4, 0.6 and 0.8 M_s were found to be 51.7, 48.99, 22.86 and 4.90 emu/g, respectively. These values are less than the corresponding values reported by Karanjkar M. *et al.* for bulk Cd-Ni ferrites [54]. The nanoparticles revealed almost zero retentivity and low coercivity. The saturation magnetization in magnetic materials depends on combination of several factors as discussed in below sections.

1) Very small particles in the distribution and/or pinning of spins on the surface of the particles,

2) Cation redistributions of metallic ions among octahedral and tetrahedral sites,

3) Canted spins within the sub lattice and,

4) Curie temperature of samples [55].

According to Neel's ferrimagnetic model, total magnetic moments of nanoparticles basically depends on cations distribution over two sub lattices (tetrahedral (A) site and octahedral (B) site) which have opposite alignments [56]. Therefore, the total magnetic moments (n_B) is given as;

$$n_B = M_{oct} - M_{tet} \tag{13}$$

Where, M_{oct} and M_{tet} are the magnetic moment on octahedral and tetrahedral sites, respectively. According to Neel's theory the magnetic moment is zero for Cd^{2+}, 2 μB for Ni^{2+} and 5 μB for Fe^{3+} ions.

Our main objective of present investigation is to study the effect of non magnetic Cd^{2+} substitution on magnetic properties of $Cd_xNi_{1-x}Fe_2O_4$ (x = 0.2, 0.4, 0.6 and 0.8) spinel ferrites. Magnetic properties study showed that substitution of Cd^{2+} and replacement of magnetic Ni^{2+}can exhibit supermagnetism. It can be seen from Table 3 that the saturation magnetization of the synthesized nanocrystals has declined continuously with the increase in Cd^{2+} ions. These results agree with FTIR analysis, Reitveld XRD analysis, and cation distributions. In general $NiFe_2O_4$ ferrite having mixed spinel structure and $CdFe_2O_4$ ferrites having normal spinel structure; where Cd^{2+} ions with zero magnetic moment prefers tetrahedral sites while magnetic Ni^{2+} and Fe^{3+} ions prefer octahedral sites [57].

According to the Reitveld analysis and magnetic study (Table 3), magnetic Ni^{2+} occupied on octahedral site which affects the presence of Fe ions causes transfer of Fe^{3+} ions from octahedral to the tetrahedral site. This increase concentrations of Fe^{3+} ions in tetrahedral sites hence reduction in magnetization at the octahedral sub-lattice. This induces disorientation in the magnetization of the nanocrystals. This could be explained on the basis of net magnetic spin moment due to two sites interactions. The magnetization of B sub-lattice decreases, while for the A sub-lattice increases with substitution of higher magnetic moment of Fe^{3+} (5 μB) as compare to Cd^{2+} (0 μB). Consequently, this results in decease of the net overall magnetization due to antiferromagnetic coupling. Since magneto crystalline anisotropy affects the coercivity (H_c) of magnetic particles. The SQR (M_r/M_s) values for all samples are listed in Table 3; all samples possess the SQR < 0.5 confirmed the formation of multi domain structure. Thus these nanoparticles have shown superparamagnetic behavior (at 300 K) [58].

Desai *et al.* [58], investigated the high field saturating behavior of bulk and nano $CdFe_2O_4$ ferrite indicates the presence of two components, (a) superparamagnetic and (b) antiferromagnetic. The Mössbauer spectroscopy investigation of $CdFe_2O_4$ showed superparamagnetic behavior. In this study characteristically be described as a transformation from multi-domain to the single-domain structure [25, 38, 78]. MNPs having single domain structure and superparamgnetic characteristics at room temperature are very useful for biomedical applications as targeted drug delivery [41], bio-molecule separation [42] and magnetic resonance imaging systems [43] etc.

6. Heat generation with managing magnetism: specific absorption ratio (SAR)

The specific absorption rate (SAR) or specific loss power (SLP) is defined as the amount of energy dissipated by the MNPs per unit mass of the particles per unit time. Therefore, measurement of specific loss power indicates an estimation of increase in temperature of the NPs in response to the applied magnetic field and hence to simultaneously observe the efficiency of the NPs. There are several physical parameters like the magnetic field strength, the magnetic field gradient, hysteresis squareness ratio and dimensional magnetic properties of the nanoparticles etc. affect effectiveness of therapy [59]. In actual practice, oscillatory magnetic field generates magnetic field gradient and that modifies the main magnetic force exerted on NPs.

6.1 Magnetic hyperthermia measurement by calorimetric method

In order to compare different materials, it is necessary to define a parameter that describes how these materials respond to external AC magnetic fields. As such, most hyperthermia experiments use a parameter called the specific absorption ratio (SAR). The SAR can be described as the absorption of electromagnetic energy per unit mass when exposed to a particular field frequency [60].

The SAR can be mathematically defined as: to evaluate the hyperthermia properties of magnetic nanoparticles, "Calorimetric Effect" is the most commonly adopted method. In this technique, a high frequency AC magnetic field of specific amplitude is applied externally for a specific time interval and finally it raises temperature of the sample as shown in Fig. 15. The magnetic induction heating system consists of water coolant to control external temperature within the coil. A high power radio frequency (200 to 400 KHz) is supplied through helical spring consist of water coolant coil, called as magnetic induction heating system. A fiber optical temperature probe is typically used to measure accurate temperature of the sample holder. A thermally insulated container is used to keep a sample to avoid heat loss to the environment during the measurement, and the heating efficiency can be measure as Specific Loss Power (SLP) is directly calculated from the temperature derivative over time at instant $t = 0$ as,

$$SAR = \frac{CV_s}{m}\frac{dT}{dt} \tag{14}$$

Where, 'C' is defined as the volumetric specific heat capacity of the sample solution, 'm' is the mass of the given sample, 'Vs' is the volume of the sample, and 'dT/dt' is the change in temperature versus time curve's slope.

In our experiment, 20 mg/ml ferrite concentration was used. From the initial slope of calorimetric method measurement, SAR has been calculated. Fig. 16 represents the magnetic induction heating curves of $Cd_xNi_{1-x}Fe_2O_4$ ($0.0 \leq x \leq 0.8$) spinel ferrite samples. It is observed from Fig.16 that that magnetic induction heating curves are not linearly proportional to an applied magnetic field (H), in fact initially it increases fast with time and then after some time it increases very slowly. The variation of SAR with field amplitude changes are observed from linear to second order. The staking curves in increasing ordered which may be credited to changing in the composition of ferrofluids [61]. It is very important to note that for superparamagnetic nanoparticles, SAR result is totally depending upon the relaxation process. These processes in ferrofluids are either due to reorientation of the particle's dipole moment or due to frictional losses. To find the time verses temperature relation, the relaxation time is given as, the Brownian relaxation time:

$$\tau_B = \frac{3V_{hyd}}{k_BT}\eta \tag{15}$$

and, Néel relaxation time,

$$\tau_N = \tau_o e^{\frac{KV}{kT}} \tag{16}$$

Where (η) is the viscosity of carrier liquid solution, (r) is the hydrodynamic radius of particle, (k) is the Boltzmann's constant, (τ_o) is the time constant, 10^{-9} second, (V) is the particle volume, and (K) is the anisotropy constant. The specific loss power (P) corresponding to relaxation i.e. Néel and Brownian and is approximately given by,

$$P = \frac{(mH\omega\tau)^2}{[2\tau k_T \rho V(1+\omega^2\tau^2)]} \tag{17}$$

Where, (m) is the particle magnetic moment, (ω) is the angular frequency, (H) is the AC field amplitude and, (ρ) is the density of the ferrite. At the resonant condition, P will reach maximum when, $\omega\tau = 10^5$. In our experimental system given particle size in the range of $10 - 40$ nm and an applied AC magnetic frequency $f = 300$ kHz. Therefore the contribution due to Brownian losses is very small as also confirmed by the polyacrylic gel experiment investigated by Thiesen *et al.* [62]. If the relaxation time depends on volume of sample (V) and anisotropic constant (K) then total loss factor contributed by Néel relaxation in given SAR value [62].

Here, the particle size is obtained from XRD calculation and magnetic moment values are from VSM measurement (Table 3). If we look out over the varying particle size with Cd^{2+} concentration (x) then the variation of power factor (P) follows the magnetic dipole moment. The variation of power factor with Cd^{2+} content (x) should be the same as magnetic moment until the anisotropy constant (K) does not change with Cd^{2+} concentration. The magnetic heating induction graphs (Fig. 16) show the rapid drop of SAR value around $x = 0.8$ may be due to variation in anisotropy (K) with Cd^{2+} concentration (x) are analogous to the results obtained by Penoyer for the variation of (K) with Mn – concentration. The SAR value shows the variation in magnetization and anisotropy together with different compositions. It is interesting to note that SAR is enhanced by 20 % in $Cd_xNi_{1-x}Fe_2O_4$ $(x = 0.0)$ compared to Fe_3O_4 [63].

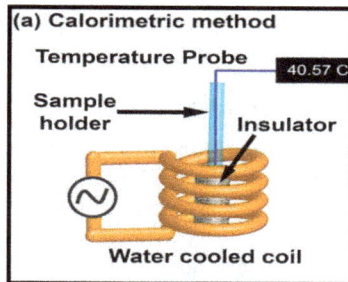

Fig. 15 *Schematic diagram of the calorimetric method used to evaluate the heating efficiency of magnetic nanoparticles.*

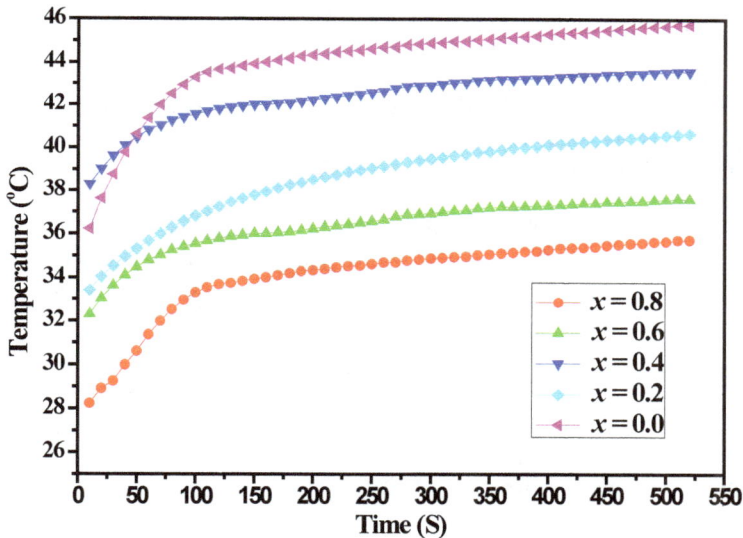

Fig. 16 *Magnetic induction heating curves of $Cd_xNi_{1-x}Fe_2O_4$ $(0.0 \leq x \leq 0.8)$ spinel ferrite samples.*

7. Applications of magnetic nanoparticles in medical field

Magnetic nanoparticles (MNP's) are exhaustively used in many biomedical applications such as targeted drug delivery as internal and external magnetism can be allied to the target, magnetic contrast agent can be introduced in forming the images by magnetic resonator.

7.1   Hyperthermia using magnetic nanoparticles

There is no involvement of any drug in this type of approach to treating cancer cells; rather heat is induced in MNP's under the influence of external applied magnetic field. This type of approaches destroys the cancer cells as they are temperature sensitive. The general principle involved in hyperthermia is specific selection of tumor body part for selecting nanostructure targeting agents influenced by external magnetic AC-field, to increases the temperature to above 40 °C emanating tumor dissemination [79].

7.2   Drug delivery

Drug delivery means spreading targeted drugs in proximity to tumor. Various materials are developed for different targets such as dendrimers, plasmon nanoparticles, liposomes, polymer capsules, micelles etc. [64-70]. In totality MNP's are not at par above its counterpart. MNP's can be guided using an external applied magnetic field in addition to diverging drugs at the infected site and the whole process can be recorded using magnetic resonance imaging (MRI) technique [71-72].

7.3   Targeting multifunctional carriers

The number of analeptic structures can be attached to the magnetic nanoparticles. While performing delivery to the target magnetic behaviors of nanoparticles provide journey through MRI [55].

7.4   MRI contrast agents

Superparamagnetic nanoparticles are used as contrast agent in MRI. Darkening of weight images is an outcome prime spin-spine relaxation time (T_2) of protons in the water in presence of external magnetic field.

7.5   Cardiovascular disease imaging

The composition of nanoparticles by plaques [11, 16] has been helpless to record the lesion-prone arterial sites. The clinical reports on MRI using SPION's are found to be helpful in calculating the risk of acute chlorosis [74-75] from VCAM-1 targeted peptide series, the specific binding of nanoparticles and MRI contrast showed an enhancement of lesion in mice as well human carotid arteries patches [76-77]. In MRI, as contrast agents magnetic nanoparticles have been proposed for several clinical applications in cardiovascular medicine including myocardial injury, atherosclerosis and, another vascular disease [72-73].

7.6 Molecular imaging

All molecular and cellular MRI techniques seemed to be non-invasive in vivo characterization and quantify biological process [75]. Molecular imaging allowed highly sensitive and to the point monitoring targets and their active response in carcinogenesis. MRI of micro phase in lymph nodes MNP's, allowed to record at millimeter level resolution in meta-stables in non-enlarged lymph nodes [14], which is out of reach of detection by other imaging process.

7.7 Conventional chemotherapeutic agents

Conventional drug line doxorubicin and methotrexate etc. can also be encapsulated in MNP's for treatment in rheunmatiod arthritis, breast tumor and malignant posted cancers. The drug loading capacities and its release profile can now be designed and monitored by fixing structural and chemical bonding with the nanoparticles.

8. Limitations of SPIONs for drug delivery

Every new technique had faced various challenges, as well some limitations in actual practice. The challenges associated with the exploitation of magnetic nanoparticles for biomedical treatment as obtain control over in vivo behavior and effects of penetration of the external magnetic field deep into the body. The drug delivery applications contain as part of (i) improving biocompatibility with human body, (ii) achieving control over bio-elimination, (iii) improving drugs targeting on specific body parts, (iv) reduce the pollydispersity of the nano particles in body parts.

The difficulties in drug delivery administration is systematic distribution of therapeutic drugs, lack of specific drug availability towards pathological sides, a specific concentration of drugs, non-toxicity and any advance side effect.

Conclusions

Cadmium doped spinel ferrites with composition $Cd_xNi_{1-x}Fe_2O_4$ (x = 0.0, 0.2, 0.4, 0.6, 0.8, 1.0) were synthesized using the sol-gel process. The structural and magnetic parameters were studied; lattice parameter and crystallite size were found to increase, whereas saturation magnetization values were observed to decrease with cadmium substitution. The magnetic analysis of x = 0.2, 0.4, 0.6, 0.8 compositions show that formed ferrites are magnetically soft and possess multi-domain structure.

References

[1] P.L. Alan and J. Youtie Jan, How interdisciplinary is nanotechnology, Journal of Nanoparticles Research, 11 (5) (2009) 1023-1041. https://doi.org/10.1007/s11051-009-9607-0

[2] E. Aghion and B. Bronfin, Magnesium alloys development towards the 21st century, Materials Science Forum. Vol. 350. (2000) 74-80. https://doi.org/10.4028/www.scientific.net/MSF.350-351.1

[3] F. Haldar, O. Arabinda, and A. Adeyeye, Deterministic control of magnetization dynamics in reconfigurable nanomagnetic networks for logic applications, ACS Nano, 10 (1) (2016) 1690-1698. https://doi.org/10.1021/acsnano.5b07849

[4] E. Wernsdorfer and S. Wolfgang, from micro-to nano-SQUIDs: applications to nanomagnetism, Superconductor Science and Technology, 22 (6) (2009) 064013-064018. https://doi.org/10.1007/s10853-006-6564-1

[5] E. Chen, J. Ching, Y. Haik, and J. Chatterjee, Development of nanotechnology for biomedical applications, Emerging Information Technology Conference, IEEE, 5 (2005) 564-570. https://doi.org/10.1109/EITC.2005.1544329

[6] H. S. Ahamad, N. S. Meshram, S. B. Bankar, S. J. Dhoble and K. G. Rewatkar, Structural properties of $Cu_xNi_{1-x}Fe_2O_4$ nano ferrites prepared by urea gel microwave auto combustion method, Ferroelectrics, 516 (2017) 167–173. https://doi.org/10.1080/00150193.2017.1362285

[7] A. Bulte, W. M. Jeff, T. Douglas, S. Mann, R. B. Frankel, B. M. Moskowitz, R. A. Brooks, C. D. Baumgarner, J. Vymazal, M. Strub, and Joseph A. Frank, Magnetoferritin: characterization of a novel superparamagnetic MR contrast agent, Journal of Magnetic Resonance Imaging, 4 (3) (1994) 497-505. https://doi.org/10.1002/jmri.1880040343

[8] E. Song, R. Qing, and Z. John Zhang, Correlation between spin orbital coupling and the superparamagnetic properties in magnetite and cobalt ferrite spinel nanocrystals, The Journal of Physical Chemistry B, 110 (23) (2006) 11205-11209. https://doi.org/10.1021/jp060577

[9] A. Laurent, G. Sophie, S. Dutz, U. O. Häfeli, and Morteza Mahmoudi, Magnetic fluid hyperthermia: focus on superparamagnetic iron oxide nanoparticles, Advances in Colloid and Interface Science, 166 (1-2) (2011) 8-23. https://doi.org/10.1016/j.cis.2011.04.003

Materials Research Forum LLC
https://doi.org/10.21741/9781644900970-9

[10] S. N. Sable, K. G. Rewatkar, V. M. Nanoti, Structural and magnetic behavioral improvisation of nanocalcium hexaferrites, Materials Science and Engineering B, 168 (2010) 156–160. https://doi.org/10.1016/j.mseb.2009.10.034

[11] F. Gutiérrez, J. Tomy, and V. A. Alvarez, Nanoparticles for Hyperthermia Applications, Handbook of Nanomaterials for Industrial Applications, Elsevier, (2018) 563-576. https://doi.org/10.1016/B978-0-12-813351-4.00032-8

[12] A. K. Nandanwar, D. S. Choudhary, N. N. Sarkar, K. G. Rewatkar, Effect of Ni^{+2} Substitutions on Structural and Electrical Behavior of Nano-Size Cadmium Ferrites, Materials Today: Proceedings 5/10P3, (2017) 22669–22674. https://doi.org/10. 1016/j. matpr.2018.06.643

[13] S. N. Sable, K. G. Rewatkar and V. M. Nanoti, Structural and magnetic behavioral improvisation of nanocalcium hexaferrites, Materials Science and Engineering: B, 168.1-3 (2010) 156-160. https://doi.org/10.1016/j.mseb.2009.10.034

[14] R. Medal, R. K. Gilchrist, W. D. Shorey, R. C. Hanselman, J. C. Parrott and C.B. Taylor, Selective inductive heating of lymph nodes, Annals of Surgery, 146(4) (1957) p.596. https://doi.org/10.1097/00000658-195710000-00007

[15] H. S. Ahamad, N. S. Meshram, S. B. Bankar, S. J. Dhoble, and K. G. Rewatkar, Structural properties of $Cu_xNi_{1-x}Fe_2O_4$ nano ferrites prepared by urea-gel microwave auto combustion method, Ferroelectrics, 516, 1 (2017) 67-73. https://doi.org/ 10. 1080/ 001 50193.2017.1362285

[16] M. W. Freeman, A. Arrott, and J. H. L. Watson, Magnetism in medicine, Journal of Applied Physics, 31(5) (1960) S404-S405. https://doi.org/10.1063/1.1984765

[17] G. Fischer, B. J. Wagner, M. Schmitt and Rolf Hempelmann, Tuning the relaxation behaviour by changing the content of cobalt in $Co_xFe_{3-x}O_4$ ferrofluids, Journal of Physics: Condensed Matter, 17 (50) (2005) 7875.

[18] M. Ansari, A. Bigham, S. M. Tabrizi, and H. A. Ahangar, Copper-substituted spinel Zn-Mg ferrite nanoparticles as potential heating agents for hyperthermia, Journal of the American Ceramic Society, 101 (8) (2018) 3649-3661.https://doi.org/10.1111 /jace.15510

[19] S. Amiri, and H. Shokrollahi, The role of cobalt ferrite magnetic nanoparticles in medical science, Materials Science and Engineering: C, 33(1) (2013) 1-8. https://doi.org/ 10.1016/j.msec.2012.09.003

[20] A. Usadel, D. Klaus, Dynamics of magnetic nanoparticles in a viscous fluid driven by rotating magnetic fields, Physical Review B, 95.10 (2017) 104430.

[21] R. Kötitz, P. C. Fannin, and L. Trahms, Time domain study of Brownian and Neel relaxation in ferrofluids, Journal of Magnetism and Magnetic Materials 149.1-2 (1995) 42-46. https://doi.org/10.1016/0304-8853(95)00333-9

[22] M. Ming, Y. W. J. Zhou, Y. Sun, Y. Zhang, and Ning Gu, Size dependence of specific power absorption of Fe_3O_4 particles in AC magnetic field, Journal of Magnetism and Magnetic Materials, 268 (1-2) (2004) 33-39. https://doi.org/10.1016/S0304-8853(03)00426-8

[23] G. Abenojar, C. Eric, S. Wickramasinghe, J. B. Concepcion, and A. C. S. Samia, Structural effects on the magnetic hyperthermia properties of iron oxide nanoparticles, Progress in Natural Science: Materials International, 26 (5) (2016) 440 -448. https://doi.org/10.1016/j.pnsc.2016.09.004

[24] H. Rudolf, W. Andra, C. G. d'Ambly, I. Hilger, W. A. Kaiser, U. Richter, and H-G. Schmidt, Physical limits of hyperthermia using magnetite fine particles, IEEE Transactions on Magnetics, 34 (5) (1998) 3745-3754. https://doi.org/10.1109/20.718537

[25] F. G. Vallejo and K. O'Grady, Effect of the distribution of anisotropy constants on hysteresis losses for magnetic hyperthermia applications, Applied Physics Letters, 103(14) (2013) 142417. https://doi.org/10.1063/1.4824649.

[26] P. Bender, J. Fock, C. Frandsen, M. F. Hansen, C. Balceris, F. Ludwig, O. Posth, Relating magnetic properties and high hyperthermia performance of iron oxide nanoflowers, The Journal of Physical Chemistry C, 122 (5) (2018) 3068-3077. https://doi.org/10.1021/acs.jpcc.7b11255.

[27] A. T. Cayless, S. R. Hoon, B. K. Tanner, R. W. Chantrell, and M. Kilner, High sensitivity measurements of néel relaxation in fine particle ferromagnetic systems, Journal of Magnetism and Magnetic Materials 30, no. 3 (1983) 303-311. https://doi.org /10.1016/0304-8853(83)90068-9

[28] R. Sergiu, R. Chantrell, and O. Hovorka, Unified model of hyperthermia via hysteresis heating in systems of interacting magnetic nanoparticles, Scientific Reports, 5 (2015) 9090. https://doi.org/10.1038/srep09090.

[29] G. B. Stillwagon, E. O. Stanley, C. Guse, S. A. Leibel, S. O. Asbell, J. L. Klein, and P. K. Leichner, Prognostic factors in unresectable hepatocellular cancer: Radiation

Magnetic Oxides and Composites II   Materials Research Forum LLC
Materials Research Foundations **83** (2020) 193-232   https://doi.org/10.21741/9781644900970-9

Therapy Oncology Group study 83-01, International Journal of Radiation Oncology Biology Physics, 20 (1) (1991) 65-71.

[30] A. Hanini, K. Kacem, J. Gavard, H. Abdelmelek, and S. Ammar, Ferrite nanoparticles for cancer hyperthermia therapy, In Handbook of Nanomaterials for Industrial Applications, Elsevier, 2018 pp. 638-661. https://doi.org/10.1016/B978-0-12-813351-4.00036-5

[31] A. C. Blanco, F. J. Teran, and D. Ortega, Current outlook and perspectives on nanoparticle-mediated magnetic hyperthermia, Iron Oxide Nanoparticles for Biomedical Applications, (2018) 197-245. https://doi.org/10.1016/B978-0-08-101925-2.00007-3

[32] K. S. Martirosyan, Thermosensitive magnetic nanoparticles for self-controlled hyperthermia cancer treatment, Journal of Nanomedicine and Nanotecholnology, 3 (6) (2012) 1-2. https://doi.org/10.4172/2157-7439.1000e112

[33] I. Apostolova and J. M. Wesselinowa, Possible low-TC nanoparticles for use in magnetic hyperthermia treatments, Solid State Communications, 149 (25-26) (2009) 986-990. https://doi.org/10.1016/j.ssc.2009.04.015

[34] K. G. Rewatkar, N. M. Patil, and S. R. Gawali, Synthesis and magnetic study of Co-Al substituted calcium hexaferrites, Bulletin of Materials Science, 28 (6) (2005) 585-587. https://doi.org/10.1007/BF02706346

[35] D. T. Edward, H. B. Lawrence and R. E. Watson, Extension of the Bloch $T^{3/2}$ law to magnetic nanostructures: Bose-Einstein condensation, Physical Review Letters, 94 (14) (2005) 147210. https://doi.org/10.1103/PhysRevLett.94.147210

[36] K. Mandal, S. Mitra, and P. A. Kumar, Deviation from Bloch $T^{3/2}$ law in ferrite nanoparticles, EPL (Europhysics Letters), 75(4) (2006) 618- 623.

[37] R. W. Chantrell, S. R. Hoon, and B. K. Tanner, Time-dependent magnetization in fine-particle ferromagnetic systems, Journal of Magnetism and Magnetic Materials, 38(2) (1983) 133-141. https://doi.org/10.1016/0304-8853(83)90037-9

[38] E. Obaidat, I. M Bashar, and Yousef Haik, Magnetic properties of magnetic nanoparticles for efficient hyperthermia, Nanomaterials, 5(1) (2015) 63-89. https://doi.org /10.3390/nano5010063

[39] T. Nattermann, V. Pokrovsky, and V. M. Vinokur, Hysteretic dynamics of domain walls at finite temperatures, Physical Review Letters, 87 (19) (2001) 197005. https://doi.org/10.1103/PhysRevLett.87.197005

[40] M. A. Hakim, S. K. Nath, S. S. Sikder, K. H. Maria, Cation distribution and electromagnetic properties of spinel type Ni-Cd ferrites, Journal of Physics and Chemistry of Solids, 74 (2013) 1316-1321. https://doi.org/10.1016/j.jpcs.2013.04.011

[41] M. Rahimi, M. Eshraghi, P. Kameli, structural and magnetic characterization of Cd substituted nickel ferrite nanoparticles, Ceramic International, 40 (2014) 15569 15575. https://doi.org/10.1016/j.ceramint.2014.07.033

[42] A. Hanini, K. Kacem, J. Gavard, H. Abdelmelek, and S. Ammar, Ferrite Nanoparticles for Cancer Hyperthermia Therapy, In Handbook of Nanomaterials for Industrial Applications, pp. 638-661. Elsevier, 2018.

[43] S. Bid and S. K. Pradhan, Characterization of crystalline structure of ball-milled nano-Ni–Zn-ferrite by Rietveld method, Materials Chemistry and Physics, 84(2-3) (2004) 291-301. https://doi.org/10.1016/j.matchemphys.2003.08.012

[44] S. F. Tehrani, V. Daadmehr, A. T. Rezakhani, R. Hosseini Akbarnejad, and S. Gholipour, Structural, magnetic, and optical properties of zinc-and copper-substituted nickel ferrite nanocrystals, Journal of Superconductivity and Novel Magnetism, 25 (7) (2012) 2443-2455. https://doi.org/10.1007/s10948-012-1655-5

[45] S.A.V. Prasad, M. Deepty, P. N. Ramesh, G. Prasad, K. Srinivasarao, Ch Srinivas, K. Vijaya Babu, E. Ranjith Kumar, N. Krisha Mohan, and D. L. Sastry, Synthesis of MFe_2O_4 (M = Mg^{2+}, Zn^{2+}, Mn^{2+}) spinel ferrites and their structural, elastic and electron magnetic resonance properties, Ceramics International, 44 (9) (2018) 10517-10524. https://doi.org/10.1016/j.ceramint.2018.03.070

[46] B. L. Konecky, J. M. Russell, T. C. Johnson, E. T. Brown, M. A. Berke, J. P. Werne, and Y. Huang, Atmospheric circulation patterns during late Pleistocene climate changes at Lake Malawi, Africa, Earth and Planetary Science Letters, 312 (3-4) (2011) 318-326. https://doi.org/10.1016/j.materresbull.2012.12.039

[47] N. Sharma, P. Aghamkar, S. Kumar, M. Bansal, Anju and R. P. Tondon, Study of structural and magnetic properties of Nd doped zinc ferrites, Journal of Magnetism and Magnetic Materials, 369 (2014) 162-167. https://doi.org/10.1016/j.jmmm.2014.05.042

[48] J. N. Christy, K. G. Rewatkar, and P. S. Sawadh, Structural and magnetic behavior of M-type Co-Zr substituted calcium hexaferrites, Materials Today: Proceedings 4.11 (2017) 11857-11865. https://doi.org/10.1016/j.matpr.2017.09.104

[49] S. Sabale, V. Jadhav, V. Khot, X. Zhu, M. Xin, and H. Chen, Superparamagnetic MFe_2O_4 (M = Ni, Co, Zn, Mn) nanoparticles: Synthesis, characterization, induction heating and cell viability studies for cancer hyperthermia applications, Journal of Materials Science: Materials in Medicine, 26 (3) (2015) 127. https://doi.org /10.1007/ s10856-015-5466-7

[50] V. A. M. Brabers, Infrared spectra of cubic and tetragonal manganese ferrites, Physica Status Solidi (b), 33(2) (1969) 563-572. https://doi.org/10.1002/pssb.19690330209

[51] M. G. Naseri , E. B. Saion, H. A. Ahangar, A. H. Shaari, Fabrication, characterization, and magnetic properties of copper ferrite nanoparticles prepared by a simple, thermal-treatment method, Materials Research Bulletin, 48 (2013) 1439–1446.https://doi.org/10.1016/j.materresbull.2012.12.039

[52] A. Abdulaziz, W. H. Abdelraheem, C. Han, M. N. Nadagouda, L. Sygellou, M. K. Arfanis, P. Falaras, V. K. Sharma, and D. D. Dionysiou, Cobalt ferrite nanoparticles with controlled composition-peroxymonosulfate mediated degradation of 2-phenylbenzimidazole-5-sulfonic acid, Applied Catalysis B: Environmental, 221 (2018) 266-279. https://doi.org/10.1016/j.apcatb.2017.08.054

[53] P. Nordblad, R. Mohan, and Samrat Mukherjee, Structural, magnetic and hyperfine characterizations of nanocrystalline Zn-Cd doped nickel ferrites, Journal of Magnetism and Magnetic Materials, 441 (2017) 710-717. https://doi.org/ 10.1016/ j.jmmm. 2017.06.040

[54] M. M. Karanjkar, N. L. Tarwal, A. S. Vaigankar, and P. S. Patil, Structural, Mössbauer and electrical properties of nickel cadmium ferrites, Ceramics International, 39(2) (2013) 1757-1764. https://doi.org/10.1016/j.ceramint.2012.08.022

[55] K. Lawrence, P. Kumar, and M. Kar, Cation distribution by Rietveld technique and magnetocrystalline anisotropy of Zn substituted nanocrystalline cobalt ferrite, Journal of Alloys and Compounds, 551 (2013) 72-81. https://doi.org /10.1016 /j.jallcom.2012.10.009

[56] F. Köseoğlu, A. Yüksel, Structural and magnetic properties of Cr doped Ni-Zn-ferrite nanoparticles prepared by surfactant assisted hydrothermal technique, Ceramics International 41.5 (2015) 6417-6423. https://doi.org/ 10.1016/ j.ceramint.2015.01.079

[57] K. G. Rewatkar, N. M. Patil, S. Jaykumar, D. S. Bhowmick, M. N. Giriya, and C. L. Khobragade, Synthesis and the magnetic characterization of iridium–cobalt substituted calcium hexaferrites, Journal of Magnetism and Magnetic Materials 316, no. 1 (2007) 19-22. https://doi.org/10.1016/j.jmmm.2007.03.192

[58] R. Desai, R. V. Mehta, R. V. Upadhyay, A. Gupta, A. Praneet, K.V. Rao. Bulk magnetic properties of $CdFe_2O_4$ in nano-regime. Bulletin of Materials Science, 30(3) (2007) 197-203. https://doi.org/10.1007/s12034-007-0035-4

[59] M. C. Boubeta, K. Simeonidis, A. Makridis, M. Angelakeris, O. Iglesias, P. Guardia, Andreu Cabot, Learning from nature to improve the heat generation of iron-oxide nanoparticles for magnetic hyperthermia applications, Scientific Reports 3 (2013) 1652. https://doi.org/10.1002/adfm.201101243

[60] M.Gharibshahian, M. S. Nourbakhsh, and O. Mirzaee, Evaluation of the superparamagnetic and biological properties of microwave assisted synthesized Zn & Cd doped $CoFe_2O_4$ nanoparticles via Pechini sol–gel method, Journal of Sol-Gel Science and Technology, 85(3)(2018) 684-692. https://doi.org/10.1007/s10971-017-4570-1

[61] A. Hanini, L. Lartigue, J. Gavard, K. Kacem, C. Wilhelm, F. Gazeau, F. Chau, and S. d Ammar, Zinc substituted ferrite nanoparticles with $Zn_{0.9}Fe_{2.1}O_4$ formula used as heating agents for in vitro hyperthermia assay on glioma cells, Journal of Magnetism and Magnetic Materials, 416 (2016) 315-320. https://doi.org/10.1016/j.jmmm.2016.05.016

[62] B. Thiesen and A. Jordan, Clinical applications of magnetic nanoparticles forhypertherm, International journal of hyperthermia, 24 (6) (2008) 467-474.

[63] M. Mahmoudi, S. Sant, B. Wang, S. Laurent, and T. Sen, Superparamagnetic iron oxide nanoparticles (SPIONs): development, surface modification and applications in chemotherapy, Advanced drug delivery reviews, 63 (1-2) (2011) 24-46. https://doi.org/ 10.1016 /j.addr.2010.05.006

[64] M. Rai and R. Shegokar, Metal Nanoparticles in Pharma. Springer, 2017

[65] R. Kötitz, P. C. Fannin, and L. Trahms, Time domain study of Brownian and Néel relaxation in ferrofluids, Journal of Magnetism and Magnetic Materials, 149 (1-2) (1995) 42-46. https://doi.org/10.1016/0304-8853 (95)00333-9.

[66] R. Valenzuela, Novel applications of ferrites, Physics Research International (2012). https://doi.org/10.1063/1.1855131

[67] J. Giri, P. Pradhan, T. Sriharsha, and D. Bahadur, Preparation and investigation of potentiality of different soft ferrites for hyperthermia applications, Journal of Applied Physics, 97 (10) (2005) 10Q916. https://doi.org/10.1063/1.1855131

[68] P. Nordblad, R. Mohan, and S. Mukherjee, Structural, magnetic and hyperfine characterizations of nanocrystalline Zn-Cd doped nickel ferrites, Journal of Magnetism and Magnetic Materials, 441 (2017) 710-717. https://doi.org/ 10.1016 /j.jmmm.2017.06.040

[69] S. Naik, A. Parveez, A. Chaudhuri, and S. A. Khader, Structural, dielectric and electrical properties of $Ni(Cd,Zn)Fe_2O_4$ by auto combustion method, Materials Today: Proceedings, 4 (11) (2017) 12103-12108. https://doi.org/ 10.10 16/ j.matpr. 2017.09.137

[70] P. Raja, T. Yadavalli, D. Ravi, H. A. Therese, C. Ramasamy, and Y. Hayakawa, Synthesis and magnetic properties of gadolinium substituted zinc ferrites, Materials Letters, 188 (2017) 406-408. https://doi.org/10.1016/j.matlet.2016.11.083

[71] R. A. Bohara, D. T. Nanasaheb, A. K. Chaurasia, and S. H. Pawar, Cancer cell extinction through a magnetic fluid hyperthermia treatment produced by superparamagnetic Co–Zn ferrite nanoparticles, RSC Advances, 5 (58) (2015) 47225-47234. https://doi.org/10.1039/C5RA04553C

[72] H. Köseoğlu and K. Yüksel, Structural and magnetic properties of Cr doped NiZn-ferrite nanoparticles prepared by surfactant assisted hydrothermal technique, Ceramics International 41.5 (2015) 6417-6423. https://doi.org/10.1016/ j.ceramint. 2015 .01.079

[73] M. F. Valan, A. Manikandan, and S. A. Antony, Microwave Combustion Synthesis and Characterization Studies of Magnetic $Zn_{1-x}Cd_xFe_2O_4$ ($0 \leq x \leq 0.5$) Nanoparticles, Journal of Nanoscience and Nanotechnology, 15 (6) (2015) 4543-4551. https://doi.org/10.1166/jnn.2015.9801

[74] W. Xiaojuan, Z. Wei, L. Zhang, C. Zhang, H. Yang, and J. Jiang, Synthesis and characterization of Fe and Ni co-doped ZnO nanorods synthesized by a hydrothermal method, Ceramics International, 40 (9) (2014) 14635-14640. https://doi.org/10.1016/j.ceramint.2014.08.022

[75] T. Tatarchuk, M. Bououdina, J. J. Vijaya, and L. J. Kennedy, Spinel ferrite nanoparticles: synthesis, crystal structure, properties, and perspective applications, In International Conference on Nanotechnology and Nanomaterials, Springer, Cham, (2016) pp. 305-325. https://doi.org/10.1007/978-3-319-56422-7_22

[76] J. I. Martın, J. Nogues, K. Liu, J. L. Vicent, and I. K. Schuller, Ordered magnetic nanostructures: fabrication and properties, Journal of Magnetism and Magnetic Materials, 256 (1-3) (2003) 449-501. https://doi.org/10.1016/S0304-8853(02)00898-3

[77] N. Y. Lanje, D. K. Kulkarni, and K. G. Rewatkar, Synthesis, transport and magnetic study of $CaLaCr_{11}O_{19}$, Materials Letters, 47 (3) (2001) 125-127. https://doi.org/ 10.1016/ S0167-577X(00)00222-6

[78] A. K. Nandanwar, N. S. Meshram, V. B. Korde, D. S. Choudhary, and K. G. Rewatkar, Effects of Ni^{2+}-substitution on structural, magnetic and electrical properties of cadmium spinel ferrite nanoparticles via chemical route, Integrated Ferroelectrics, 203 (1) (2019) 12-18. https://doi.org/10.1080/10584587.2019.1674955

[79] S. N. Kamde, A. K. Nandanwar, P. G. Agone, and K. G. Rewatkar, Effect of Cr^{3+} doped on structural, magnetic and electrical properties of sol-gel synthesized $SrFe_{12}O_{19}$ hexaferrite nanoparticle, Integrated Ferroelectrics, 203 (1) (2019) 150-155. https://doi.org/10.1080/10584587.2019.1674959

Chapter 10

Investigation on Structural, Electrical and Magnetic Properties of Copper Substituted Nickel Nanoferrites

Rapolu Sridhar[1, a], D. Ravinder[2, b], K. Vijaya Kumar[3, c], G. Helen Ruth Joice[4, d]

[1]Department of Basic Science and Humanities, Vignan Institute of Technology and Science, Deshmukhi (V), Yadadri-Bhuvanagiri Dist. 508 284, Telangana, India

[2]Department of Physics, Osmania University, Hyderabad 500 007, Telangana, India

[3]Department of Physics, JNTUH College of Engineering Sultanpur, Sultanpur (V), Pulkal (M), Sangareddy Dist. 502 293, Telangana, India,

[4]Department of Physics, Thiru Kollanjiappar Arts College, Vriddhachalam 606 001, Cuddalore Dist, Tamilnadu, India

[a]rapolu31@gmail.com, [b]ravindergupta28@rediffmail.com, [c]kvkphd@gmail.com, [d]helenruthjoice@gmail.com

Abstract

$Ni_{1-x}Cu_xFe_2O_4$ $(0.0 \leq x \leq 1.0)$ nanoferrites were prepared using citrate-gel auto combustion technique. The effect of Cu substitution on the magnetic properties, dielectric properties as well as structural parameters like lattice constant, hopping length and X-ray density were reported and discussed. The surface morphology and elemental analysis of prepared compositions were examined using SEM and EDAX analysis. Electrical properties such as dc resistivity, drift mobility and Seebeck coefficient were reported as function of temperature and composition, it is explained with hopping mechanism between $Fe^{2+} \Leftrightarrow Fe^{3+}$ ions and the activation energy was measured in the ferrimagnetic and paramagnetic regions.

Keywords

Nanoferrites, Citrate-Gel Auto Combustion Technique, Structural Parameters, Magnetic Properties, Dielectric Properties

Contents

1. Introduction

Spinel ferrites are playing a major role in scientific research and technology because of their remarkable magnetic and electrical properties [1, 2]. Amending microstructure as nanostructure, improved the structural, electrical and magnetic properties of ferrites, from past decade researchers are attentive on the preparation of nano size spinel ferrites with ultra-fine size and remarkable magnetism. Nano structured materials illustrate novel properties over those exhibited by bulk materials of same compositions, so these materials are most significant in the field of science and technology [3, 4]. When particle size reduces into nano meter scale, the materials properties can be modify and they are useful for various applications [5, 6].

Conventional physical methods are used to prepare the spinel ferrites, but they have few disadvantages like sintering temperature is high and time period is more etc. So, from last few years for preparing nano materials, different chemical methods like sol-gel, co-precipitation, spray dying and citrate gel methods are used because they are easy, eco-friendly, require low temperature and low cost [7, 11].

In the recent years, among spinel ferrites, nano size nickel ferrite and substituted nickel nano ferrites have attracted researchers more because of their high electrical resistance, high mechanical hardness, low preparation cost and low eddy current loss etc. [12]. These materials have found interesting applications in magnetic refrigerators, high density recording devices, colour imaging and microwave devices [13].

By considering significance of nickel nano ferrites' applications and their novel modified properties by doping with divalent ions, we decided to prepare copper substituted nickel nano ferrites, $Ni_{1-x}Cu_xFe_2O_4$ (x = 0.0, 0.2, 0.4, 0.5, 0.6, 0.8 and 1.0) using citrate gel auto combustion technique and to study the effect of copper substitution on the structural, magnetic, electrical and dielectric properties of nickel ferrites in a systematic manner.

2. Experimental procedure

2.1 Sample preparation

Copper substituted nickel ferrites with chemical composition $Ni_{1-x}Cu_xFe_2O_4$ ($0.0 \leq x \leq 1.0$) were synthesized using citrate gel auto combustion technique. Stoicheometric amount of AR grade nickel nitrate ($Ni(NO_3)_2 \cdot 6H_2O$), copper nitrate ($Cu(NO_3)_2 \cdot 6H_2O$), ferric nitrate ($Fe(NO_3)_3 \cdot 9H_2O$), citric acid ($C_6H_8O_7 \cdot H_2O$) and ammonia solution were taken as starting materials. An appropriate amount of metal nitrates were dissolved in deionized double distilled water in separate beakers and then mixed together. As prepared mixture was added in to citric acid (1:3 molar ratio of nitrate to citric acid) and kept on a magnetic stirrer to obtain a homogeneous solution. The ammonia solution was added drop wise to adjust the pH of solution to 7 and then heated at 80 °C until it converts into viscous gel. Further, as obtained viscous gel was heated at 180 °C. The auto combustion reaction started in the gel and large amounts of gaseous ash propagated from the bottom to the top, like the eruption of volcano, giving rise to dark grey ash. Finally the as-burnt ash was ground well then calcined in a muffle furnace at 700 °C for 5 h. As recovered ferrite powders were again ground well. The prepared powder samples were compressed in to pellet using the KBr hydraulic press machine (*Model: M-15*).

2.2 Characterization

FT-IR spectra were recorded on the FT-IR spectrometer (Spectrum 100, Perkin Elmer, USA) in wave number range of 350-4000 cm^{-1} using KBr pellet method. Obtained IR spectra analyzed to get structural information about the prepared ferrites system.

X-ray diffraction patterns of all samples were recorded at room temperature using X-ray diffractomerter (Bruker Karlsruhe, German, D8 Advanced System) with Cu-kα radiation (λ = 1.5405Å). X-ray pattern of each sample was recorded from 20° to 80°, with scan step 0.04°/sec. The crystallite size was estimated from the most intense (311) Bragg peak using the Scherer's formula as given in an equation 1 [14].

$$D_{xrd} = \frac{0.91\,\lambda}{\beta \cos\theta} \tag{1}$$

Where, λ - X-ray wavelength, θ - angle of Bragg diffraction and, β - full width at half maxima of the peak.

Lattice parameter value is measured from d-spacing value corresponding Miller indices (h k l) with the following equation 2 [15].

$$a = \frac{d_{hkl}}{\sqrt{h^2 + k^2 + l^2}} \tag{2}$$

Hopping length was calculated at tetrahedral (A-site) d_A and octahedral (B-site) d_B using equations 3 and, 4 respectively.

$$\text{A site (Tetrahedral) hopping length } d_A = 0.25a\sqrt{3} \tag{3}$$

$$\text{B site (Octahedral) hopping length } d_B = 0.25a\sqrt{2} \tag{4}$$

X-ray density (d_x) was calculated using the equation 5 [16].

$$d_x = \frac{8M_W}{N_A a^3} \left(\frac{g}{cm^3}\right) \tag{5}$$

Materials Research Forum LLC
https://doi.org/10.21741/9781644900970-10

Where, M_w - Composition Molecular weight and N_A - Avogadro's number (6.02×10^{23}/mole).

Bulk density (d_B) was calculated using the equation 6 [17].

$$d_B = \frac{m}{\pi r^2 h} \tag{6}$$

Where, m - mass of the sample, r - radius of the sample and h - thickness of the sample

Porosity (P) of the sample was measured using the equation 7 [18].

$$P = \left(1 - \frac{d_B}{d_x}\right) \times 100 \text{ \%} \tag{7}$$

Where, (d_B) - bulk density and (d_x) - X-ray density

The microstructure and surface morphology of prepared samples were examined by SEM (Hitachi-S520, Japan). The morphology and crystallite size were examined by TEM (Tecnai-12, FEI, Netherlands) images.

Magnetic hysteresis loops of all samples were recorded at room temperature using a VSM (GMW Magnet System, Model 3473) instrument. Magnetic parameters such as saturation magnetization (M_s) and coercivity (H_C) were obtained from hysteresis loops and Magnetic moment (μ_B) calculated using the equation 8 [19].

$$\mu_B = \frac{M_w \times M_s}{5585} \tag{8}$$

Where, M_w – Composition Molecular weight and M_S - saturation magnetization.

Temperature and composition dependent DC electrical measurements were carried out using two probe method [20]. The systematical procedure can be explained as follows.

Clean the sample holder by polish paper. Keep the sample in the sample holder properly, put the sample holder in the furnace and make all necessary connections. Switch on the furnace, increase the temperature to heat the sample and cool the sample up to room temperature. Again heat the sample upto desired temperature level, then switch off furnace and take the reading on multi meter at the time of cooling at regular interval of temperature in the step of 10°C up to room temperature. The resistivity (ρ) and temperature (T) Kelvin relationship expressed as Arrhenius relation [21].

$$\rho = \rho_0 e^{\Delta E/k_B T} \tag{9}$$

The Arrhenius plots graph $\ln(\rho T)$ versus $10^3/T$. It was observed that a change at a point on plotted graph, it is known as curie temperature and it indicates a change of magnetic ordering and dividing the curve into two regions, resultant to ferrimagentic region and paramagnetic region. From plots graphs extract the slope and find the activation energy (ΔE) of each sample was calculated using the equation 10.

$$\Delta E = (2.303)\, k_B 10^3 \times slope \;\; (eV) \tag{10}$$

Where, ρ_o - Resistivity at room temperature, K_B - Boltzmann constant (8.617×10^{-5}eV K^{-1}) and ΔE - Activation energy

Drift mobility (μ_d) of charge carriers was calculated using the equation 11 [22].

$$\mu_d = \frac{1}{\eta e \rho} \tag{11}$$

Where η - Number of charge carriers, e - Electron charge and ρ - Resistivity at a given temperature.

Charge carriers concentration was estimated using equation 12 [23].

$$\eta = \frac{N_A d_B P_{Fe}}{M_w} \tag{12}$$

Where, P_{Fe} - Number of iron atoms in ferrite composition.

Thermoelectric power was studied as a function of temperature and composition by differential method [24-25]. The experimental arrangement consists of two electrodes, the top one works as hot junction and the lower one works as cold junction and sample is inserted in sample holder. The hot and cold electrodes are connected to electro meter for measuring thermo emf. The temperature of the hot junction is increased by keeping the cold junction at room temperature, however cold junction temperature goes on rising due to thermal conductivity of the sample which is low. Thermoelectric power study was carried out using this method. Temperature difference between two junctions is maintained constant (about 10 °C) and is used for the purpose of calculation of thermo emf generated by the sample.

Seebeck coefficient (*S*) was calculated using the following relation.

$$S = \left(\frac{\Delta V}{\Delta T}\right) \tag{13}$$

Where, ΔV - Thermo emf and ΔT - Temperature gradient.

Dielectric measurements were carried out at room temperature in frequency range from 20 Hz to 2 MHz using an Agilent E4980A precession LCR meter. LCR meter bridge instrument directly measure the capacitance of the sample material (pellet), capacitance of material (with air) and dielectric loss tangent (tanδ). By using this data different dielectric parameters such as dielectric constant (real part ε′) and ac conductivity were calculated using equations 14 and

15, respectively [26, 27].

$$(\varepsilon') = \frac{Cd}{\varepsilon_0 A} \tag{14}$$

$$(\sigma_{ac}) = \omega \varepsilon_0 \varepsilon'(\tan \delta) \tag{15}$$

Where, ε_0 is permittivity of free space (8.854×10^{-12} F/m), ε' is real dielectric constant and $\tan \delta$ is dielectric loss tangent.

3. Results and discussion

3.1 Structural characterization

3.1.1 FT-IR analysis

FT-IR spectra of $Ni_{1-x}Cu_xFe_2O_4$ ($0.0 \leq x \leq 1.0$) ferrites were recorded in wave number range of 350-4000 cm^{-1}. As recorded FT-IR spectra are indicated in Fig. 1 and corresponding band positions are presented in Table 1. It is observed form Fig. 1 and Table 1 that position of absorption bands v_1 and v_2 are found in between 570-584 cm^{-1} and 405-414 cm^{-1}, due to vibration of ions in the crystal lattices [28], which are in good agreement with ferrite vibration mentioned by Waldron [29] and Hafner [30]. It shows formation of the single-phase spinel structure with two sub lattices [31]. The high frequency band (v_1) represents Fe^{+3} - O^{-2} stretching vibrations at tetrahedral (A) site, while the low frequency band (v_2) represents M^{+2} - O^{-2} vibrations at octahedral (B) site.

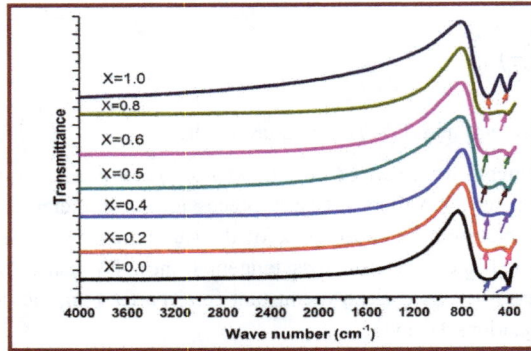

Fig. 1 FT-IR spectra of $Ni_{1-x}Cu_xFe_2O_4$ (0.0 ≤ x ≤ 1.0) ferrites.

Table 1 FT-IR parameters of $Ni_{1-x}Cu_xFe_2O_4$ (0.0 ≤ x ≤ 1.0) nano ferrites.

Cu content (x)	v_1 (cm^{-1})	v_2 (cm^{-1})
0.0	569.83	414.19
0.2	572.23	411.28
0.4	574.42	408.55
0.5	577.45	404.69
0.6	577.04	407.21
0.8	581.56	409.43
1.0	584.24	404.89

3.1.2 XRD analysis

XRD patterns of $Ni_{1-x}Cu_xFe_2O_4$ (0.0 ≤ x ≤ 1.0) ferrites, heated at 700 °C for 5 h. are shown in Fig. 2; it demonstrates sharp reflection peaks, confirmed that formed samples are well crystalline and obtained XRD peaks were indexed using ICDD card no (PDF# 86-2267) and (PDF # 73-2314) for Ni and Cu ferrites, respectively. The reflections from (111), (220), (311), (222), (400), (511) and (440) planes confirmed a cubic unit cell and (311) plane shows spinel phase structure. So, these allowed planes confirmed formation of single phase having a cubic spinel structure [32]. The crystallite size was found in the range from 36.70 nm to 58.91 nm.

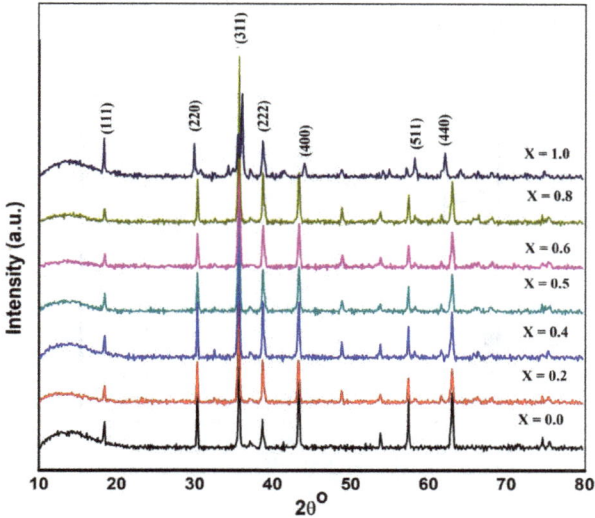

Fig. 2 X-ray diffraction pattrens of $Ni_{1-x}Cu_xFe_2O_4$ (0.0 $\leq$ x $\leq$ 1.0) ferrites, heated at 700 °C for 5 h.

Table 2 Structural parameters of $Ni_{1-x}Cu_xFe_2O_4$ (0.0 $\leq$ x $\leq$ 1.0) ferrites.

Cu content (x)	Crystallite size (nm)	Lattice parameter a (Å)	Hopping length		X-ray density d_x (gram/cm³)	Bulk density d_B (gram/cm³)	Porosity P (%)
			(A-site) d_A (Å)	(B-site) d_B (Å)			
0.0	58.9	8.35	2.950	3.614	5.35	5.07	5.34
0.2	46.3	8.34	2.947	3.610	5.39	4.98	7.78
0.4	49.2	8.33	2.944	3.607	5.43	4.94	8.95
0.5	37.3	8.32	2.942	3.604	5.45	4.93	9.45
0.6	36.9	8.31	2.939	3.600	5.48	4.89	10.73
0.8	43.3	8.29	2.932	3.590	5.54	4.86	12.28
1.0	36.7	8.24	2.914	3.57	5.67	4.84	14.55

Fig. 3 shows variation of lattice parameter 'a' with Cu content (x); there is decrease in lattice parameter with Cu content (x). The average value of lattice parameter is 8.313 Å. It is in good agreement with the reported work [33]. The hopping length values are found to decrease at tetrahedral and octahedral sites with increase of Cu^{+2} content (x) in prepared ferrites (Table 2).

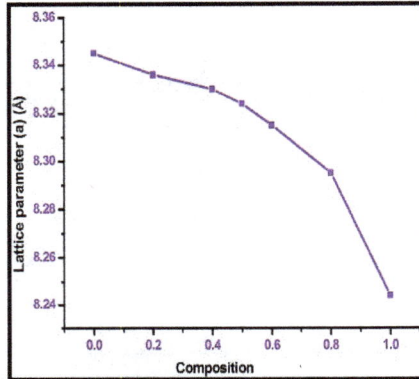

Fig. 3 Variation of lattice parameter 'a' with Cu composition (x).

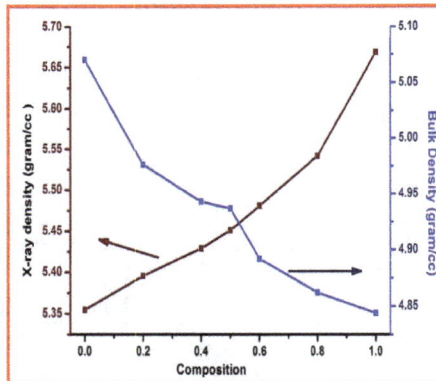

Fig. 4 Variations of X-ray density and bulk density with Cu composition (x).

Fig.4 illustrates the variation of X-ray density and bulk density as a function of Cu composition (x). It is clear from Table 2 that the X-ray density (d_x) rises from 5.35 g/cm^3 to 5.67 g/cm^3; while the bulk density (d_B) reduces from 5.07 g./cm^3 to 4.84 g./cm^3 with Cu^{+2} composition (x). The increase of the X-ray density is partially due to the higher atomic weight of copper (63.546 g/mole), compared with nickel (53.933 g/mole). The X-ray density of prepared ferrite is in good agreement with the previously reported value of 5.38 g/cm^3[34]. The X-ray density is higher than the bulk density due to the presence of pores. The porosity value raises from 5.34 % to 14.55% with increase Cu^{+2} composition in prepared ferrites; similar trend was observed by other researchers [35, 36].

3.2   Morphological studies

SEM images and the EDS spectra of Ni$_{1-x}$Cu$_x$Fe$_2$O$_4$ ($0.0 \leq x \leq 1.0$) ferrites are shown in Fig. 5. The surface morphology of all samples indicated spongy like structure with porous grains. The EDS results show the presence of Ni, Cu, Fe and O elements.

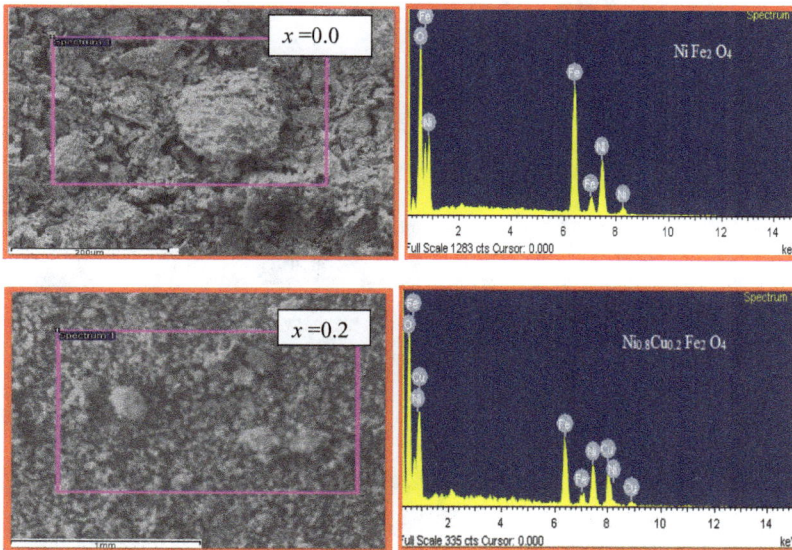

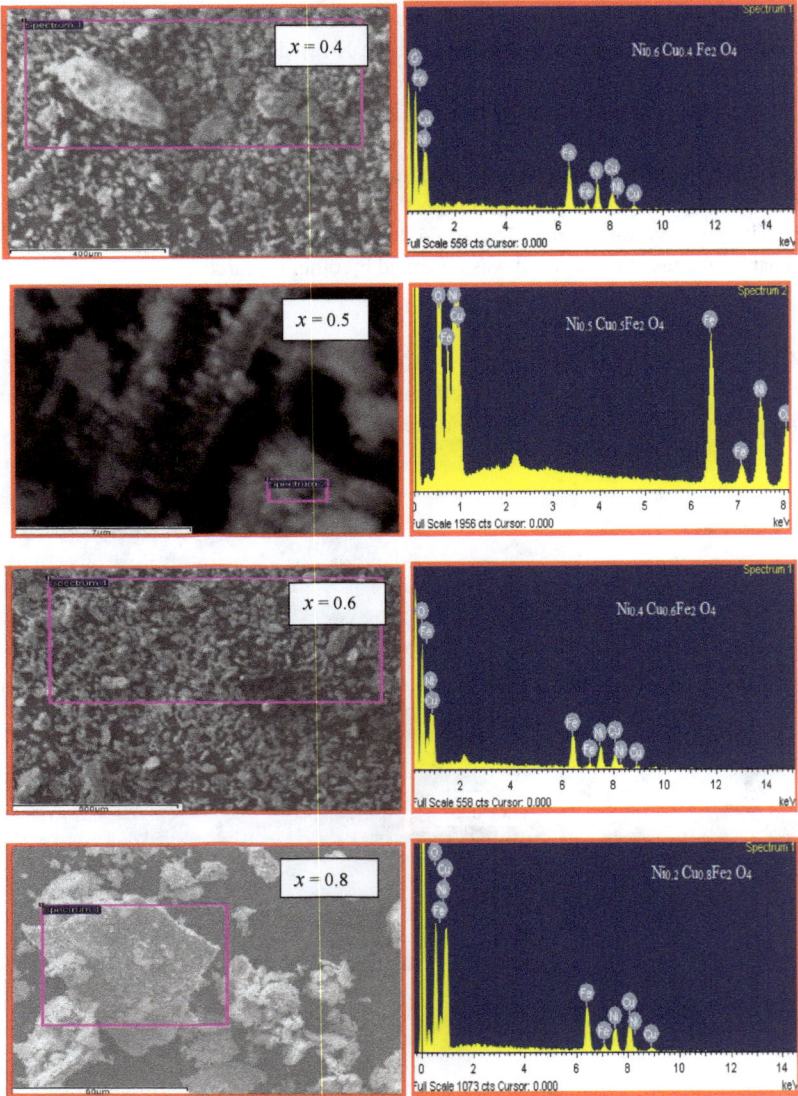

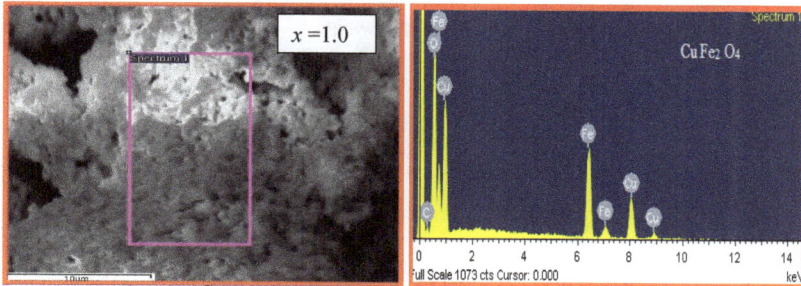

Materials Research Forum LLC
https://doi.org/10.21741/9781644900970-10

Fig. 5 SEM images and EDS spectra of $Ni_{1-x}Cu_xFe_2O_4$ (0.0 ≤ x ≤ 1.0) nano ferrites.

Fig. 6(a) and Fig. 6(b) represent the TEM images of $NiFe_2O_4$ and $Ni_{0.2}Cu_{0.8}Fe_2O_4$ nano ferrites. It is clear from Fig. 6 (a, b) that formed particles possess platelet shape.

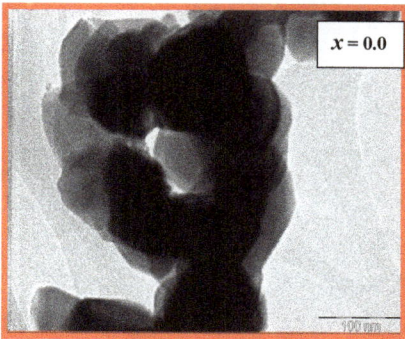

Fig. 6(a). TEM image of $NiFe_2O_4$ ferrite

Fig. 6(b). TEM image of $Ni_{0.2}Cu_{0.8}Fe_2O_4$ ferrite

3.3 Magnetic properties

Fig. 7 illustrates magnetic M-H loops of $Ni_{1-x}Cu_xFe_2O_4$ *(0.0 ≤ x ≤ 1.0)* ferrites, all compositions show soft magnetic behavior, except x =1.0 composition. Nano size nickel ferrite with x =0.0 sample shows saturation magnetization value of 33.75emu/gr, which is less than reported bulk nickel ferrite value (50 emu/gr.) [37]. In the case of small size nano ferrite particles, surface to volume ratio is large; as a result of the canting and disorder of the surface layer spins, super exchange interaction occurs among those atoms; these factors reduce saturation magnetization.

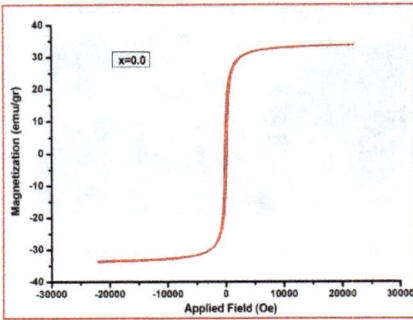
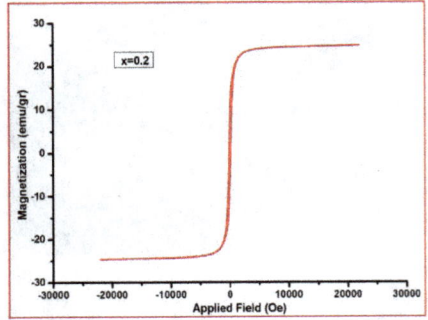

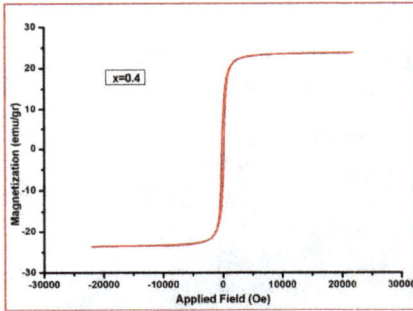
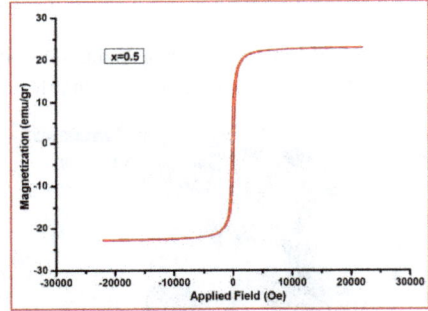

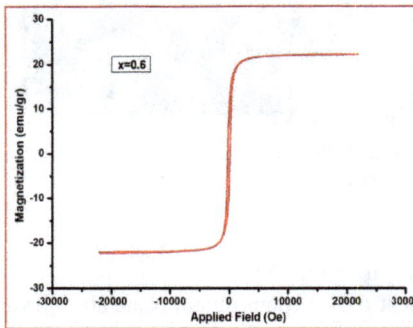
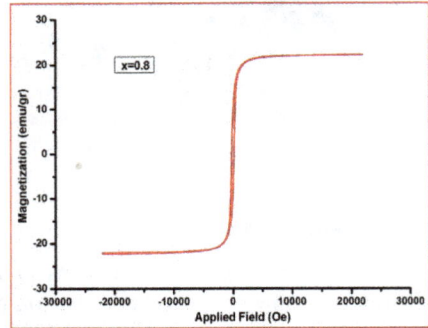

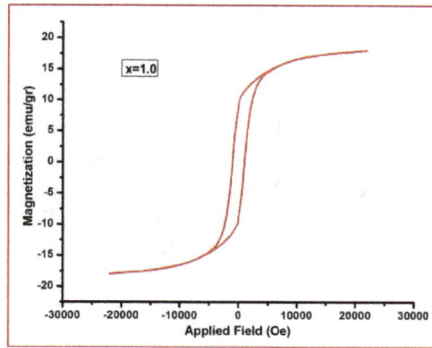

Fig. 7 Magnetic hysteresis loops of $Ni_{1-x}Cu_xFe_2O_4$ (0.0 ≤ x ≤ 1.0) ferrites.

Table 3 Magnetic parameters of $Ni_{1-x}Cu_xFe_2O_4$ (0.0 ≤ x ≤ 1.0) ferrites.

Cu content (x)	Saturation Magnetization M_S (emu/g.)	Coercive Field H_C (O_e)	Magnetic moment μ_B
0.0	33.75	165.49	1.416
0.2	24.75	170.17	1.042
0.4	23.70	179.13	1.002
0.5	22.85	184.28	0.969
0.6	22.20	188.79	0.944
0.8	21.15	188.89	0.902
1.0	18.00	992.02	0.771

The magnetic parameters calculated from hysteresis loops are listed in Table 3. It is clear from Table 3 that saturation magnetization (M_S) decreases from 33.75 emu/g. to 18.00 emu/g, and the magnetic moment decreases from 1.416 μ_B to 0.771 μ_B with Cu^{2+} content (x). As Cu^{2+} concentration increases, the ratio between Fe^{+3} ions on octahedral to tetrahedral sites decreases, as a result, an A-B interaction also decreases [38], and hence net magnetization decreases. Similar result was reported earlier by other researchers [39].

The coercivity (H_c) values are found in between 165 Oe -992 Oe, increasing with the increase of dopant Cu^{+2} content (x) in prepared ferrites, due to the anisotropy field decreases. These nanoferrites are multi-domain particle systems. In the present study

$CuFe_2O_4$ composition has high coercivity and the crystallite size (36.7 nm) is less (Table 2).

It is clear from Table 3 that x =0.0 sample shows high saturation magnetization value (33.75 emu/g) with small coercive field (165.487 Oe); therefore, $NiFe_2O_4$ is considered as 'soft' magnetic material, while x = 1.0 sample shows low the saturation magnetization value (18.00 emu/g) with large (992.02 Oe) coercive field and, $CuFe_2O_4$ is considered as 'hard' magnetic material. So $CuFe_2O_4$ is very useful for application in recording heads, magnetic shielding, microwave devices and also in fabrication of hard permanent magnets.

3.4   Electrical properties

3.4.1 Temperature dependent dc conductivity

Fig. 8 illustrates variation of the DC electrical resistivity with temperature for $Ni_{1-x}Cu_xFe_2O_4$ $(0.0 \leq x \leq 1.0)$ ferrites. It is clear from Fig. 8, that the resistivity (ρ) decreases with increase in temperature for each sample, which is a normal behavior of semiconductor [41].

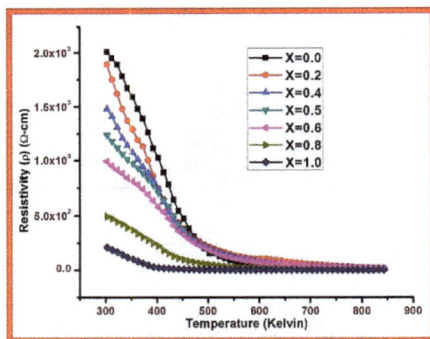

Fig. 8 Variation of dc resistivity with temperature of $Ni_{1-x}Cu_xFe_2O_4$ $(0.0 \leq x \leq 1.0)$ ferrites.

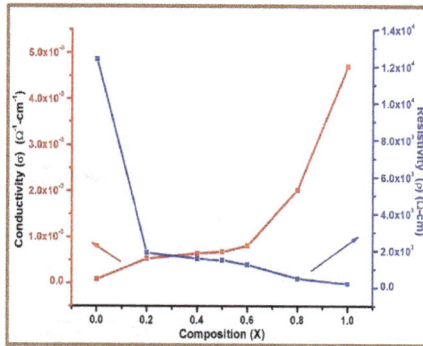

Fig. 9 Variation of dc resistivity and ac conductivity with Cu^{2+} concentration (x) of Ni_{1-x}
$Cu_xFe_2O_4$ (0.0 ≤ x ≤ 1.0) ferrites.

Fig. 9 represents the variation of dc resistivity and ac conductivity with Cu^{2+} concentration (x) for $Ni_{1-x}Cu_xFe_2O_4$ (0.0 ≤ x ≤ 1.0) ferrites. At room temperature, dc resistivity decreases from 1.24×10^4 Ω-cm to 2.12×10^2 Ω-cm and ac conductivity increases from 8.04×10^{-5} Ω-cm to 4.71×10^{-3} Ω-cm, because of the resistivity of Cu (1.7×10^{-6} Ω-cm) is smaller than that of Ni(7.0×10^{-6} Ω-cm) [42]. It can be explained well with Verway's hopping mechanism [43], When Cu^{2+} ion doped for Ni^{2+}, Cu^{2+} ions partially occupy on the A sites then some Fe ions are shifted from A to B sites, hence Ni^{2+} ions decrease on B sites. Consequently, the number of $Fe^{2+} \Leftrightarrow Fe^{3+}$ ions increase on the B sites. As a result, the electrical resistivity decreases and ac conductivity increases with Cu^{2+} substitution (x). Similar behavior was observed and reported by other researchers [44, 45]. Fe^{2+} was interpreted based on the cation distribution of Ni-Cu nano-ferrites.

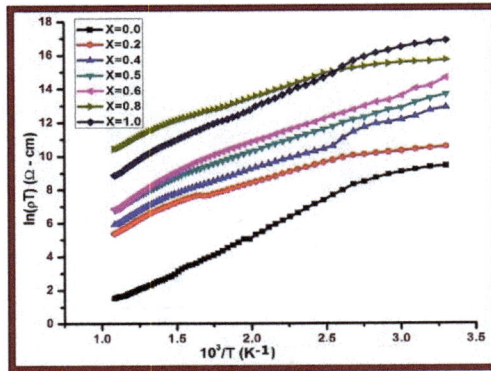

Fig. 10 Variation of ln(ρT) with inverse temperature of $Ni_{1-x}Cu_xFe_2O_4$ (0.0 ≤ x ≤ 1.0) ferrites.

Fig. 10 shown the Arrhenius plots ln(ρT) versus $10^3/T$ of Ni-Cu nano ferrites. It was observed that a change at a point indicates a change of magnetic ordering and dividing the curve into two regions, resultant to ferrimagentic region and paramagnetic region.

Table 4 Activation energy values of $Ni_{1-x}Cu_xFe_2O_4$ (0.0 ≤ x ≤ 1.0).

Cu content (x)	Para Region (E_P) eV	Ferri Region (E_F) eV	Activation Energy (ΔE) eV
0.0	0.698	0.369	0.329
0.2	0.717	0.402	0.315
04	0.750	0.454	0.296
0.5	0.788	0.502	0.286
0.6	0.836	0.566	0.270
0.8	0.859	0.609	0.250
1.0	0.954	0.731	0.223

According to magnetic semiconductor theory; the ferrimagnetic region is an ordered one while the paramagnetic region is disordered one [46]. Therefore, for the conduction in the paramagnetic region, more energy is required compared with the ferrimagnetic region. Hence, the activation energy in the paramagnetic region (E_P) is found greater than in ferrimagnetic region (E_F). Similar results were reported by others in Zn-Ni ferrites [47].

From Table 4, it is clear that the activation energy (ΔE) is decreases from 0.329 eV to 0.223 eV with increase in Cu^{2+} content (x). As low conductivity of the ferrites goes hand with high activation energy, vice versa. It may also be acceptable activation energy behaves as similar way as that of dc electrical resistivity.

3.4.2 Temperature dependent drift mobility

Fig. 11 illustrates variation of drift mobility with inverse temperature for different compositions. It is clear from Fig. 11 that as the temperature increases, the drift mobility also increases in all compositions. There is no change in carrier concentration but there is a change in charge carrier mobility with increase in temperature. As a result the change in charge carrier mobility rather than the change in carrier concentration and it is accountable for the drift mobility variation with temperature. Hence, it is clear that increase in drift mobility with increases in temperature due to charge carriers' start hopping from one site to another site as reported by others [48].

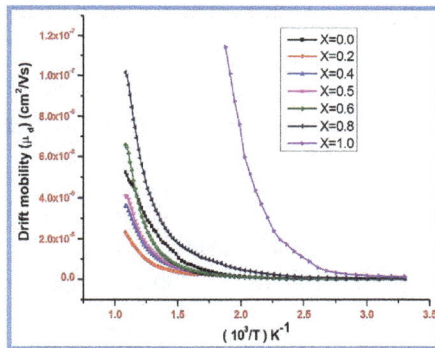

Fig. 11 Variation of drift mobility with inverse temperature for $Ni_{1-x}Cu_xFe_2O_4$ ($0.0 \leq x \leq 1.0$) ferrites.

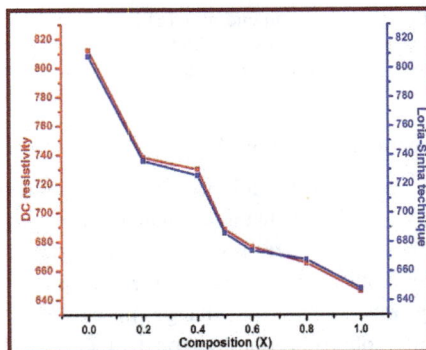

Fig. 12 Variation of Curie temperature with copper composition (x) using dc resisitivity measurements and Loria-Sinha method for $Ni_{1-x}Cu_xFe_2O_4$ (0.0 ≤ x ≤ 1.0) ferrites.

Fig. 12 shows variation of Curie temperature with Cu content for different compositions. From Loria-Sinha method, Curie temperature decreases from 808 K *to* 648 K, while from dc resistivity measurements; it decreases from 812.08 K *to* 646.20 K on increasing Cu content (*x*). The increase in Cu^{2+} concentration reduces Fe^{3+} concentration on A-sublattice, that leads to a reduction in the number of Fe_A^{3+}– O– Fe_B^{3+} linkage, this linkage determines the magnitude of the Curie temperature and hence decreases A-B interactions. So there is decrease in Curie temperature with increase in Cu^{2+} ion concentration [49]. Similar results were observed by Ravinder *et al.* [50].

3.4.3 Temperature dependence thermoelectric power

Fig. 13 illustrates variation of Seebeck coefficient (*S*) with copper content (*x*) for different compositions. The negative value of Seebeck coefficient indicates that majority charge carriers are electrons. Hence, the prepared ferrites possess n-type semiconductor nature and Seebeck coefficient varies from -150 µV/K to – 40 µV/K with copper content (*x*). It explains that conduction mechanism is mainly due to hopping of electrons between $Fe^{2+} \Leftrightarrow Fe^{3+}$ ions [51]. Similar results were reported by L.G. Van Uitert *et al.* [52].

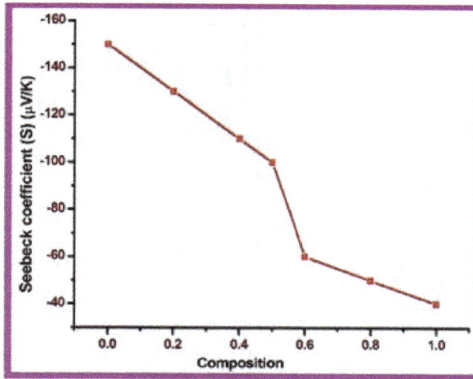

Fig. 13 Variation of Seebeck coefficient (S) with copper composition (x) for Ni$_{1-}$$_xCu_xFe_2O_4$ (0.0 ≤ x ≤ 1.0) ferrites.

Fig. 14 illustrates variation of Seebeck coefficient (S) with temperature for different compositions. It explains that when temperature increases, the hot surface losses electrons and it possesses positive charges (holes) and the cold surface accepts the electrons from hot surface and it possesses negative charges (electrons). This conduction mechanism between $Fe^{2+} \Leftrightarrow Fe^{3+}$ results, Seebeck coefficient increases up to a certain temperature, it represents as transition temperature T_S (°K). Further, temperature increases; Seebeck coefficient starts to decrease because of oxygen vacancies and migration of ions from one site to other, thus reducing the charge carrier's mobility [53]. A similar tendency was reported by other researchers [54, 55].

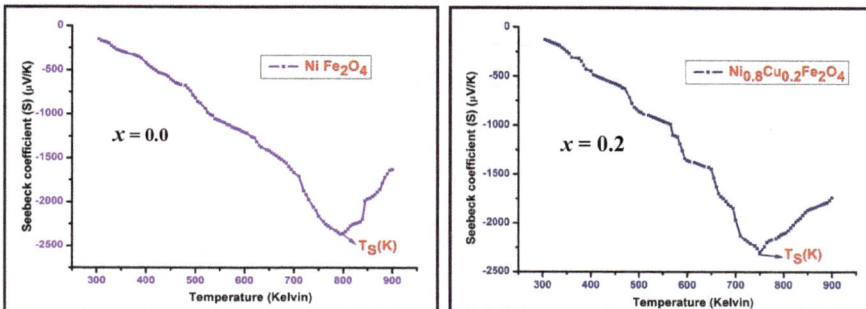

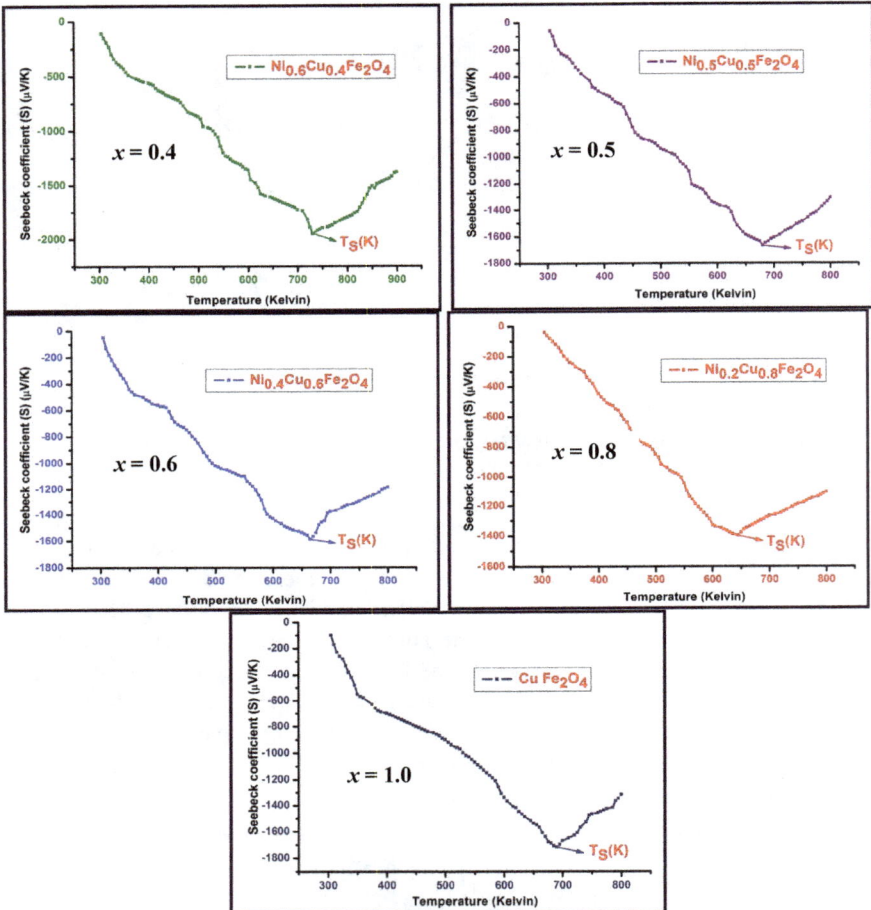

Fig. 14 Variation of thermo electric power with temperature for $Ni_{1-x}Cu_xFe_2O_4$ (0.0 ≤ x ≤ 1.0) ferrites.

Magnetic Oxides and Composites II   Materials Research Forum LLC
Materials Research Foundations **83** (2020) 233-265   https://doi.org/10.21741/9781644900970-10

3.5   Dielectric measurements

3.5.1   Frequency dependence dielectric constant (real-ε')

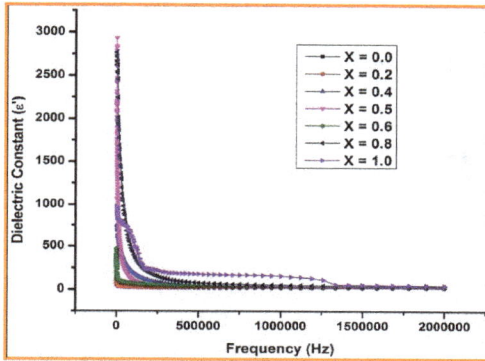

Fig. 15 Variation of dielectric constant (ε') as a function of frequency for $Ni_{1-x}Cu_xFe_2O_4$ (0.0 $\leq x \leq$ 1.0) ferrites.

Fig. 15 represents variation of dielectric constant (real-ε') as a function of frequency for different compositions. It is clear from Fig.15 that at lower frequency the dielectric constant is high, with increasing frequency dielectric constant decreasing and at high frequency, it shows normal behavior of ferrimagnetic materials [56, 57]. Similar trend has been observed by several other investigations [58]. It is due to the lagging of hopping electrons between $Fe^{2+} \Leftrightarrow Fe^{3+}$ ions at localized sites. At high temperature, the Fe^{2+} ions are formed in the samples, these ions are responsible for polarization behind the applied frequency [59, 60].

3.5.2 Frequency dependence dielectric loss tangent (*tan δ*)

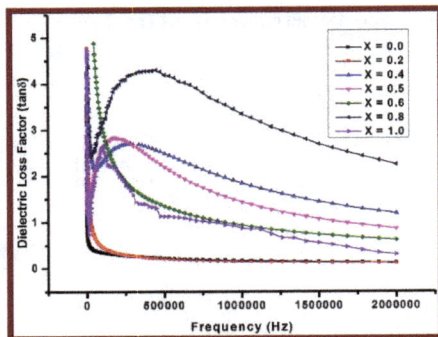

Fig. 16 Variation of dielectric loss tangent (tan δ) as a function of frequency of Ni$_{1-}$$_xCu_xFe_2O_4$ (0.0 ≤ x ≤ 1.0) ferrites.

From Fig. 16, it is clear that for x = 0.0, 0.2, 0.6 compositions, the value of *tan δ* linearly decreases with increasing the frequency, it shows normal behavior of any spinal ferrite materials and for x = 0.4, 0.5, 0.8 and 1.0 compositions, the value of *tan δ* increases initially and exhibits the loss factor that is maximum between frequency from 3×10^5 Hz to 6×10^5 Hz and further decreases by increasing the frequency. It is observed when the jumping frequency between $Fe^{2+} \Leftrightarrow Fe^{3+}$ ions is exactly equal to the applied field frequency [61]. Similar results were reported by others [62].

From Fig. 16, it is clear that for x = 0.0, 0.2, 0.6 compositions, the value of *tan δ* linearly decreases with increasing the frequency, it shows normal behavior of any spinal ferrite materials and for x = 0.4, 0.5, 0.8 and 1.0 compositions, the value of *tan δ* increases initially and exhibits the loss factor that is maximum between frequency from 3×10^5 Hz to 6×10^5 Hz and further decreases by increasing the frequency. It is observed when the jumping frequency between $Fe^{2+} \Leftrightarrow Fe^{3+}$ ions is exactly equal to the applied field frequency [61]. Similar results were reported by others [62].

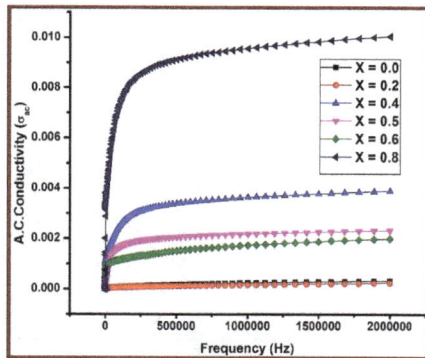

Fig. 17　Variation of ac conductivity (σ_{ac}) as a function of frequency of $Ni_{1-x}Cu_xFe_2O_4$
(0.0 $\leq x \leq$ 1.0) ferrites.

3.5.3　Frequency dependence ac conductivity (σ_{ac})

Fig.17 shows frequency dependence ac conductivity for all compositions. It is clear from Fig.17 that the ac conductivity increases linearly at lower frequency, that confirms the polaron type of conduction and in the higher frequency region almost independent behavior for all the prepared samples, this behavior can be explain using Maxwell-Wagner's polarization model that is in agreement with Koop's theory [63, 65]. The frequency dependent conduction is attributed to small polarons [66]. These results are in good agreement with the other reports [67, 68].

3.5.4　Composition dependence dielectric parameters (ε', *tan δ* and σ_{ac})

Table 5 illustrates the dielectric parameters at 2 MHz. It is clear from Table 5 that dielectric constant is increases with increase of copper concentration, whereas ac conductivity and the dielectric loss factor values are increased up to $x = 0.8$, thereafter it decreased, this can be explain by considering the polarization mechanism that is similar to the conduction mechanism in ferrite system. When Ni^{2+} ions are replaced by Cu^{2+} ions, the number of Fe^{3+} ions increases at octahedral (B) site as well as it will increase the number of Fe^{2+} and Fe^{3+} ion pairs at octahedral sites. Therefore the resistance of grains decreases, and hence increases the probability of electrons to reach the grain boundary, which increases the polarization. Therefore, the dielectric constant increases with an increasing copper concentration. The similar trend has been observed in different compositions, reported by others [69, 70]. The dielectric loss factor increases from $x =$

0.0 to 0.8, thereafter it decreases with increase of copper concentration due to more dispersion. As these samples have low ac conductivity, so they are useful for microwave applications [71].

Table 5 Dielectric parameters of $Ni_{1-x}Cu_xFe_2O_4$ $(0.0 \leq x \leq 1.0)$ nano ferrites at 2 MHz frequency.

Cu content (x)	Dielectric parameters at frequency of 2 MHz		
	ε'	$tan\ \delta$	σ_{ac}
0.0	17.771	0.125	3.32×10^{-04}
0.2	19.109	0.115	2.44×10^{-04}
0.4	23.477	1.189	3.90×10^{-03}
0.5	28.670	1.861	3.32×10^{-03}
0.6	33.036	2.618	3.00×10^{-03}
0.8	40.265	3.245	1.06×10^{-03}
1.0	43.595	0.305	4.94×10^{-01}

Conclusions

➢ $Ni_{1-x}Cu_xFe_2O_4$ $(0.0 \leq x \leq 1.0)$ ferrites were prepared using citrate-gel auto combustion method.

➢ XRD analysis confirms the formation of single cubic phase spinel structure with crystallite size is in between 36.7 nm - 58.91 nm.

➢ The lattice constants and at tetrahedral and octahedral sites hopping length are decreases, X-ray density, porosity increases and decrease in bulk density with substitution of Cu^{2+} content.

➢ SEM images show that formed grains are inhomogeneous and grain size distributions is broad. EDS spectra show presence of Ni, Cu, Fe and O elements and TEM images confirm the crystallites formation is in nano size.

➢ Saturation magnetization and magnetic moment are decrease with increase Cu^{2+} ion concentration. The saturation magnetization decreased from 33.75 emu/gr. to 18.00 emu/gr and the coercivity increased from 165 Oe to 992 Oe with copper substitution.

All compositions show soft magnetic behavior except $x = 1.0$, which shows magnetically hard nature.

➢ Temperature dependent dc resistivity shows that the prepared samples have semiconductor behavior. Prepared samples resistivity is high; these are desirable as core materials in electronic applications like electronic inductors, power transformers, electromagnets and in telecommunication applications.

➢ The activation energy (ΔE) variation with Cu^{2+} ion concentration increases, it represents low conductivity of the ferrite goes hand with high activation energy.

➢ Prepared Ni-Cu nano ferrites thermoelectric power value originates as negative (-Ve) in entire temperature range, it reflects natively charged electrons are majority charge carriers so the prepared materials behave as n-type semiconductors.

➢ In prepared ferrite system, at higher frequencies dielectric loss is low and ac conductivity value is low.

References

[1]   M.K. Nazemi, S. Sheibani, F. Rashchi, V.M. Gonzalez-DelaCruz, A. Caballero, Preparation of nanostructured nickel aluminate spinel powder from spent NiO/Al_2O_3 catalyst by mechano-chemical synthesis, Advanced Powder Technology, 23 (2012) 833–838. https://doi.org/10.1016/j.apt.2011.11.004

[2]   M. M. Rashad, R. M. Mohamed, M. A. Ibrahim, L.F.M. Ismail, E. A. Abdel Aal, Magnetic and catalytic properties of cubic copper ferrite nanopowders synthesized from secondary resources, Advanced Powder Technology, 23 (2012) 315–323. https://doi.org/10.1016/j.apt.2011.04.005

[3]   Xiaoguang Pan, Aimin Sun, Yingqiang Han, Wei Zhang and Xiqian Zhao, Effects of different sintering temperature on structural and magnetic properties of Ni-Cu-Co ferrite nanoparticles, Modern Physics Letters B, 32 (27) (2018) 1850321-1850334. https://doi.org/10.1142/S0217984918503219

[4]   Q Chen, Z. Zhang, Size-dependent superparamagnetic properties of $MgFe_2O_4$ spinel ferrite nanocrystallites, Journal of Applied Physics Letters, 73 (1998) 3156-3158. https://doi.org/10.1063/1.122704

[5]   N. M. Deraz, A. Alarifi, Controlled synthesis, physicochemical and magnetic properties of nano-crystalline Mn ferrite system, International Journal of Electrochemical Science, 7 (2012) 5534-5543.

[6]   H. A. Begum, N. Khatun, S. Islam, N. A. Ahmed, M. S. Hossain, M. A. Gafur and A. Siddika, Synthesis and characterization of structural, magnetic and electrical properties of Ni-Mn-Zn ferrite, International Journal of Nanoelectronics and Materials, 11(1) (2018) 15-24.

[7]   Ebrahim Roohani, Hadi Arabi and Reza Sarhaddi, Influence of nickel substitution on crystal structure and magnetic properties of strontium ferrite preparation via sol-gel auto-combustion route, International Journal of Modern Physics B, 31 (2017) 1750271-1750282. https://doi.org/10.1142/S021797921750271X

[8]   T. Tatarchuk, M. Bououdina, W. Macyk, O. Shyichuk, N. Paliychuk, I. Yaremiy, B. A. Najar and M. Pacia, Structural, optical, and magnetic properties of Zn-doped $CoFe_2O_4$ nanoparticles, Nanoscale Research Letters, 12 (2017) 141-152. https://doi.org/10.1186/s11671-017-1899-x

[9]   D. Sharma and N. Khare, Tailoring the optical bandgap and magnetization of cobalt ferrite thin films through controlled zinc doping, AIP Advances, 6 (2016) 8. https://doi.org/10.1063/1.4960989

[10]   M. T. Jamila, J. Ahmad, S. H. Bukhari, T. Sultan, M. Y. Akhter, H. Ahmad and G. Murtaza, Effect on structural and optical properties of Zn-substituted cobalt ferrite $CoFe_2O_4$, Journal of Organic Research, 13 (2017) (1) 45 to 53.

[11]   A. Chatterjee, D. Das, S. K. Pradhan and D. Chakravarty, Synthesis of nanocrystalline Ni-Zn ferrite by the sol-gel method, Journal of Magnetism and Magnetic Materials, 127 (1-2) (1993) 214-218. https://doi.org/10.1016/0304-8853(93)90217-P

[12]   M. Kaiser, Electrical conductivity and complex electric modulus of titanium doped nickel–zinc ferrites, Physica B, 407 (2012) 606-613. https://doi.org/10.1016/j.physb.2011.11.043

[13]   K. Maaz, S. Karim, A. Mashiatullah, J. Liu, M. D. Hou, Y. M. Sun, J. L. Duan, H. J. Yao, D. Mo, Y.F. Chen, Structural analysis of nickel doped cobalt ferrite nanoparticles prepared by co precipitation route, Physica B: Condensed Matter, 404 (2009) 3947-3951. https://doi.org/10.1016/j.physb.2009.07.134

[14]   Von Avdenne M, Das Elektronen-Rastermikroskop, Zeitschrift für Physik 109 (1938) 553-572. https://doi.org/10.1007/BF01341584

[15] B. D. Cullity, Elements of X-ray Diffraction, 2nd edition, Addison-Wesley, London (1978) 102.

[16] B. Viswanathan and V.R.K. Murthy, Ferrite Materials Science and Technology, Narosa Publishing House, New Delhi (1990).

[17] R.C. Kambale, P.A. Shaikh, S.S. Kamble, Y.D. Kolekar, Effect of cobalt substitution on structural, magnetic and electric properties of nickel ferrite, Journal of Alloys and Compounds, 478 (2009) 599- 603. https://doi.org/10.1016/j.jallcom.2008.11.101

[18] Muhammad Javed Iqbal, Zahoor Ahmad, Yevgen Melikhov, Ikenna Cajetan Nlebedim, Effect of Cu–Cr co-substitution on magnetic properties of nanocrystalline magnesium ferrite, Journal of Magnetism and Magnetic Materials, 324 (2012) 1088-1094. https://doi.org/10.1016/j.jmmm.2011.10.030

[19] D. R. Mane, D. D. Birajdar, Sagar E. Shirsath, R. A. Telugu, R. H. Kadam, Structural and magnetic characterizations of $Ni_{0.7-x}Mn_xZn_{0.3}Fe_2O_4$ ferrite nanoparticles, Physica Status Solidi (a), 207 (2010) 2355-.2363. https://doi.org/10.1002/pssa.201026079

[20] Iqbal, M. J., Ashiq. M. N. Gomez. P. H., Synthesis, physical, magnetic and electrical properties of Al–Ga substituted co-precipitated nanocrystalline strontium hexaferrite, Journal of Magnetism and Magnetic Materials, 320 (2008) 881-886. https://doi.org/10.1016/j.jmmm.2007.09.005

[21] C.B.Kolekar, P.N.Kumble and S.G.Kulkarni, Effect of Gd3+ substitution on dielectric behavior of copper-cadmium ferrites, Journal of Materials Science, 30 (22) (1995) 5784-5788. https://doi.org/10.1007/BF00356721

[22] J. Smit, H.P.J. Wijn, Ferrites, John Wiley, New York (1959) 226-233.

[23] N. F. Mott, E. A. Davis, Electron, IC Processes in Non-crystalline Material, Oxford, London (1979).

[24] S. M. Hoque, Md. A. Choudhury, Md. F. Islam, Characterization of Ni-Cu mixed spinel ferrite, Journal of Magnetism and Magnetic Materials, 251 (2002) 292-303. https://doi.org/10.1016/S0304-8853(02)00700-X

[25] Z. Simsa, Sensitivity and curie point of Li-Zn ferrites, Journal of Physics B, 16 (1996) 919-921. https://doi.org/10.1007/BF01699826

[26] C. Surig, K. A.Hempel, D. Bonnenborg, Hexaferrite particles prepared by sol-gel technique IEEE Transactions on Magnetics 30 (1994) 4092-4094. https://doi.org/10.1109/20.333999

[27] V.V. Kresin, Collective resonances and response properties of electrons in metal clusters, Physics Reports, 220 (1992) 1. https://doi.org/10.1016/0370-1573(92)90056-6

[28] M.Siva Ram Prasad, B.B.V.S.V Prasad, B.Rajesh, K.H.Rao , K.V.Ramesh, Magnetic properties and DC electrical resistivity studies on cadmium substituted nickel–zinc ferrite system, Journal of Magnetism and Magnetic Materials, 323 (2011) 2115-2121. https://doi.org/10.1016/j.jmmm.2011.02.029

[29] R. D. Waldron, Infrared Spectra of Ferrites, Physical Review, 99 (1955) 1727-1735. https://doi.org/10.1103/PhysRev.99.1727

[30] S. T. Hafner, Z. Crystallogr. Ordung/unordung and Ultrabsorption IV Die Adrsorbtion einiger Minerale mit spinel, 115 (1961) 331-340. https://doi.org/10.1524/zkri.1961.115.5-6.331

[31] V. A. M. Brabers, Infrared spectra of cubic and tetragonal manganese ferrites, physica status solidi(b), 33 (1969) 563-572. https://doi.org/10.1002/pssb.19690330209

[32] M.H.R. Khan, A.K.M. Akther Hossain, Reentrant spin glass behavior and large initial permeability of $Co_{0.5-x}Mn_xZn_{0.5}Fe_2O_4$, Journal of Magnetism and Magnetic Materials, 324 (2012) 550-558. https://doi.org/10.1016/j.jmmm.2011.08.039

[33] A. Maqsood, K. Khan, M. Anis-ur-Rehman, M.A. Malik, Structural and Electrical Properties of Ni-Co Nanoferrites Prepared by Co-precipitation Route Journal of Superconductivity and Novel Magnetism, 24 (2011) 617–622. https://doi.org/10.1007/s10948-010-0956-9

[34] S.A. Shahid, B. Shahzad , A. Naz, J.N. Ramiza, M. Yaseen, Convenient synthesis of ni-zn ferrites from metal chlorides, Journal of Faculty of Engineering & Technology 19 (2012) 43-50

[35] Ahmed Faraz, Mudasara Saqib, Nasir M. Ahmad, Synthesis, structural, and magnetic characterization of $Mn_{1-x}Ni_xFe_2O_4$ spinel nanoferrites, Journal of Superconductivity and Novel Magnetism, 25 (2012) 91-100. https://doi.org/10.1007/s10948-011-1212-7

[36] A. M. Abdeen, O.M. Hemeda, E.E.Assem, M.M.El-Sehly, Structural, electrical and transport phenomena of Co ferrite substituted by Cd, Journal of Magnetism and Magnetic Materials, 238 (2002) 75-83. https://doi.org/10.1016/S0304-8853(01)00465-6

[37] S. S. Bellad, R. B. Pujar, B. K. Chougule, structural and magnetic-properties of some mixed Li-Cd ferrites, Materials Chemistry and Physics, 52 (1998) 166-169. https://doi.org/10.1016/S0254-0584(98)80019-9

[38] Xiwei Qi, Ji Zhou, Zhenxing Yue, Zhilum Gui, Longtu Li, Effect of Mn substitution on the magnetic properties of MgCuZn ferrites, Journal of Magnetism and Magnetic Mateials, 251 (202) 316-322. https://doi.org/10.1016/S0304-8853(02)00854-5

[39] I. H. Gul, A. Maqsood, Structural, magnetic and electrical properties of cobalt ferrites prepared by the sol–gel route, Journal of Alloys and Compounds, 465 (2008) 227-231. https://doi.org/10.1016/j.jallcom.2007.11.006

[40] Q. Zeng, I. Baker, V. McCreary, Zh. Yan, Soft ferromagnetism in nanostructured mechanical alloying Fe-Co based powers, Journal of Magnetism and Magnetic Materials, 318 (2007) 28-38. https://doi.org/10.1016/j.jmmm.2007.04.037

[41] V. D. Reddy, M. A. Malik and P. V. Reddy, Electrical transport properties of manganese-magnesium mixed Ferrites, Materials Science and Engineering B, 8 (4) (1991) 295-301. https://doi.org/10.1016/0921-5107(91)90050-6

[42] M. U. Islam, I. Ahmad, T. Abbas, M. A. Chaudhry, R. Nazmeen, Advanced Materials, Proceedings of the 6th International Symposium, (1999) 155-158

[43] A. M. Sankpal, S. R. Sawant, A. S. Vaingankar, Relaxation time studies on $Ni_{0.7}Zn_{0.3}Al_xFe_{2-x}O_4$ and $Ni_{0.7}Zn_{0.3}Cr_xFe_{2-x}O_4$, Indian Journal of Pure and Applied Physics, 26 (1988) 459-462

[44] J. Bijal, S. Phanjouban, D. Kothari, C. Prakash, P. Kishan, Hyperfine interactions and magnetic studies of Li-Mg ferrites, Solid State Communication, 83 (1992) 679-682. https://doi.org/10.1016/0038-1098(92)90144-X

[45] S. Manjura Hoque, Md.Amanullah Choudhury, Md. Fakhrul Islam, Characterization of Ni-Cu Mixed Spinel Ferrite, Journal of Magnetism and Magnetic Materials, 251 (2002) 292-303. https://doi.org/10.1016/S0304-8853(02)00700-X

[46] D. Ravinder, K.Vijaya Kumar and B.S. Boyanor, Elastic behaviour of Cu–Zn ferrites, Materials Letters, 38 (1999) 22-27. https://doi.org/10.1016/S0167-577X(98)00126-8

[47] M. El-Shabasy, DC electrical properties of Zn-Ni ferrites, Journal of Magnetism and Magnetic Materials, 172 (1997) 188-192. https://doi.org/10.1016/S0304-8853(97)00014-0

[48] M. U. Islam, M. A. Chaudhry, T. Abbas, M. Umar, Temperature dependent electrical resistivity of Co–Zn–Fe–O system, Materials Chemistry and Physics, 48 (1997) 227-229. https://doi.org/10.1016/S0254-0584(96)01890-1

[49] Y. Irkhin, E.Turnov, Physics, Journal of Experimental and Theoretical Physics, 33 (1957) 673.

[50] D. Ravinder, K. Latha, Electrical conductivity of Mn–Zn ferrites, Journal of Applied Physics, 75 (1994) 6118-6120. https://doi.org/10.1063/1.355479

[51] T. E. Whal, N. Salerno, Y. G Projkova, K. A. Mizza and S. Mazen, The electrical conductivity and thermoelectric power of lithium ferrite in the vicinity of the order-disorder transition temperature, journal of philosophical magazine B, 53 (5) (1986) 107-113. https://doi.org/10.1080/13642818608240648

[52] L.G. Van Uitert, DC resistivity in the nickel and nickel zinc ferrite system, Journal of Chemical Physics, 23 (1955) 1883-1887. https://doi.org/10.1063/1.1740598.

[53] T. T. Srinivasan, P. Ravindranathan, L.E. Cross, R. Roy, R.E. Newman, S.G. Sankar, Studies on high-density nickel zinc ferrite and its magnetic properties using novel hydrazine precursors, Journal of Applied Physics, 63 (1988) 3789-3791. https://doi.org/10.1063/1.340615

[54] M. I. Klinger, Electron conduction in magnetite and ferrites, Physica Status Solidi (B), 79 (1) (1977) 9-48. https://doi.org/10.1002/pssb.2220790102

[55] T. Gron, S. Mazur, H. Duba, J. Krok-Kowalski and E. Maciazek, Effect of double exchange on thermoelectric power of $Cu_xCo_yCr_zSe_4$, Journal of Alloys and Compounds, 467 (1-2) (2009) 112-119. https://doi.org/10.1016/j.jallcom.2008.01.001

[56] P. K. Gupta, C.T. Hung, Magnetically controlled targeted micro-carrier systems Life Sciences, 44 (1989) 175-186. https://doi.org/10.1016/0024-3205(89)90593-6

[57] Chandra Babu B, Naresh V, Jayaprakash B, Buddhudu S, Structural, thermal and dielectric properties of lithium zinc silicate ceramic powders by sol-gel method, journal of ferroelectric letters section , 38 (2011) 124-130. https://doi.org/10.1080/07315171.2011.623610

[58] M.R. Anantharaman, S. Sindhu, S. Jagatheesan, K.A. Malini, P. Kurian, Dielectric properties of rubber ferrite composites containing mixed ferrites, Journal of Physics. D, 32 (1999) 1801-1810. https://doi.org/10.1088/0022-3727/32/15/307

[59] R.V. Mangalaraja, S. Ananthakumar, P. Manohar, Magnetic, electrical and dielectric behaviour of $Ni_{0.8}Zn_{0.2}Fe_2O_4$ prepared through flash combustion

technique, Journal of Magnetism and Magnetic Materials, 253 (2002) 56-64.
https://doi.org/10.1016/S0304-8853(02)00413-4

[60] S. Sindhu, M.R. Antharaman, B.P. Thampi, K.A. Malini, Dielectric properties of
      rubber ferrite composites, Bulletin of Materials Science, 24 (2001) 623-631.
      https://doi.org/10.1007/BF02704011

[61] A. A. Zaky and R. Hawley, Dielectric solids, Routledge and Kegan Paul Ltd.
      London (1990) 21-27.

[62] N. Rezlescu and E. Rezlescu, Dielectric properties of copper containing ferrites,
      physica status solidi (a) 23 (1974) 575. https://doi.org/10.1002/pssa.2210230229

[63] S.F. Mansour, Frequency and composition dependence on the dielectric properties
      for Mg-Zn Ferrite, Journal of Solids, 28 (2) (2005) 263-273

[64] J. C. Maxwell, Electricity and Magnetism, Oxford University, Press, London,
      (1973).

[65] W. A. Yager, The distribution of relaxation times in typical dielectrics. Journal of
      Applied Physics, 7(12) (1936) 434–450. https://doi.org/10.1063/1.1745355

[66] K. Iwauchi, Dielectric properties of fine particles of Fe_3O_4 and some ferrites,
      Journal of Applied Physics, 10 (1971) 1520 – 1523.
      https://doi.org/10.1143/JJAP.10.1520

[67] D. Alder and J. Fienleib., Electrical and optical properties of narrow-band
      materials, Physics Revision B, 2 (1970) 3112-3134.
      https://doi.org/10.1103/PhysRevB.2.3112

[68] K. M. Batoo, S. Kumar, C. G. Lee, A. Current, Influence of Al doping on electrical
      properties of Ni–Cd nano ferrites, Current Applied Physics, 9 (2009) 826-832.
      https://doi.org/10.1016/j.cap.2008.08.001

[69] M. Ajmal and A. Maqsood, conductivity, density related and magnetic properties of
      $Ni_{1-x}Zn_xFe_2O_4$ ferrites with the variation of zinc concentration Materials Letters, 62
      (2008) 2077-2080. https://doi.org/10.1016/j.matlet.2007.11.019

[70] A. K. Singh, T.C. Goel, R.G. Mendiratta, Dielectric properties of Mn-substituted
      Ni–Zn ferrites, Journal of Applied Physics, 91 (2002) 6626-6629.
      https://doi.org/10.1063/1.1470256

[71] J. Azadmanjiri, H. K. Salehani, M.R. Barati, F. Farzan, Preparation and
      electromagnetic properties of $Ni_{1-x}Cu_xFe_2O_4$ nanoparticle ferrites by sol–gel auto-
      combustion method, Materials Letters, 61 (2007) 84-87.
      https://doi.org/10.1016/j.matlet.2006.04.011

Keyword Index

About the Editors

Dr. Rajshree B. Jotania is a professor of Physics, Department of Physics, Electronics and Space science, University School of Sciences at Gujarat University, Ahmedabad, India. She obtained her B.Sc., M.Sc. and Ph.D from Saurashtra University, Rajkot, India. She was Junior Research Fellow (DAE-BRNS project) during 1987 to 1989 at Physics Department, Saurashtra University, Rajkot, India. She obtained a few regional, and national awards for contribution toward scientific research. She worked at National Chemical Laboratory, Pune, India for two months as a Summer Visiting Teacher Fellow in 2005 and as a Visiting Scientist fellow in 2011. She possesses 30 years of teaching experience at UG and PG level. She is a member of Board of studies at few Universities of Gujarat, India and a Mentor of DST-INSPIRE (Department of Science and Technology-Innovation in Science Pursuit for Inspired Research) program. She has published more than 100 papers in various research journals and conference proceedings. She has delivered more than 20 invited talks at various DST-INSPIRE Internship science camp in India. She has edited three books entitled 'Ferrites and ceramic composites' (Vol. I & II, Trans Tech Publisher, Switzerland) and Magnetic Oxides and Ceramic Composites (MRF, USA). She has visited Singapore, Malaysia, New York and North Africa for research work. She has attended more than 50 international, national conferences/symposiums/seminars/ academy meeting and worked as a chair person as well as delivered invited talks in a few international and national conferences. She possesses a life membership of eight professional bodies and she has guided six Ph. D, ten M. Phil students. At present few more students are working under her guidance for M.Phil and Ph. D. To date she has completed five research projects of various agencies. She has worked as deputy co-coordinator, DRS (SAP-I) program. She is an active member of Indian Association of Physics Teacher.

S. H. Mahmood obtained his B.Sc. degree in Physics from The University of Jordan, Amman in 1978, and his PhD degree in Physics from Michigan State University, East Lansing, Michigan, USA in 1986. Between 1986 and 2010, he was a faculty member at Yarmouk University, Irbid, Jordan, and since 2010, he is a professor of physics at The University of Jordan, Amman. During his academic career, he was involved in teaching, research, graduate work supervision, and administration. He held the positions of Director of the Center for Theoretical and Applied Physical Sciences, Chairman of Physics, Dean of Science, Dean of Scientific research and Graduate Studies, and Vice President. He published more than 130 articles in peer-reviewed international journals, participated in tens of regional and international conferences, and supervised tens of M.Sc. and PhD theses. He also received several national, regional and international Awards and Honors for Academic excellence and contribution to science. Also, he participated in the management and execution of nationally and internationally funded projects concerned with establishing long-term research programs, new academic programs, capacity building, and curricular development. Additionally, he actively participated as a scientific advisor, and a member of scientific committees and councils of Scientific Research Funds in Jordan, and of editorial boards of international journals.

About the Authors

Dr. Khaled Snini is working as a Ph.D student in Physics, Department of Physics at Sfax University, Tunisia. His research work focuses on magnetic, electrical and dielectric properties of manganite perovskites, magnetic materials and energy materials. He has published 3 research papers in International Journals Dr. Khaled visited many countries for collaborative research work and for presentation of papers and invited talks.

Prof. Dr. Mohamed Ellouze is full professor in physics, in magnetic, crystallographic field at Sfax University, Tunisia. He is a member of the International Center of Diffraction Data (ICCD) in metallic and alloys. He is a former president of the Maghreb Alexander Von Humboldt Alumni. He works in Physics department, Faculty of Sciences of Sfax, Sfax University, Tunisia. He is a reviewer in some international journals with impact factor. He has published more than hundred research papers and two book chapters. He has guided 12 Ph. D. students. He has attended more than forty international conferences and worked as a chair person in four international conferences.

Prof. Ibrahim Bsoul is working as a professor in Physics, Department of Physics, Al al-Bayt University, Mafraq, Jordan since 19 years. He obtained B.Sc. in Physics and Ph.D. in Solid State Physics (Ferroelectric Materials) degrees from Dnepropetrovsk State University, Ukraine, and his research work focuses on Ferroelectrics and magnetic materials. He is working on the synthesis, structural and magnetic properties of various types of hexaferrites, spinel ferrites, and garnets, with focus on the nanocrystalline systems. To his credit, published 63 research papers (h-index 18, > 1000 citations) in International Journals. In addition, he published two chapters in Materials Research Foundations book series.

Prof. Ebtesam E. Ateia is a Former Vice Dean for Education and Students Affairs, Faculty of Science, Cairo University, Egypt. She is working as a Professor in Physics, Department of Physics, Cairo University, Egypt since 14 years. Her research work focuses on Ferrites and Multiferroic materials, Polymer Blends, Swelling properties of polymer composites, Physical properties (Electrical, Mechanical, Magnetic, Structural, Optical and Thermodynamics Properties of Solids), Nanotechnology, Nano composites, Graphene Science and Technology, and Nano-materials (Nanotechnology). She has published more than 78 research papers in different International Journals. She is an Editorial Board member of Material Science Research-India, American Journal of Modern Physics and Advanced Journal of Chemistry. She has supervised 25 M. Sc. and 20 Ph. D. dissertations. She has been awarded the International Publication award from Cairo University, Egypt in 2007 and 2019. Prof. Ebtesam Ateia also visited France, China, and Spain, for Collaborative Research and conference papers presentation and invited talks.

Mohamed Farag Shokry is a researcher at Egypt Nanotechnology Center (EGNC), Cairo University, El-Sheikh Zayed, Giza, Egypt. He graduated from Physics department, Faculty of science, Ain shams university, Egypt in 2009. He was awarded the master degree in March 2020. He has published two research papers in Applied Physics A Materials Science & Processing and Journal of Inorganic and Organometallic Polymers and Materials in 2018 and 2020 respectively.

Dr. Rapolu Sridhar is an Associate Professor of Physics, Department of Basic Science & Humanities at Vignan Institute of Technology and Science, Hyderabad, India. His area of research is nanomaterials, material science and magnetic materials. He obtained his M.Sc. (Physics) from Osmania University, Hyderabad and Ph.D. from Jawaharlal Nehru Technological University, Hyderabad, Telengana India. He is involved in teaching since last 21 years and possesses total 10 years of research experience. He has published eleven research papers in peer-reviewed SCOPUS

and UGC approved journals, two chapters in peer reviewed books as well as two text books. He has attended and presented more than thirty five research papers at national and international conferences, organized Academic and Research events. He has worked as a reviewer for Engineering Physics text book published by Tata McGraw – Hill publication. He possesses life membership in Magnetic Society of India (MSI), Senior Membership (Life Time) in the Asia Society of Researchers (ASIR), Profession Membership (Life time) in the Institution for Engineering Research and publications (*IFERP*).

Prof. Ravinder Dachepalli is working as a professor in Physics, Department of Physics, Osmania University, Hyderabad, India since thirty six years and his research work on magnetic and electrical properties of nano-ferrites,magnetic materials, ferrites, Energy materials, thin films, GMR materials, Cu-Co alloy thin films and nano-materials (Nanotechnology) by pulsed laser deposition, sol-gel, citrate precursor method and electro deposition. To his credit he has published 199 research papers in various International Journals. He has been awarded Young Scientist award received by Dr. Abdual Kalam (former president of India), for outstanding contributions in the field of science and Technology, UGC career award, Boyscast fellowship by DST (Department of science and technology, Government of India, JSPS fellowship Japan and Royal Society fellowship, UK. Prof. Ravinder also visited USA, UK, Sweden, Ireland, Canada, Singapore and Japan for collaborative research, conference papers presentation and invited talks. He has completed five research projects and guided ten Ph.D. students. At present eight more students are working under his guidance for Ph.D.

Dr. K. Vijaya Kumar is Professor of Physics and Coordinator, Directorate of University Academic Audit Cell, Jawaharlal Nehru Technological University Hyderabad, Telangana State, India. He is having more than twenty three years of teaching, admistation and research experience. He works on magnetic and electrical properties of nano-ferrites, magnetic materials, thin films and nano-materials by sol-gel, citrate precursor method and citrate gel auto combustion method. He has published more than 60 research papers in peer-reviewed SCOPUS and UGC approved journals. He has

attended and published more than 80 research papers in national and international conferences. He authored good number of books for different disciplines of undergraduates. He also organized Academic & Research events such as UGC sponsored orientation courses, short term courses, refresher courses, workshops. He is reviewer and member in editorial boards for reputed journals. Two research projects are ongoing under TEQIP-III, JNTUH, Hyderabad. Under his guidance; seven students are received Ph.D degree and six more students are pursuing the Ph.D. research work. He is life time member in several Professional Bodies. He has received many awards: Adarsh Vidya Saraswathi Rasthriya Puraskar from Glacier Journal of Research Foundation, Ahmedabad, India; Best Young Faculty Award for the Futuristic and Outstanding best practices in the field of education From DK International Research Foundation, Perambalur, Chennai; Associate Fellow Award for Contributions to Science and Technology, Telangana Academy of Sciences, Osmania University, Hyderabad, Telangana, India; Best Citizens Of India, In Recognition of Exceptional Caliber and Outstanding Performance in the Chosen Area by International Publishing House, New Delhi, India; Vishista Seva Puraskar for outstanding services towards academic and administrative activities by JNTUH College of Engineering Jagitial, India.

Dr. G. Helen Ruth Joice is an Assistance Professor of Physics, department of Physics at Thiru Kolanjiappar Govt. arts college, Vridhachalam, Tamil Nadu, India. She obtained M.Sc (Physics) degree from Bharathidasan University, Turuchirapalli, Tamil Nadu, India and Ph.D from Manonmaniam Sundaranar University, Tirunelveli, India. Her interested area of research is nanomaterials, material science and magnetic materials. She has total 20 years teaching experience along with 10 years research experience. She has published 7 research papers in peer-reviewed journals and authored two text books. She has attended and presented more than 13 research papers at national and international conferences and organized two research events sponsored by Tamil Nadu State Council for Higher Education, India.

Dr. Amar Keshao Nandanwar is currently working as a research student in Department of Physics, Dr. Ambedkar College Deeksha Bhoomi Nagpur, India affiliate to RTM Nagpur University, India. He has worthy academics and teaching experience to UG and PG students. He is involved in research work on magnetic and electrical properties of nano-ferrites, magnetic materials, ferrites, Energy materials, GMR materials, Sensors, Dielectric loss, Biomedical Application, Hyperthermia, Super-paramagnetism, Spinel Ferrites, Hexaferrites and nano-materials (Nanotechnology) by sol-gel auto combustion method and co-precipitation method. He is interested in various characterizations techniques. He has published many research papers on material science in reputed journals and presented his research work in many international conferences. He has been awarded Rastrasant Tukdoji Maharaj Nagpur University memorial research fellowship.

Dr. Kishor Govindrao Rewatkar is an Associate Professor in Department of Physics at Dr. Ambedkar College, Nagpur (DACN), India and teaching to UG and PG and Post PG students including Nanoscience and Nanotechnology diploma. He is also faculty and Coordinator in the Nanoscience and Nanotechnology, DACN. He was faculty member of Board of studies in Physics of RTM Nagpur University, Indai and was Research Recognition committee member of Gondwana University, Gadchiroli, India during 2010-2015 and 2013-2018 respectively. He is also member of various academic, professional and administrative organizations. He is president of Marathi Vidnyan Parishad, Nawargaon Chapter, Dist. Chandrapur and Founder secretary of Society for Technologically Advanced Materials of India (STAMI).

Dr. Kishor Rewatkar is an author of six books for UG students of RTM Nagpur University and published chapters in Internationally published book and also to his credit more than 127 national and international research papers and received awards including Marquis Who's Who of America 1999. His research is focused on the development of nanoferrites, magnetic and electrical properties of nano-ferrites, magnetic materials and ferrites for various technological applications. He is also reviewer of several International journals. Twenty seven students have been received Ph. D degree under his guidance and ten more students are pursuing their Ph. D degree registered at RTM Nagpur University, India.

Dr. Kishor has completed four major and minor research projects. Four Indian government patents are on his credit. He has successfully organized national and International conferences, seminars and workshops in and outside of India. He has visited Sri Lanka, Thailand, Malaysia, Singapore, Uzbekistan, France, Germany, Belgium, Netherland, Italy, England, wherein he visited internationally acclaimed universities and academic institutions.

Dr. Ratiram Gomaji Chaudhary is currently working as an Assistant Professor and Head of Post Graduate Department of Chemistry, S. K. Porwal College of Arts, Science and Commerce, Kamptee, India. He did his Ph.D. in Chemistry under the faculty of Science and Technology, RTM, Nagpur University, Nagpur. His research areas are bionanomaterials, carbon-based nanomaterials, metal oxide nanoparticles, coordination polymers, antimicrobial assay, catalysis and photocatalytical performances. He has completed a 'Major Research Project' funded by SERB/DST, New Delhi, India. He is the RTM Nagpur University, Nagpur's recognized Ph.D supervisor, and under his guidance two students have been awarded Ph.D, four students are working for Ph. D degree at present. Moreover, thirty five M.Sc Projects dissertations have been completed under his guidance. To his credit he has two books and published more than seventy two research articles in peer-reviewed SCI and Scopus indexing journals. He is a recipient of three prestigious awards viz. *'Rajiv Gandhi National Fellowship Award'* as JRF, *'Young Scientist Award'* and *'Award of Appreciation.* He is a reviewer of twenty reputed journals and reviewed more than fifty research articles. He is a review editor and managing guest editor of two reputed journals. He has organized one national conference (NCSCA-2019) in Nagpur, India; he delivered five Invited talks, he is a member of several scientific bodies.

Mr. Ajay Potbhare is a Research Scholar in the Department of Chemistry, S. K. Porwal College of Arts, Science and Commerce, Kamptee, India. His research focuses on the Nanostructured bionanomaterials, carbon-based nanocomposites, metal oxide/phosphide nanoparticles, polymer nanocomposites, antimicrobial assay, Li-ion battery, Supercapacitor, Electrocatalysis, energy storage and photocatalytic performances. He has published 15 peer-reviewed journal articles, one book chapter. He has attended several National and International the conferences and symposiums. He is a recipient of two prestigious

awards viz. *'CSIR-UGC'* as JRF, and *Award of Appreciation.* He is a reviewer of one reputed journal, and reviewed more than 5 research articles.

Mr. Prashant Bhanudasji Chouke is an Assistant Professor of Chemistry in Government Polytechnic, Bramhapuri and Head of the Department, Science and Humanities Department, Government Polytechnic, Bramhapuri, India. He has 15 year teaching experience in education at various reputed institutes. He has completed his M. Phil. in Chemistry under the faculty of Science from Rashtrasant Tukadoji Maharaj, Nagpur University Nagpur, India under the supervision of Dr. V. N. Ingle, Ex-Head Department of Chemistry RTM, and Nagpur University Nagpur. His research is focus on synthesis, Catalyst, Schiff-bases, Green synthesis, Antimicrobial activity, Photocatalytic activity. He is a recipient of. *"Rajiv Gandhi National Fellowship Award"* as JRF. He has published 10 peer-reviewed journal articles and reviewed two research articles, and attended several national and international conferences.

Dr. Alok Ramkesh Rai is currently working as an Assistant Professor and Head of Post Graduate Department of Microbiology, S. K. Porwal College of Arts, Science and Commerce, Kamptee, Indai. He did his PhD. in Microbiology under the faculty of Science and Technology, RTM, Nagpur University, Nagpur, India. His research areas are bionanomaterials, Biological application of nanoparticles, metal oxide nanoparticles, antimicrobial assay, microbial diversity and Plant microbial interaction. He has completed one 'Major Research Project' funded by UGC, New Delhi, India. He is the RTM Nagpur University, Nagpur's recognized Ph. D supervisor, and under his guidance, two research students are working. Fifty M.Sc Projects dissertations have been completed under his guidance. To his credit, he has two books and published more than twenty five research articles in peer-reviewed SCI and Scopus indexing journals. He has worked experience of more than eight years as an Senior Research fellow, Junior Research fellow in different reputed projects funded by Indian Council of Agriculture Research (ICAR). He is a reviewer of reputed journals and reviewed more than fifteen research articles. He organized one International conference (Future Tech of Life Sciences Feb.2020), he delivered five Invited talks, and is a member of several scientific bodies.

Raghvendra Kumar Mishra is currently working as an EPSRC funded researcher at Cranfield University, United Kingdom. He was working as Research Assistant at the IMDEA Materiales, C/ Eric Kandel, 2, Tecnogetafe, 28906 Getafe, Madrid - Spain from the year 2018-2019. He has widely studied the processing of blends, in-situ generation micro and nano fibrillar composites, the electromagnetic shielding effect of nanocomposites, decorating and alignment of carbon nanotubes and thermal, dynamic mechanical and structural relationship in polymer blends and nanocomposites. He has received several awards from different organizations and technology events such as best presentation, best reviewers (Vacuum, Elsevier). He is a reviewer in many international journals e.g. Environmental Chemistry Letters (Springer), Cellulose, Vacuum and so on. He has worked as well research experience in Mechanical engineering, Materials science and Technology, and Nanoscience and Nanotechnology. He has published eleven books and more than seventy research articles in peer-reviewed SCI and Scopus indexing journals.

Prof. Dr. Martin Federico Desimone was born in Buenos Aires-Argentina. He graduated from Pharmacy, Biochemistry and received his Ph. D degree from the University of Buenos Aires, Argentina. After a postdoctoral period in the University of the Basque Country, Spain, he joined in 2008 the Laboratoire de Chimie de la Matière Condensée de Paris. Currently, he is a CONICET researcher and Professor in the Faculty of Pharmacy and Biochemistry at the same University. His current focus is on the design of functional organic, inorganic and biological hybrid materials and the study of interactions on the nano- and microscale for biomedical and biotechnological applications. He was recipient of awards and distinctions including the participation in the 2016 Emerging Investigator themed issue of the Journal of Materials Chemistry B, the Award Innovar in the category applied research received in two consecutive years (2016 and 2017) from the Ministry of Science, Technology and Productive Innovation (MINCYT), the distinction "Dr. José A. Balseiro" XV edition, received in The Argentine Senate, which is the upper house of the Argentine National Congress and the "Academic Excellence" UBA.

Dr. Ahmed Abdala is an Associate Professor in the Chemical Engineering Program at Texas A&M University at Qatar. He received his Ph.D. in Chemical Engineering from North Carolina State University in 2003. His research focuses on polymers, nanomaterials, and nanocomposites including polymer nanocomposites, functionalized 2D nanomaterials, anticorrosion coatings, polymeric membranes for gas separation and water treatment, and nanohybirds of metal/metal oxides and 2D carbon nanomaterials for catalysis, adsorption, energy storage, and thermoelectric materials. He published 68 peer-reviewed journal articles, 7 book chapters, and 22 refereed conference papers (H-index= 31, total citations = 13500). He is also the inventor of 7 US patents. Dr. Abdala is the recipient of Distinguished Researcher Award from the Indian Association of Solid-State Chemists & Allied Scientists (ISCAS) in 2019, and Qatar Excellence in Innovation Award from Thomson Reuters in 2016, "Science Lantern Faculty Award" from Abu Dhabi National Oil Company in 2014.

Sílvia L. Soreto Teixeira, is a postdoctoral researcher at the Department of Physics-I3N, University of Aveiro, Portugal. At this moment, she is dedicated to the investigation of materials namely ferrites and composites containing nanoparticles for energy storage which can also be applied in biomedical applications. Her work focuses, essentially, on the synthesis of materials and their characterization (structural, morphological, electrical and magnetic).

She is the author and co-author of twenty one peer-reviewed papers and nine book chapters. She has also reviewed articles in international scientific journals. She has made more than thirty oral/poster communications and has participated in several scientific projects such as "Pessoa Program", "Research and Development Cooperation Slovakia-Portugal", "Agreement FCT Portugal-CNRST Morocco" and "Research and Development Cooperation Morocco-Portugal". She received an "Oral Presentation Award-First Rank" award at the "ISyDMA2018" conference that took place in April 2018. She has experience in organizing conferences and has participated in the organization of two events.

Manuel Pedro Fernandes Graça, is working as a Professor in the Physics Department of the University of Aveiro, Portugal and member of the Institute of nanostructures, nanomodelling, and nanofabrication (I3N). He is graduated in Physics Engineering, Master degree in Science and Materials Engineering and completed the Ph.D. degree in Physics at University of Aveiro (UA) in Portugal and UFC-LOCEM in Brasil. After the Ph.D. he worked as a director of the Energy Department of Prirev-Surface Technology S.A. and after as Principal Researcher at UA. He is specialized in experimental techniques in the field of solid state physics, with major focus in electrical and magnetic properties of solid materials and in micro and nano materials processing techniques. MPG is the author of one book, twenty one chapter books and two hundred twenty articles in international peer-review journals, plus twelve in conference proceedings, more than three thousand five hundred citations and h-index of 29. He was chief Editor of the book "Electrical Measurements: Introduction, Concepts, and Applications" published by the Nova Science Publishers. MPG gave more than fifty oral communications including international plenary sessions. He is a co-author of more than two hundred posters in international scientific meetings. He was supervisor of fourteen post-doc projects, eight Ph.D. students (five students have already finished) and five research fellows with industrial cooperation. MPG completed the supervision of sixteen final course theses and twenty two master theses. In 2010 he was elected member of Aveiro University Physics Department Council and between 2013 and 2016 he was a member of the Scientific Council of Aveiro University, Portugal.

Luís Cadillon Costa is working as a Professor in the Physics Department of the University of Aveiro, Portugal and member of the research laboratory I3N, classified as Outstanding. His research activity has been dedicated to the synthesis and characterization of materials for applications in electronics and electrical engineering. He is member of the editorial board of three scientific journals. He is co-author of about one hundred sixty papers in journals of Science Citation Index (SCI), seven books, sixteen book chapters and referee in forty five international scientific journals. He has about four hundred communications in conferences, being fourteen invited plenary talks, about two thousand citations and h-factor 23. He participated in forty six I&D projects, being coordinator of thirteen. He is responsible for the supervising of fifty three students, of Bachelor, Master, Ph.D and

Post-Doc. He is director of the Ph.D program in Physics Engineering, member of the Scientific Council of the University of Aveiro and member of the Installation Committee of the Faculty of Sciences of the University of Timor Leste.

Manuel Almeida Valente (13/05/1955) is working as an associate professor in Physics, Department of Physics, University of Aveiro, Portugal since 39 years with scientific interests on: Magnetic properties of nano-ferrites, magnetic materials, ferrites; and Electric Properties, Dielectric spectroscopy; Ferroelectric materials; multiferroics; Non-crystalline materials. He is the author or co-author of: 1 Books; 10- Chapters of books; 240 Articles International journal with referee; 13 conference Proceedings; 6 Invited oral communications in congress; 30 Oral communications; > 250 "posters" Communications in Conferences;

He has also: > 2800 citations (ISI Web of Knowledge/Scopus); H factor – 30; Referee in > 50 International Scientific Journals; 49 projects, being coordinator of 9 projects; 3 Ph.D; 11 Post-docs supervisings; 45 Master / project supervising. International Collaboration with 13 Universities and Institutes. Coordinator of the Research Group "Materials with enhanced electrical and magnetic properties".

Dr. M. Raghasudha is working as an Assistant Professor in Chemistry, Department of Chemistry, National Institute of Technology, Warangal, India since 2018. She has a total of 23 years of teaching experience with various positions. She has been awarded a Post Doctoral Fellowship by UGC, New Delhi at Department of Chemistry, University College of science, Osmania University, Hyderabad, India for three years till March 2018. Her research interests include Nanomaterials in broad and synthesis of metal oxide nano particles, magnetic materials, nanoferrites, their electrical and magnetic properties, and their biomedical applications, catalytic and photocatalytic applications for environmental remediation, Synthesis of nanomaterials using medicinal plant extracts by Sol-gel method, Hydro thermal method and their biological applications. To her credit she has published 35 research papers in reputed international Journals She is an Associate Fellow of Telangana Academy of

Science, Life member of Indian Science Congress Association, Life member of Magnetic Society of India, Life member of Materials Research Society of India. She is a Peer Reviewer of many international Journals.

Dr. Aniruddha Mondal has now been working Postdoctoral Research Fellow in the Department of Chemical Engineering and Biotechnology, Tatung University, Taipei, Taiwan since 2019. He received his PhD from the Department of Inorganic Materials and Catalysis Division (IMCD), Council of Scientific and Industrial Research-Central Salt and Marine Chemicals Research Institute (CSIR-CSMCRI), Bhavnagar, India, in October 2018. Additionally, Dr Mondal has received Post Graduate Diploma in Patents Law from Nalsar Law University, 2015. His main research focuses on the synthesis of different dimensional based nanostructured materials, nanocomposites of functionalised 2D nanomaterials and polymer-based, synthesis of corrosion inhibitor, Nanostructured material of metal/metal oxides/metal sulfide/metal selenide/carbide for Li-ion Battery, Supercapacitor, Electrocatalysis, Photocatalysis, Photoelectrochemical cell, Urea electrolysis, Methanol Oxidation, Heterogeneous catalysis in simple organic transformation. He published 27 peer-reviewed journal articles, one book chapter, and inventor of one Patent, attended several National and International conferences and symposium. Dr Aniruddha Mondal is the honourable member of the "Catalysis Society Taiwan".

Mr. S K Tarik Aziz is now pursuing his *PhD* degree in the Bar Ilan Institute for Nanotechnology and Advanced Materials (BINA), Chemistry, Bar Ilan University, Israel with prestigious MILGA fellowship (Presidential Scholarship). Additionally, Tarik has worked on synthesis of in-situ polymer-supported catalysis for C-C coupling organic reaction at Birla Institute of Technology, Mesra, Ranchi, India. At present, his research focuses on the synthesis of transition metal phosphides, characterization and its electrochemical applications. He has published eight research articles in peer-reviewed SCI and Scopus indexing journals. He awarded best poster-speech in conferences Israel Vacuum Society (IVS-MRS-2019) and Catalysis (2019) in Technion (Israel).

Dr. Sudip Mondal is presently working as an Assistant Professor in the Post Graduate Department of Chemistry, S. K. Porwal College of Arts, Science and Commerce, Kamptee, Nagpur, Maharashtra, India. He has qualified CSIR-NET examination conducted by UGC-CSIR, New Delhi. He perused Ph.D. in Chemistry from, RTM, Nagpur University, Nagpur. His area of research includes solution thermodynamics, materials science, surface chemistry, ionic liquids, molecular simulation and cyclic voltammeter. He has guided more than 22 students to complete their PG dissertations. He has published two books and more than twelve research articles in peer-reviewed SCI and Scopus indexing journals. He is an editorial board member of American Journal of Physical Chemistry. He also acted as guest editor of Advanced Journal of Chemistry (Special Issue) Science Publishing Group. He is the reviewer of two reputed journals and reviewed more than three research articles. He organized several academic events which includes National Conference, International Webinar and National Webinar. He is a life member of several scientific bodies. He delivered guest lecturers for many reputed educational institutes.

Dr. Trimurti L. Lambat is an Assistant Professor in the Department of Chemistry, Manoharbhai Patel College, Deori dist- Gondia affiliated to R. T. M. Nagpur University, Nagpur, India. He received his Ph.D. in synthetic organic and medicinal chemistry from Government Institute of Science, Nagpur, India affiliated to R. T. M. Nagpur University, Nagpur, India in 2017. His research focuses on Green Chemistry, development of novel methodologies for multifunctional bioactive molecules, nanocatalyst, stereochemistry and medicinal activities of synthetic scaffolds. He published twenty eight peer-reviewed journal articles, three book chapters, and six refereed conference papers (H-index 06, total citations-80). Dr. Trimurti Lambat is the recipient of "INSPIRE FELLOW" Award from the Department of Science and Technology, Ministry of Science, New Delhi, India in 2012.

www.ingramcontent.com/pod-product-compliance
Lightning Source LLC
Chambersburg PA
CBHW071333210326
41597CB00015B/1435

* 9 7 8 1 6 4 4 9 0 0 9 6 3 *